BRIDGET BEHE

Postharvest Technology of Horticultural Crops

ADEL A. KADER
ROBERT F. KASMIRE
F. GORDON MITCHELL
MICHAEL S. REID
NOEL F. SOMMER
JAMES F. THOMPSON

Cooperative Extension
University of California
Division of Agriculture
and Natural Resources

1985

Authors

Author	Mailing address	Telephone number (area code 916)
ADEL A. KADER Professor and Pomologist Agricultural Experiment Station Cooperative Extension	Department of Pomology University of California Davis, CA 95616	752-0909/2-0122
ROBERT F. KASMIRE Vegetable Marketing Specialist Cooperative Extension	Mann Laboratory University of California Davis, CA 95616	752-1412/2-1410
F. GORDON MITCHELL Pomologist Cooperative Extension	Department of Pomology University of California Davis, CA 95616	752-0508/2-0506
MICHAEL S. REID Professor and Postharvest Physiologist Agricultural Experiment Station Cooperative Extension	Department of Environmental Horticulture University of California Davis, CA 95616	752-7060/2-0130
NOEL F. SOMMER Lecturer and Postharvest Pathologist	Department of Pomology University of California Davis, CA 95616	752-0908/2-0122
JAMES F. THOMPSON Agricultural Engineer and Lecturer Cooperative Extension	Department of Agricultural Engineering University of California Davis, CA 95616	752-6167/2-0120

©1985 by The Regents of the University of California,
Division of Agriculture and Natural Resources.
All rights reserved. Published 1985.

Printed in the United States of America

Library of Congress Catalog Number: 85-70729
International Standard Book Number: 0-931876-72-9

Contents

	Preface	v
1	Sources of Information Related to Postharvest Biology and Technology ADEL A. KADER	1
2	Postharvest Biology and Technology: An Overview ADEL A. KADER	3
3	Product Maturation and Maturity Indices MICHAEL S. REID	8
4	Harvesting Systems JAMES F. THOMPSON	12

5	Preparation for Fresh Market
	   I. Fruits. F. GORDON MITCHELL ....
	   II. Vegetables. ROBERT F. KASMIRE ....
	   III. Automation Trends in Packinghouse Operations. P. CHEN
	   IV. Cull Utilization. JAMES F. THOMPSON ....
		14
		22
		24
		26

6	Packages for Horticultural Crops F. GORDON MITCHELL	28
7	Cooling Horticultural Commodities F. GORDON MITCHELL	35
8	Cooling Operations: Evaluation of Efficiency ROBERT F. KASMIRE	44
9	Storage Systems JAMES F. THOMPSON	49
10	Psychrometrics and Perishable Commodities JAMES F. THOMPSON	54
11	Modified Atmospheres and Low-pressure Systems during Transport and Storage ADEL A. KADER	58
12	Methods of Gas Mixing, Sampling, and Analysis ADEL A. KADER	65
13	Ethylene in Postharvest Technology MICHAEL S. REID	68
14	Principles of Disease Suppression by Handling Practices NOEL F. SOMMER	75
15	Strategies for Control of Postharvest Diseases of Selected Commodities NOEL F. SOMMER	83
16	Postharvest Treatments for Insect Control F. GORDON MITCHELL AND ADEL A. KADER	100
17	Transportation of Horticultural Commodities ROBERT F. KASMIRE	104
18	Handling of Horticultural Crops at Destination Markets ROBERT F. KASMIRE	111
19	Energy Use in Postharvest Technology Procedures JAMES F. THOMPSON	115

Contents

20	Quality Factors: Definition and Evaluation for Fresh Horticultural Crops ADEL A. KADER	118
21	Standardization and Inspection of Fresh Fruits and Vegetables ADEL A. KADER	122
22	Postharvest Handling Systems: Leafy, Root, and Stem Vegetables ROBERT F. KASMIRE	131
23	Postharvest Handling Systems: Fruit Vegetables ROBERT F. KASMIRE	139
24	Postharvest Handling Systems: Temperate Fruits F. GORDON MITCHELL	143
25	Postharvest Handling Systems: Table Grapes F. GORDON MITCHELL	149
26	Postharvest Handling Systems: Subtropical Fruits ADEL A. KADER	152
27	Postharvest Handling Systems: Tropical Fruits NOEL F. SOMMER	157
28	Postharvest Handling Systems: Tree Nuts ADEL A. KADER	170
29	Postharvest Handling Systems: Ornamentals MICHAEL S. REID	174

30	The Extension Link: University Research and the Horticultural Crops Industry in California
	    I. Cooperative Extension Methods. ROBERT F. KASMIRE
	    II. Extension and the California Fruit Industry.
	        F. GORDON MITCHELL
		179
		
		182

	Index	185

Published by
Agriculture and Natural Resources Publications
Division of Agriculture and Natural Resources
University of California, Berkeley 94720

For information about ordering this publication, write to:
Publications
Division of Agriculture and Natural Resources
University of California
6701 San Pablo Avenue
Oakland, CA 94608-1239

Preface

Postharvest Technology of Horticultural Crops is the outcome of a syllabus that was developed for a short course initiated in 1979 and offered annually since then in two modes: (1) as a regular University of California, Davis course (Plant Science 196) for advanced undergraduate and graduate students interested in postharvest biology and technology of horticultural crops, and (2) as a short course organized through University Extension for participants who are not current UCD students. The latter group usually includes research and extension workers, quality control personnel, and other persons concerned with postharvest handling of fresh horticultural crops.

Emphasis is on current postharvest technology procedures for fresh fruits, vegetables, and ornamentals in California. However, all the principles discussed are applicable to postharvest handling of fresh horticultural crops worldwide.

Thirty chapters are included, of which 20 present various aspects of postharvest technology of horticultural commodities and eight briefly cover postharvest handling systems for certain commodities or commodity groups. It was not possible to include every horticultural crop in these eight chapters and keep the book to a reasonable length. The remaining two chapters deal with sources of information and extension efforts. We are continuously working on improving all aspects of this book and so we welcome comments and suggestions for incorporation into future editions.

On behalf of the authors, I wish to thank all those individuals who assisted us. I especially want to acknowledge the tireless efforts of Pamela Moyer (Department of Pomology, UC Davis) in compiling the subject index, proofreading, and organizing production of this book. Thanks are also due Barbara Weatherly, Kathi Miller, and Ann Altamirano for their excellent job in typing the manuscript. Thanks are also due Don Edwards (Department of Pomology, UC Davis) and the staff of Cooperative Extension's Visual Media for their help with illustrations. Furthermore, we greatly appreciate the assistance and cooperation of Franz Baumhackl, Jim Coats, Susan Gardner, Louis Perica, and other University of California Agriculture and Natural Resources Publications staff members who participated in the production of this book.

ADEL A. KADER
TECHNICAL EDITOR

1
Sources of Information Related to Postharvest Biology and Technology

ADEL A. KADER

There are numerous sources of information related to postharvest biology and technology of horticultural crops that can be used to supplement and expand the information in this book. The purpose of this chapter is to point out the library resources available.

A Starting Point

A list of textbooks and general references related to postharvest biology and technology of horticultural crops is included at the end of this chapter. In addition, a list of references for further reading is included in each subsequent chapter of this book. Consult the following reference, which provides a good starting point for further development of background information.

Kader, A. A., L. L. Morris, and M. Cantwell. 1983. *Postharvest handling and physiology of horticultural crops—a list of selected references.* 3d ed., Vegetable Crops Series 169, University of California, Davis.

Additional library resources

For a review of published literature about a specific topic or a given commodity, either use a computerized information service if one is available, or consult one or more of the following abstracting journals:

Horticultural Abstracts

Food Science Abstracts

Biological Abstracts

Chemical Abstracts

Bibliography of Agriculture (titles only)

Review of Plant Pathology

Several review journals that periodically contain some chapters dealing with postharvest biology and technology:

Horticultural Reviews

Advances in Food Research

Critical Reviews in Food Science and Nutrition

Annual Review of Plant Physiology

Annual Review of Phytopathology

To follow current publications, on a regular basis, use the weekly publication *Current Contents: Agriculture, Biology & Environmental Sciences* (published by Institute for Scientific Information, 3501 Market St., Philadelphia, PA, 19104), which includes tables of contents from a large number of periodicals, including most of those listed below. It also includes listings of new publications from the U.S. Department of Agriculture and major U.S. Agricultural Experiment Stations.

There is no one scientific journal that specializes in postharvest biology and technology of horticultural crops. Research reports are published in a wide range of scientific journals, which include the following:

Journal of the American Society for Horticultural Science

HortScience

Scientia Horticulturae

Journal of Horticultural Science

Proceedings of the Florida State Society for Horticultural Science

Proceedings of the Tropical Region of the American Society for Horticultural Science

Fruits d'Outre Mer

Tropical Agriculture

Tropical Science

American Potato Journal

Economic Botany

Plant Physiology

Phytochemistry

Journal of Food Science

Food Technology

Journal of Food and Agricultural Chemistry

Journal of the Science of Food and Agriculture

Journal of Food Biochemistry

Journal of Food Quality

Journal of Textural Studies

Transactions of the American Society of Agricultural Engineers

Agricultural Engineering

Phytopathology

Plant Disease

Several semitechnical and popular periodicals that are also available:

> *California Agriculture*
>
> *Western Grower and Shipper*
>
> *American Fruit Grower*
>
> *American Vegetable Grower*
>
> *The Goodfruit Grower*
>
> *Florida Grower and Rancher*
>
> *Florists Review*
>
> *Citrus and Vegetables Magazine*
>
> *Modern Packaging*
>
> *Outlook* (United Fresh Fruit & Vegetable Association)
>
> *Produce Marketing Association Magazine*
>
> *The Packer* (a weekly newspaper)

A few newsletters are published periodically by Cooperative Extension Specialists in postharvest biology and technology of horticultural crops in California, Florida, and other locations, including:

> *Perishables Handling* (available from: F. G. Mitchell, Department of Pomology, University of California, Davis, CA, 95616)
>
> *Packinghouse Newsletter* (available from: W. Wardowski, University of Florida, AREC, 700 Experiment Station Rd., Lake Alfred, FL, 33850)
>
> *Handling Florida Vegetables* (available from: M. Sherman, University of Florida, Department of Vegetable Crops, 1217 HSPP Building, Gainesville, FL, 32611)
>
> *Postharvest Pomology Newsletter* (available from: E. Kupferman, Tree Fruit Research Center, 1100 N. Western Ave., Wenatchee, WA, 98801)

Publications of the U.S. Department of Agriculture; Agricultural Experiment Stations, and Cooperative Extension in California, Florida, New York, Michigan, Oregon, Washington, and other states; International Institute of Refrigeration (Paris, France); Tropical Development and Research Institute (London, England); and other organizations.

Publications of industry organizations such as United Fresh Fruit and Vegetable Association (UFFVA) and Produce Marketing Association (PMA).

Visual aids

To find out about available audiovisual programs (slide sets, film strips, 16 mm movies, and so on) dealing with various aspects of postharvest technology of horticultural crops, consult the directory published by UFFVA (North Washington at Madison, Alexandria, VA, 22314) and the listing of slide programs available from the American Society for Horticultural Science (701 North Saint Asaph St., Alexandria, VA, 22314).

References

1. ASHRAE. 1982. *ASHRAE handbook and product directory, applications volume*. Am. Soc. Heating, Refrigeration and Air conditioning Engineers. Atlanta, GA.

2. Burton, W. G. 1982. *Postharvest physiology of food crops*. London and New York: Longman. 339 pp.

3. Debney, H. G., K. J. Blacker, B. J. Redding, and J. B. Watkins. 1980. *Handling and storage practices for fresh fruits and vegetables-product manual*. Aust. United Fresh Fruit & Veg. Assn. Brisbane, Australia.

4. Dennis, C. 1983. *Postharvest pathology of fruits and vegetables*. London: Academic Press. 264 pp.

5. Duckworth, R. B. 1966. *Fruits and vegetables*. Oxford: Pergammon Press. 306 pp.

6. Friend, J., and M. J. C. Rhodes, eds. 1981. *Recent advances in the biochemistry of fruit and vegetables*. New York: Academic Press. 278 pp.

7. Haard, N. F., and D. K. Salunkhe, eds. 1975. *Symposium: Postharvest biology and handling of fruits and vegetables*. Westport, CT: AVI Publ. Co. 193 pp.

8. Hulme, A. C., ed. 1970. *The biochemistry of fruits and their products*. Vol. 1. New York: Academic Press. 620 pp.

9. _____. 1971. *The biochemistry of fruits and their products*. Vol. 2. New York: Academic Press. 788 pp.

10. Hultin, H. O., and M. Milner, eds. 1978. *Postharvest biology and biotechnology*. Westport, CT: Food and Nutrition Press. Westport, CT. 460 pp.

11. Lieberman, M., ed. 1983. *Postharvest physiology and crop preservation*. New York: Plenum. 575 pp.

12. Lutz, J. M., and R. E. Hardenburg. 1968. *The commercial storage of fruits, vegetables, and florist and nursery stocks*. USDA Agric. Handb. No. 66. 94 pp.

13. Pantastico, Er. B., ed. 1975. *Postharvest physiology, handling and utilization of tropical and subtropical fruits and vegetables*. Westport, CT: AVI Publ. Co. 560 pp.

14. Ryall, A. L., and W. J. Lipton. 1979. *Handling, transportation and storage of fruits and vegetables*. Vol. 1, *Vegetables and melons*. 2d ed. Westport, CT: AVI Publ. Co. 588 pp.

15. Ryall, A. L., and W. T. Pentzer. 1982. *Handling, transportation and storage of fruits and vegetables*. Vol. 2, *Fruits and tree nuts*. Westport, CT: AVI Publ. Co. 610 pp.

16. Salunkhe, D. K. 1975. *Storage, processing, and nutritional quality of fruits and vegetables*. Boca Raton, FL: CRC Press. 166 pp.

17. Salunkhe, D. K., and B. B. Desai. 1984a. *Postharvest biotechnology of fruits*. Vol. I. Boca Raton, FL: CRC Press. 184 pp.

18. _____. 1984b. *Postharvest biotechnology of fruits*. Vol. II. Boca Raton, FL: CRC Press. 168 pp.

19. _____. 1984c. *Postharvest biotechnology of vegetables*. Vol. I. Boca Raton, FL: CRC Press. 232 pp.

20. _____. 1984d. *Postharvest biotechnology of vegetables*. Vol. II. Boca Raton, FL: CRC Press. 288 pp.

21. Seelig, R. A., ed. Various dates. *Fruit and vegetables facts and pointers*. A series of reports on each of 81 commodities. United Fresh Fruit & Veg. Assn. Alexandria, VA.

22. Wills, R. H. H., T. H. Lee, D. Graham, W. B. McGlasson, and E. G. Hall. 1981. *Postharvest: An introduction to the physiology and handling of fruit and vegetables*. Westport, CT: AVI Publ. Co. 163 pp.

2
Postharvest Biology and Technology: An Overview

ADEL A. KADER

Losses in quantity and quality occur in horticultural crops between harvest and consumption. The magnitude of postharvest losses in fresh fruits and vegetables is estimated to be 5 percent to 25 percent in developed countries and 20 percent to 50 percent in developing countries, depending upon the commodity. Our aim is to reduce these losses and to do so, we must: (1) understand the biological and environmental factors involved in deterioration, and (2) use those postharvest technology procedures which will delay senescence and maintain the best possible quality. This chapter includes a brief discussion of item (1) and an introduction to item (2), which will be covered in detail in subsequent chapters of this book.

Fresh fruits, vegetables, and ornamentals are living tissues which are subject to continuous change after harvest. While some of these changes are desirable, most are not desirable from the consumer's standpoint. Postharvest changes in fresh produce cannot be stopped, but can be slowed down within certain limits. Senescence is the final stage in development of plant organs during which a series of essentially irreversible events leads to cellular breakdown and death.

Fresh horticultural crops are diverse in morphological structure (roots, stems, leaves, flowers, fruits, and so on), in composition, and in general physiology. Thus, commodity requirements and recommendations for maximum postharvest life vary among various groups of commodities. All fresh horticultural crops are high in water content and thus are subject to desiccation (wilting, shriveling) and to mechanical injury. They are also susceptible to attack by bacteria and fungi resulting in pathological breakdown.

Biological Factors Involved in Deterioration

Respiration

Respiration is the overall process by which stored organic materials (carbohydrates, proteins, fats) are broken into simple end products with a release of energy. Oxygen (O_2) is used in this process and carbon dioxide (CO_2) is produced by the commodity. The loss of stored food reserves in the commodity during respiration means: (1) senescence is hastened as the reserves which provide energy for maintaining the living status of the commodity are exhausted, (2) loss of food value (energy value) for the consumer, (3) reduced flavor quality, especially sweetness, and (4) loss of salable dry weight (especially important for commodities destined for dehydration). The energy released as heat, known as vital heat, is very important in postharvest technology considerations, such as the estimation of refrigeration and ventilation requirements.

The rate of deterioration (perishability) of harvested commodities is generally proportional to their respiration rate. A classification of horticultural commodities according to their respiration rates is shown in table 2.1. On the basis of their respiratory and ethylene production patterns during maturation and ripening, fruits can be placed in two groups: climacteric fruits and nonclimacteric fruits (table 2.2). Climacteric fruits exhibit a large increase in CO_2 and ethylene (C_2H_4) production rates coincident with their ripening, while nonclimacteric fruits exhibit no change in their generally low CO_2 and C_2H_4 production rates during ripening.

Ethylene production

Ethylene is the simplest of the organic compounds that have an effect on physiological processes of plants. It is a natural product of plant metabolism and is produced by all tissues of higher plants and by some microorganisms. Ethylene is considered the natural aging and ripening hormone and is physiologically active in trace amounts (less

Table 2.1. Classification of horticultural commodities according to their respiration rates

Class	Range at 5°C (41°F) (mg CO_2/Kg-hr)*	Commodities
Very low	<5	Nuts, dates, dried fruits and vegetables
Low	5-10	Apple, citrus, grape, kiwifruit, garlic, onion, potato (mature), sweet potato
Moderate	10-20	Apricot, banana, cherry, peach, nectarine, pear, plum, fig (fresh), cabbage, carrot, lettuce, pepper, tomato, potato (immature)
High	20-40	Strawberry, blackberry, raspberry, cauliflower, lima bean, avocado
Very high	40-60	Artichoke, snap bean, green onion, brussels sprouts, cut flowers
Extremely high	>60	Asparagus, broccoli, mushroom, pea, spinach, sweet corn

*Vital heat (Btu/ton/24 hrs) = mg CO_2/Kg-hr × 220.

than 0.1 ppm). It also plays a major role in abscission of plant organs.

A classification of horticultural commodities according to their ethylene production rates is included in table 2.3. There is no consistent relationship between the C_2H_4 production capacity of a given commodity and its perishability. However, exposure of most commodities to C_2H_4 accelerates their senescence.

Generally, C_2H_4 production rates increase with maturity at harvest, physical injuries, disease incidence, increased temperatures up to 30°C (86°F), and water stress. On the other hand, ethylene production rates by fresh horticultural crops are reduced by storage at the lowest safe temperature for each commodity and by reduced O_2 (less than 8%) and/or elevated CO_2 (more than 2%) levels around the commodity.

Compositional changes

Many changes in pigments take place during development and maturation of the commodity on the plant. The following changes may continue after harvest and this can be desirable or undesirable.

1. Loss of chlorophyll (green color)—desirable in fruits but not in vegetables

2. Development of carotenoids (yellow and orange colors)—desirable in fruits such as apricots, peaches, and citrus; red color development in tomatoes is due to a specific carotenoid (lycopene); beta-carotene is provitamin A and thus is very important in nutritional quality

3. Development of anthocyanins (red and blue colors)—desirable in fruits such as apples (red cultivars), cherries, strawberries, and bush berries; pigments are water-soluble and are much less stable than carotenoids

4. Changes in anthocyanins and other phenolic compounds—may result in tissue browning which is undesirable from appearance quality standpoint

Changes in carbohydrates include: (1) starch to sugar conversion (undesirable in potatoes, desirable in fruits), (2) sugar to starch conversion (undesirable in peas and sweet corn), and (3) conversion of starch and sugars to CO_2 and water through respiration. Breakdown of pectins and other polysaccharides results in softening of fruits and a consequent increase in susceptibility to mechanical injuries. Increased lignin content is responsible for toughening of asparagus spears and root vegetables.

Changes in organic acids, proteins, amino acids, and lipids can influence flavor quality of the commodity. Loss in vitamin content, especially ascorbic acid (vitamin C) is detrimental to nutritional quality. Production of flavor volatiles associated with ripening of fruits is very important to their eating quality.

Growth and development

Sprouting of potatoes, onions, garlic, and root crops greatly reduces their utilization value and accelerates deterioration. Rooting of onions and root crops is also undesirable. Asparagus spears continue to grow after harvest; elongation and curvature (if the spears are held horizontally) are accompanied with increased toughness and decreased palatability. Similar geotropic responses occur in cut gladiolus and snapdragon flowers stored horizontally. Seed germination inside fruits such as tomatoes, peppers, and lemons, is an undesirable change from the standpoint of quality.

Transpiration or water loss

Water loss can be one of the main causes of deterioration since it results in not only direct quantitative losses (loss of salable weight), but also causes losses in appearance (due to wilting and shriveling), textural quality (softening, flaccidity, limpness, loss of crispness, and juiciness), and nutritional quality.

The dermal system (outer protective coverings) plays an important role in the regulation of water loss by the

Table 2.2. Classification of some fruits according to respiratory behavior during ripening

Climacteric fruits		Nonclimacteric fruits	
Apple	Muskmelon	Blackberry	Orange
Apricot	Nectarine	Cacao	Pepper
Avocado	Papaya	Cashew apple	Pineapple
Banana	Passion fruit	Cherry	Pomegranate
Biriba	Peach	Cucumber	Raspberry
Blueberry	Pear	Eggplant	Satsuma mandarin
Breadfruit	Persimmon	Grape	Strawberry
Cherimoya	Plantain	Grapefruit	Summer squash
Feijoa	Plum	Jujube	Tamarillo
Fig	Sapote	Lemon	Tangerine
Guava	Soursop	Lime	
Jackfruit	Tomato	Loquat	
Kiwifruit	Watermelon	Lychee	
Mango		Olive	

Table 2.3. Classification of horticultural commodities according to ethylene production rates

Class	Range at 20°C (68°F) (μl C_2H_4/Kg-hr)	Commodities
Very low	Less than 0.1	Artichoke, asparagus, cauliflower, cherry, citrus, grape, jujube, strawberry, pomegranate, leafy vegetables, root vegetables, potato, most cut flowers
Low	0.1–1.0	Blueberry, cranberry, cucumber, eggplant, okra, olive, pepper, persimmon, pineapple, pumpkin, raspberry, tamarillo, watermelon
Moderate	1.0–10.0	Banana, fig, guava, honeydew melon, mango, plantain, tomato
High	10.0–100.0	Apple, apricot, avocado, cantaloupe, feijoa, kiwifruit (ripe), nectarine, papaya, peach, pear, plum
Very high	More than 100.0	Cherimoya, mammee apple, passion fruit, sapote

commodity. It includes the cuticle, epidermal cells, stomata, lenticles, and trichomes (hairs). The cuticle is composed of surface waxes, cutin embedded in wax, and a layer of mixtures of cutin, wax, and carbohydrate polymers. The thickness structure and chemical composition of the cuticle vary greatly among commodities and among developmental stages of a given commodity.

Transpiration rate is influenced by internal or commodity factors (morphological and anatomical characteristics, surface to volume ratio, surface injuries, and maturity stage) and external or environmental factors (temperature, relative humidity, air velocity, and atmospheric pressure). Transpiration (evaporation of water from the plant tissues) is a physical process that can be controlled by various treatments applied to the commodity (e.g., surface coatings and wrapping with plastic films) or by manipulation of the environment (e.g., maintenance of high relative humidity and control of air circulation rate).

Physiological breakdown

Exposure of the commodity to undesirable temperatures can result in physiological disorders.

1. Freezing injury—commodities held below their freezing temperatures
2. Chilling injury—commodities (mainly tropical and subtropical origin) held at temperatures above their freezing point and below 5° to 15°C (41° to 59°F) depending on commodity

   Physiological injury symptoms—surface and internal discoloration, pitting, water-soaked areas, uneven ripening or failure to ripen, off-flavor development, and accelerated incidence of surface molds and decay

3. Heat injury—exposure to direct sunlight or to excessively high temperatures

   Symptoms—bleaching, surface burning or scalding, uneven ripening, excessive softening, and desiccation

Certain types of physiological disorders originate from preharvest nutritional imbalances. For example, blossom-end rot of tomatoes and bitter pit of apples result from calcium deficiency. Increased calcium content in certain fruits via preharvest or postharvest treatments can reduce their susceptibility to physiological disorders.

Very low oxygen (<1%) and/or high carbon dioxide (>20%) atmospheres can result in physiological breakdown of most fresh horticultural commodities. Exposure to C_2H_4 can induce various types of physiological disorders in certain commodities. The interactions among O_2, CO_2, and C_2H_4 concentrations, temperature, and duration of storage influence the incidence and severity of physiological disorders related to atmospheric composition.

Physical damage

Various types of physical damage (surface injuries, impact bruising, vibration bruising, and so on) are major contributors to deterioration. Mechanical injuries are not only unsightly but also accelerate water loss, provide loci for fungal infection, and stimulate CO_2 and C_2H_4 production by the commodity.

Pathological breakdown

One of the most common and obvious symptoms of deterioration results from the activity of bacteria and fungi. Attack by most organisms follows physical injury or physiological breakdown of the commodity. In a few cases, pathogens can infect apparently healthy tissues and become the primary cause of deterioration. In general, harvested fruits and vegetables exhibit considerable resistance to potential pathogens during most of their postharvest life. The onset of ripening in fruits, and senescence in all commodities, results in their becoming susceptible to infection by pathogens. Stresses, such as mechanical injuries, chilling, and sunscald, lower the resistance of the commodity to pathogens.

Environmental Factors Influencing Deterioration

Temperature

Temperature is the most important environmental factor that influences the deterioration rate of harvested commodities. For each increase of 10°C (18°F) above optimum, the rate of deterioration increases by two- to three-fold (table 2.4). Exposure to undesirable temperatures results in many physiological disorders, as mentioned above. Temperature also influences how ethylene, reduced oxygen, and elevated carbon dioxide, affect the commodity. Spore germination and growth rate of pathogens are greatly influenced by temperature. Some pathogens, such as Rhizopus rot, are sensitive to low temperatures. Thus, cooling of commodities below 5°C (41°F) immediately after harvest can greatly reduce Rhizopus rot incidence. A comparison of temperature effects on postharvest responses of chilling-sensitive and nonchilling-sensitive horticultural crops is shown in table 2.5.

Relative humidity

The rate of water loss from fruits and vegetables depends upon the vapor pressure deficit between the commodity and the surrounding ambient air, which is influenced by temperature and relative humidity. At a given temperature

Table 2.4. Effect of temperature on deterioration rate of a non-chilling sensitive commodity

Temperature		Assumed Q_{10}*	Relative velocity of deterioration	Relative shelf-life	Loss per day (%)
°F	°C				
32	0	*	1.0	100	1
50	10	3.0	3.0	33	3
68	20	2.5	7.5	13	8
86	30	2.0	15.0	7	14
104	40	1.5	22.5	4	25

$$*Q_{10} = \frac{\text{Rate of deterioration at T} + 10°C}{\text{Rate of deterioration at T}}$$

Table 2.5. Classification of horticultural crops according to sensitivity to chilling injury

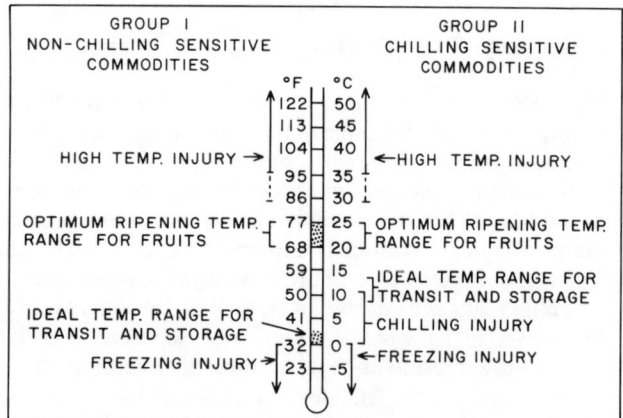

Group I (Non-chilling sensitive)		Group II (Chilling sensitive)	
Apple*	Artichoke	Avocado	Bean, snap
Apricot	Asparagus	Banana	Cucumber
Bush berry	Bean, lima	Cherimoya	Eggplant
Cherry	Beet	Citrus	Muskmelon
Fig	Broccoli	Feijoa	Okra
Grape	Brussels sprouts	Guava	Pepper
Kiwifruit	Cabbage	Jujube	Potato
Nectarine	Carrot	Mango	Pumpkin
Peach	Cauliflower	Olive	Squash
Pear	Celery	Papaya	Sweet potato
Persimmon*	Corn, sweet	Passion fruit	Tomato
Plum	Garlic	Pineapple	Watermelon
Prune	Lettuce	Plantain	
Strawberry	Onion	Pomegranate	
	Pea	Sapote	
	Radish		
	Spinach		
	Turnip		

*Some varieties are chilling sensitive.

and rate of air movement, the rate of water loss from the commodity depends on the relative humidity. At a given relative humidity, water loss increases with the increase in temperature.

Atmospheric composition

Reduction of oxygen and elevation of carbon dioxide, whether intentional (modified or controlled atmosphere storage) or unintentional, can either delay or accelerate deterioration of fresh horticultural crops. The magnitude of these effects depends upon: commodity, cultivar, physiological age, O_2 and CO_2 level, temperature, and duration of holding.

Ethylene

The effects of ethylene on harvested horticultural commodities can be desirable or undesirable. Thus, it is of major concern to all handlers of fruits, vegetables, and ornamentals. Ethylene can be used to promote faster and more uniform ripening of fruits picked at the mature-green stage. On the other hand, exposure to ethylene can be detrimental to the quality of most nonfruit vegetables and ornamentals.

Light

Exposure of potatoes to light should be avoided because it results in greening due to formation of chlorophyll and/or solanine (toxic to humans). Light-induced greening of Belgian endive is also undesirable.

Other factors

Various kinds of chemicals (e.g., fungicides, growth regulators) may be applied to the commodity to affect one or more of the biological deterioration factors.

Postharvest Technology Procedures

Temperature management procedures

Temperature management is the most important tool that we have to extend the shelf-life of fresh horticultural commodities. Proper temperature management begins with the rapid removal of field heat by using one of the following cooling methods: hydrocooling, in-package icing, top-icing, evaporative cooling, room cooling, forced-air cooling, serpentine forced-air cooling, vacuum cooling, and hydro-vacuum cooling.

Cold storage facilities should be well-engineered and adequately equipped. They should have (1) good construction and insulation, including a complete vapor barrier on the warm side (usually outside) of the insulation, (2) strong floors, (3) adequate and well-positioned doors for loading and unloading, (4) effective distribution of refrigerated air, (5) sensitive and properly located controls, (6) enough refrigerated coil surface to minimize the difference between the coil and air temperatures, and (7) adequate capacity for expected needs. Commodities should be stacked in the cold room leaving air spaces between pallets and room walls as well as among pallets to insure good air circulation. Storage rooms should not be loaded beyond their limit for proper cooling. In monitoring temperatures, commodity temperature rather than air temperature should be used.

Transit vehicles must be cooled before loading the commodity. Delays between cooling after harvest and loading into transit vehicles should be avoided. Proper temperature maintenance should be ensured throughout the handling system.

Control of relative humidity

Relative humidity can influence water loss, decay development, incidence of some physiological disorders, and uniformity of fruit ripening. Condensation of moisture on the commodity (sweating) over long periods of time is probably more important than is the relative humidity of ambient air in enhancing decay. Proper relative humidity is 85 percent to 95 percent for fruits and 90 percent to 98 percent for vegetables except dry onions and pumpkins (70% to 75%). Some root vegetables can best be held at 95 percent to 100 percent relative humidity.

Relative humidity control can be achieved by one or more of the following procedures:

1. Addition of moisture (water mist or spray, steam) to air by humidifiers
2. Regulation of air movement and ventilation in relation to produce load in cold storage room
3. Maintaining temperature of refrigeration coils to within about 1°C (2°F) of air temperature
4. Moisture barriers—insulation of storage room and transit vehicle walls, polyethylene liners in containers, plastic films for packaging
5. Wetting the floor in storage rooms
6. Crushed ice in shipping containers or in retail displays for commodities not injured by such practice
7. Sprinkling produce with water during retail marketing—use on leafy vegetables, cool-season root vegetables, and immature fruit vegetables (e.g., snap beans, peas, sweet corn, summer squash)

Supplements to temperature management

Many technological procedures are used commercially as supplements to temperature management. None of these procedures, alone or in their various combinations, can substitute for maintenance of optimum temperature and relative humidity in extending shelf-life of harvested horticultural commodities. But they can help extend shelf-life beyond what is possible using refrigeration alone.

Supplemental procedures

Treatments applied to the commodity. (1) Curing of certain root, bulb, and tuber vegetables; (2) Cleaning removal of excess surface moisture; (3) Sorting for elimination of defects; (4) Waxing and other surface coatings, film wrapping; (5) Heat treatments (hot water, vapor heat); (6) Treatment with postharvest fungicides; (7) Use of sprout inhibitors; (8) Special chemical treatments (scald inhibitors, calcium); (9) Fumigation for insect control; and (10) Ethylene treatment (degreening, ripening).

Treatments to manipulate the environment. (1) Packaging; (2) Control of air movement and circulation; (3) Control of air exchange or ventilation; (4) Exclusion and/or removal of ethylene; (5) Controlled or modified atmospheres; and (6) Sanitation procedures.

Future Trends and Developments in Perishables Handling

Research and development efforts are continually aimed at improving existing technology and testing new ideas for possible alternatives to current procedures. Some trends follow:

1. Cooling methods—improved, more energy efficient
2. Temperature monitoring and control methods—better microprocessor technology in transit vehicles and storage rooms
3. Expedited handling—rapid transportation, more efficient distribution systems, and increased direct marketing
4. Increased mechanization—in harvesting, bulk handling from field to packinghouse and during transport to destination markets, and unitization (palletization) of loads
5. Shipping container types—reduction to a few sizes, each adequate for numerous commodities
6. Slip sheets in palletization—increased use
7. Returnable plastic containers for produce distribution—increased use within an organization or geographical area
8. Modified atmospheres—expansion in consumer packages, pallet shrouds, transit vehicles, and cold storage rooms; may facilitate more shipments of mixed loads
9. Partially prepared fresh fruits and vegetables—increased marketing of cut lettuce, carrots, and cabbage, for example, for both institutional and consumer use
10. Other modifications in handling procedures—to economize in labor, materials, and energy use, and to protect the environment

References

1. United Nations Food and Agriculture Organization. 1981. *Food loss prevention in perishable crops.* FAO Agric. Serv. Bull. 43, UN Food & Agric. Org. Rome: 72 pp.

2. Grierson, W., and W. F. Wardowski. 1978. Relative humidity effects on the postharvest life of fruits and vegetables. *HortScience* 13:570-74.

3. Harvey, J. M. 1978. Reduction of losses in fresh market fruits and vegetables. *Annu. Rev. Phytopath.* 16:321-41.

4. International Institute of Refrigeration. 1979. *Recommended conditions for cold storage of perishable produce.* Int. Inst. Refrig. Bull. Sup. 148 pp.

5. Kader, A. A. 1983. Postharvest quality maintenance of fruits and vegetables in developing countries. In *Postharvest physiology and crop preservation*, ed. M. Lieberman, 520-36. New York: Plenum.

6. Kader, A. A., J. M. Lyons, and L. L. Morris. 1974. Quality and postharvest responses of vegetables to preharvest field temperature. *HortScience* 9:523-27.

7. National Academy of Sciences. 1978. *Postharvest food losses in developing countries.* Science & Technology for International Development. Washington, D.C.: NAS. 202 pp.

8. Rhodes, M. J. C. 1980a. The maturation and ripening of fruits. In *Senescence in plants*, ed. K. V. Thimann, 157-205. Boca Raton, FL: CRC Press.

9. _____. 1980b. The physiological basis for the conservation of food crops. *Prog. Food & Nutr. Sci.* 4(3-4): 11-20.

10. Tindall, H. D., and F. J. Proctor. 1980. Loss prevention of horticultural crops in the tropics. *Prog. Food & Nutr. Sci.* 4(3-4): 25-40.

3
Product Maturation and Maturity Indices

MICHAEL S. REID

The maturity of harvested perishable commodities has an important bearing on the way in which they are handled, transported, and marketed, and on their storage life and quality. A study of the concept of maturity, what it means, and how it is measured, is, therefore, central to postharvest technology. The meaning of the term *mature*, the importance of maturity determination, and some examples of approaches to determining and applying a satisfactory index of maturity, are discussed here.

Definition of Maturity

To most people "mature" and "ripe" mean the same thing when describing fruit. For example, *mature* is defined in Webster's dictionary as:

mature (fr. L *maturus* ripe): **1.** based on slow and careful consideration; **2 a (1):** having completed natural growth and development: RIPE **(2):** having undergone maturation, **b:** having attained a final or desired state; **3 a:** of or relating to a condition of full development.

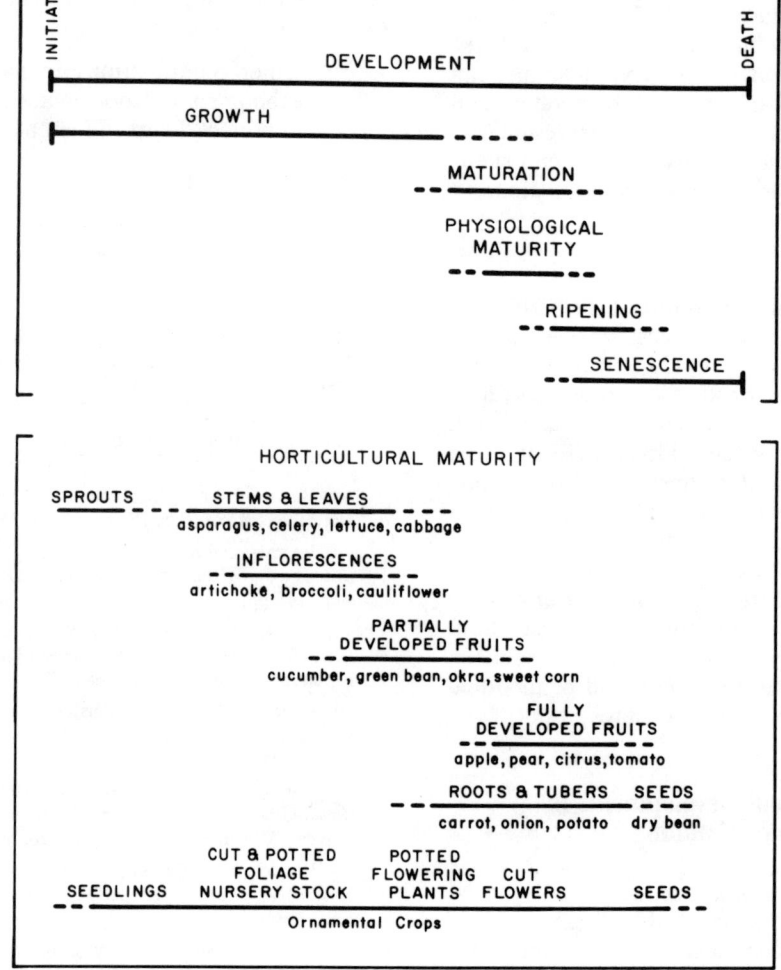

Fig. 3.1. Horticultural maturity in relation to developmental stages of the plant (Watada et al. 1984).

In postharvest physiology we consider "mature" and "ripe" as distinct terms for different stages of fruit development (fig. 3.1). *Mature* is best defined by 2 a (1) above (having completed natural growth and development), and for fruits, is defined in the U.S. Grade standards as "that stage which will ensure proper completion of the ripening process." This latter definition lacks precision in that it fails to define "proper completion of the ripening process." Most postharvest technologists consider that the definition should be "that stage at which a commodity has reached a sufficient stage of development that after harvesting and postharvest handling (including ripening, where required), its quality will be at least the minimum acceptable to the ultimate consumer."

In many fruits, for example mature (but green) bananas, the eating quality at maturity will be far from optimal. In most vegetables, optimal maturity coincides with optimal eating quality.

Horticultural maturity is the stage of development when a plant or plant part possesses the prerequisites for utilization by consumers for a particular purpose. A given commodity may be horticulturally mature at any stage of development (fig. 3.1). For example, sprouts or seedlings are horticulturally mature in the early stage of development, whereas most vegetative tissues, flowers, fruits, and underground storage organs become horticulturally mature in the midstage, and seeds and nuts in the late stage of development.

Determination of Maturity

The search for an objective determination of maturity has occupied the attention of many horticulturists, working with a wide range of commodities, for years. The number of satisfactory indices that have been suggested is nevertheless rather small, and for most commodities the search for a satisfactory maturity index continues.

Maturity indices

The definition of maturity as the stage of development giving minimum acceptable quality to the ultimate consumer implies a need for objective measures of maturity. In addition, indication of maturity is of considerable importance during the marketing chain.

Regulations. Regulations published by grower groups, marketing orders, or legally appointed authorities (such as the State Departments of Agriculture and the USDA) frequently include a statement as to the minimum (and sometimes maximum) maturity acceptable for a given commodity. Objective maturity standards are available for relatively few commodities and most regulations rely on subjective judgments related to the broad definitions quoted above.

Marketing strategy. In most markets the laws of supply and demand mean price incentives for the earliest (or sometimes the latest) shipments of any particular commodity. This encourages growers and shippers to expedite (or delay) the harvesting of their crop to take advantage of premium prices. This is, of course, the reason for the minimum maturity statement in grade standards that is mentioned above. An objective measure of maturity would enable growers to know whether a commodity could be harvested at times that would take advantage of buoyant market conditions.

Efficient use of labor and resources. With many crops the need for labor and equipment for harvesting and handling is seasonal. In order to plan operations efficiently, growers need to predict the likely starting and finishing dates for harvest of each commodity. Objective maturity indices are of paramount importance for accurate prediction of harvest dates.

Methods for Establishing Maturity Indices

Two rather different problems will be addressed here. The first is to measure maturity at harvest or at a subsequent inspection point. The second, and more complex, problem is to find some way of predicting the time at which a commodity will mature. For both problems, similar techniques may be appropriate, but the ways in which they are applied differ.

Requirements for maturity indices

Maturity measures to be made by producers, handlers, and quality control personnel must be simple, readily performed in the field or orchard, and should require relatively inexpensive equipment. The index should preferably be objective (a measurement) rather than subjective (an evaluation). The index should consistently relate to the quality and postharvest life of the commodity for all growers, districts, and years. For example, if a firmness measurement of kiwifruit with acceptable storage life and eating quality after ripening shows 15 pounds-force in 1 season and 21 pounds-force in another season, it is clear that firmness is not a satisfactory method for determining maturity of this commodity.

Many features of fruits and vegetables have been used in attempting to provide adequate estimates of maturity. Indices that have been proposed, or that are presently in use, are shown in table 3.1.

A wide range of equipment has been devised to measure many of the characteristics of fruits and vegetables noted in table 3.1. A summary of the most important of these methods, together with the nature of the determination, is given in table 3.2.

Methods for predicting maturity

Prediction of maturity is more complex than assessing the maturity at or after harvest. The basic requirement is for a measurement that can be made prior to, but which is highly correlated with, the date of maturation. The simplest prediction system uses measurements which relate to development of the fruit in a regular way through the latter part of the growing season. When the relationship

Table 3.1. Maturity indices for selected fruits and vegetables

Index	Examples
Elapsed days from full bloom to harvest	Apples, pears
Mean heat units during development	Peas, apples, sweet corn
Development of abscission layer	Some melons, apples, feijoas
Surface morphology and structure	Cuticle formation on grapes, tomatoes Netting of some melons Gloss of some fruits (development of wax)
Size	All fruits and many vegetables
Specific gravity	Cherries, watermelons, potatoes
Shape	Angularity of banana fingers Full cheeks of mangoes Compactness of broccoli and cauliflower
Solidity	Lettuce, cabbage, brussels sprouts
Textural properties:	
Firmness	Apples, pears, stone fruits
Tenderness	Peas
Color, external	All fruits and most vegetables
Internal color and structure	Formation of jelly-like material in tomato fruits Flesh color of some fruits
Compositional factors:	
Starch content	Apples, pears
Sugar content	Apples, pears, stone fruits, grapes
Acid content, sugar/acid ratio	Pomegranates, citrus, papaya, melons, kiwifruit
Juice content	Citrus fruits
Oil content	Avocados
Astringency (tannin content)	Persimmons, dates
Internal ethylene concentration	Apples, pears

Table 3.2. Methods of maturity determination

Index	Methods of determination	Subjective	Objective	Destructive	Nondest.
Elapsed days from full bloom	Computation		×		×
Mean heat units	Computation from weather data		×		×
Development of abscission layer	Visual or force of separation	×	×		×
Surface structure	Visual	×			×
Size	Various measuring devices, weight		×		×
Specific gravity	Use of density gradient solutions, flotation techniques vol/wt		×		×
Shape	Dimensions, ratio charts	×	×		×
Solidity	Feel, bulk density, γ-rays, x-rays	×	×		×
Textural properties:					
Firmness	Pressure testers, deformation		×	×	
Tenderness	Tenderometer		×	×	
Toughness	Texturometer, fibrometer (also: chemical methods for determination of polysaccharides)		×	×	
Color, external	Light reflectance		×		×
	Visual color charts	×			×
Color, internal	Light transmittance, delayed light emission		×		×
Internal structure	Light transmittance, delayed light emission		×		×
	Visual examination	×		×	
Compositional factors:					
Starch content	KI test, other chemical tests		×	×	
Sugar content	Hand refractometer, chemical tests		×	×	
Acid content	Titration, chemical tests		×	×	
Juice content	Extraction, chemical tests		×	×	
Oil content	Extraction, chemical tests		×	×	
Tannin content	Ferric chloride test, chemical tests		×	×	
Internal ethylene	Gas chromatography		×	×	

between changes in the index quantity and the quality and storage life of the commodity has been determined, an index value can be assigned for the minimal acceptable maturity. Theoretically then, once the pattern of change is established for the chosen index quantity, measurements made early in the season can be used to predict the date at which the commodity will reach minimum acceptable maturity.

The way in which this strategy might be applied can be illustrated best by the following three examples.

Apples. Literature relating to the prediction of maturity in apples is voluminous, yet no really satisfactory index of maturity has been proposed. The use of climatic data to predict the date of harvest by a modification of the "days from full bloom" index noted in table 3.1, even when adapted by using "days from the 'T' stage" has provided only general predictions of the harvest date.

In an attempt to provide a more satisfactory prediction of the maturation date, workers have examined a number of changes during fruit development. Measurement of respiration, ethylene production, sugar content, starch content, and changing firmness of the fruit each failed to meet some of the criteria outlined above for a satisfactory maturity index. The "starch pattern," an old method of determining maturity, refined by assigning scores to a range of patterns, proved to be a good index. Changes in the mean starch index score during the period prior to harvest were readily analyzed as a linear regression, and the date of minimal acceptable maturity could be predicted several weeks in advance.

Avocados. The State of California has for many years promulgated a minimum oil content as the maturity standard for avocados. This index has been unsatisfactory, inasmuch as avocados, having more than the minimum oil requirement, may well be lacking in organoleptic qualities. However, raising the minimum oil content might eliminate from the market particular avocado varieties or crops whose organoleptic quality is adequate. Recent work, using taste panel evaluations to determine quality, has shown that the pattern of avocado fruit growth can be used, not only to determine the date at which minimum acceptable maturity is achieved, but also to predict that date. A good correlation between oil content and dry weight also was found. Consequently the California minimum maturity index was changed from oil content to dry weight.

Kiwifruit. In contrast to apples and avocados, very little developmental information is available for kiwifruit. In recent studies, workers measured changes in a wide range of chemical and physical parameters during growth and development. This information was compared to storage and taste panel results to decide on possible methods for determining, and if possible predicting, the minimal acceptable maturity for this crop. When grown under New Zealand conditions, fruit reaches a minimum maturity at 6.25 percent soluble solids. The change in soluble solids in the 6 weeks prior to the normal harvest date can be used, with regression analysis, to predict the date of harvest for different orchards, seasons, and growing districts.

References

1. Arthey, V. D. 1975. *Quality of horticultural products*. New York: Halstead Press, John Wiley and Sons. 228 pp.

2. Hulme, A. C., ed. 1971. *The biochemistry of fruits and their products*. Vol. 2. New York: Academic Press. 788 pp.

3. Lewis, C. E. 1978. The maturity of avocados—a general review. *J. Sci. Food Agric.* 29(10): 857.

4. Ryall, A. L., and W. J. Lipton. 1979. *Handling, transportation and storage of fruits and vegetables*. Vol. 1, *Vegetables and melons*. 2d ed. Westport, CT: AVI Publ. Co. 588 pp.

5. Ryall, A. L., and W. T. Pentzer. 1982. *Handling, transportation and storage of fruits and vegetables*. Vol. 2, *Fruits and tree nuts*. 2d ed. Westport, CT: AVI Publ. Co. 610 pp.

6. Watada, A. E., R. C. Herner, A. A. Kader, R. J. Romani, and G. L. Staby. 1984. Terminology for the description of developmental stages of horticultural crops. *HortScience* 19:20-21.

4
Harvesting Systems

JAMES F. THOMPSON

The goals of harvesting are to gather a commodity from the field at the proper level of maturity, with a minimum of damage and loss, as rapidly as possible, and at a minimum cost. Today, as in the past, these goals are best achieved through hand harvest in most fruit, vegetable, and flower crops. Tables 4.1 and 4.2 list the level of hand harvest in various vegetable, fruit, nut, and berry crops. All flower crops are hand harvested.

Hand Harvest

Primary advantages

1. Humans can accurately select for maturity, allowing accurate grading and multiple harvest
2. Humans can handle fruit with a minimum of damage
3. Rate of harvest can be easily increased by hiring more workers
4. Hand harvest requires a minimum of capital investment (although some farmers provide housing for their employees)

Fig. 4.1. Hand harvest, as shown here for strawberries, is the primary method of harvesting fresh market horticultural commodities in the U.S. Courtesy of Mel Gagnon.

The main problems with hand harvest are centered around labor management. Labor supply is a problem for farmers who cannot offer a long employment season. Labor strikes during the harvest period can be very costly. In recent years, there has been a significant increase in costs associated with complying with government labor regulations.

In spite of these problems, quality is such an important aspect in successful marketing of fresh market commodities that hand harvest is still the dominant method of harvest. In fact, tables 4.1 and 4.2 indicate that many of the crops that are largely machine harvested are used for processing or are crops that are not easily damaged, such as nuts, roots, and tubers.

Effective use of hand labor requires very careful management. New employees must be trained to harvest the fruit quality needed at an acceptable rate. Employees must know what level of performance is expected of them, and be encouraged and trained to reach that level. Benefits such as paid vacations, insurance, and so on, will help

Table 4.1. Level of hand harvest for selected vegetable crops in the U.S.

Acreage hand harvested		Commodity		
%				
76-100	Artichoke	Asparagus	Broccoli*	Cabbage
	Cantaloupe	Cauliflower	Celery	Cucumber*
	Lettuce	Green onion	Collard greens	Cress
	Dandelion	Eggplant	Endive	Escarole
	Fennel	Kale	Kohlrabi	Mushroom*
	Okra	Pepper	Rapini	Rhubarb*
	Romaine	Sorrel	Squash	Watercress
	Cassava	Celeriac	Ginger root	Parsley root
	Parsnip	Rutabaga	Salsify	Turnip
	Taro root*	Jerusalem artichoke		
51-75	Sweet potato	Mustard greens	Parsley	Swiss chard
	Turnip greens			
26-50	Snap bean*	Dry onion	Pumpkin*	Tomato*
0-25	Carrot	White potato*	Lima bean*	Snap bean*
	Sweet corn*	Peas*	Spinach*	Horseradish*
	Red beet*	Boniato	Garlic	Brussels sprouts*
	Malanga	Radish		

*≥50% of crop processed.

ensure a return of employees who have already been trained.

With some commodities, machines have been used to aid hand harvest. Belt conveyors are used in some vegetable crops such as lettuce and melons to move a harvested commodity to a central loading or in-field handling device. Scoops with rods protruding from the end of the scoop are used by workers to comb through some berry crops. Platforms or moveable worker positioners have been used in place of ladders in crops such as dates, papayas, and bananas. Lights have been used to a limited extent for night harvest of melons in California. This allows harvest when outside temperatures are cool, which is easier on workers and can improve melon quality. Numerous other mechanical aids have been tried, but they often do not increase productivity enough to warrant their expense.

Mechanical Harvest

Main advantages of mechanical equipment

1. Potential of rapid harvest
2. Improved conditions for workers
3. Reduced problems associated with hiring and managing hand labor

Effective use of mechanical harvesting equipment requires many skills not associated with hand harvest. The equipment must be operated by dependable, well-trained people. Improper operation of the equipment results in very costly damage to the expensive machinery and quickly damages a lot of commodity. The equipment must be regularly maintained and emergency maintenance must be available. The commodity must be grown to accept mechanical harvest. For example, trees must be pruned for strength and to minimize fruit damage caused by fruit falling through the tree canopy. Maximum and uniform stand establishment is necessary for vegetable crops. Also, cropping patterns must be set up to utilize the expensive equipment as long as possible in order to pay for the high capital investment. (This can severely limit the production choices of some farmers.)

However, mechanical harvest is not presently used for most fresh market crops because machines are rarely capable of selective harvest, they tend to damage the commodity, and they are expensive. Mechanical harvest can be used with commodities which can be harvested at one time and are not sensitive to mechanical injury (roots, tubers, and nuts). Rapid processing after harvest will minimize the effects of mechanical injury. Mechanical harvesting of crops now hand harvested will probably require breeding new varieties that are more suited to mechanical harvest. This is a lengthy process and has been done only for a very few commodities.

Mechanical harvest problems

1. Damage to perennial crops (e.g., damage to bark from a tree shaker)
2. Lack of processing and handling capacity to handle the high rate of harvest
3. Technological obsolescence before equipment is paid for
4. Social impacts of lower labor requirements

The easiest crops to mechanically harvest have had equipment developed for them. Other crops will be mechanized at a slow rate compared to the advancement of mechanization in the past 40 years.

Table 4.2. Level of hand harvest for selected fruit, nut, and berry crops in the U.S.

Acreage hand harvested	Commodity			
%				
76-100	Apple	Apricot	Avocado	Banana
	Breadfruit	Sweet cherry*	Coffee*	Grape*
	Guava*	Kiwi	Kumquat*	Loquat
	Lychee	Mango	Nectarine	Peach*
	Pear*	Persimmon	Pineapple*	Pomegranate
	Quince*	Rosehip*	Wild blueberry*	Currant*
	Gooseberry*	Strawberry	Grapefruit*	Lemon*
	Lime	Orange*	Olive*	Papaya
	Passion fruit*	Tangelo*	Tangerine	Cashew
	Coconut*	Chestnut*	Jojoba*	
51-75	Red raspberry*	Macadamia		
26-50	Prune*	Blackberry*	Highbush blueberry*	Black raspberry*
	Pecan			
0-25	Tart cherry*	Date	Fig	Cranberry*
	Almond	Filbert	Peanut*	Pistachio
	Walnut			

*≥50% of crop processed.

References

1. ASAE. 1983. *Status of harvest mechanization of horticultural crops.* Am. Soc. Agric. Eng., St. Joseph, MI: 78 pp.

2. Grierson, W., and W. C. Wilson. 1983. Influence of mechanical harvesting on citrus quality: Cannery vs. fresh fruit crops. *HortScience* 18(4): 407-09.

3. Kader, A. A. 1983. Influence of harvesting methods on quality of deciduous tree fruits. *HortScience* 18(4): 409-11.

4. Kasmire, R. F. 1983. Influence of mechanical harvesting on quality of nonfruit vegetables. *HortScience* 18(4): 421-23.

5. Morris, J. R. 1983. Influence of mechanical harvesting on quality of small fruits and grapes. *HortScience* 18(4): 412-17.

6. O'Brien, M., B. F. Cargill, and R. B. Fridley. 1983. *Principles and practices for harvesting and handling of fruits and nuts.* Westport, CT: AVI Publ. Co. 636 pp.

7. Studer, H. E. 1983. Influence of mechanical harvesting on the quality of fruit vegetables. *HortScience* 18(4): 417-21.

5
Preparation for Fresh Market
I. Fruits

F. GORDON MITCHELL

Protection of fresh fruits must begin with cultural practices in the field and continue until fruits are consumed. Deterioration can result from improper pruning, thinning, fertilization, disease control, and so on, during production. Many deterioration problems are the result of cumulative insults to the fruits throughout the postharvest handling period. Thus, protection is vital, both in the field and the packinghouse to avoid immediate deterioration causes and to delay the onset of deterioration later in the distribution channels.

Table 5.1 presents data on accumulated impact bruising throughout a Bartlett pear handling operation in California. While these results represent an extreme of what might be experienced commercially, they do show the serious consequences of repeated insults to the fruit at all stages of handling.

Harvesting

Field containers

Most fresh market fruits are now harvested by hand into buckets or bags, which are then emptied into field bins for subsequent handling. Metal or plastic picking buckets are typically used for the softer fruits, while bottom-dump picking bags are used for fruits with a lower compression bruise potential. Certain delicate fruits are still transferred from buckets to field lugs (many sweet cherries), picked directly into field lugs (table grapes), or picked into buckets and packed into shipping containers directly from the bucket (some stone fruits). Very soft, delicate fruits may be harvested, sorted, and packed directly into the shipping container by the picker (strawberries, bushberries). Despite extensive past research on mechanical harvesting, there is presently no mechanical harvest of fresh market fruits in California.

In California, field bins are generally standardized at 120 cm by 120 cm (47 inch by 47 inch) outside horizontal dimensions and 61 cm (24 inch) inside depth. While most fruits will tolerate this depth, some sweet cherries are handled in 30.5 cm (12 inch) deep bins to avoid bruising. Bins are typically of 1.9 cm (3/4 inch) plywood, smooth or coated on the inside, and vented on sides and bottom to facilitate air circulation. To avoid fruit cutting, ventilation slots are normally routed so that inside edges are tapered. Ventilation requirements for cooling are discussed in chapter 7.

Bin surfaces should be clean and smooth to avoid fruit abrasion injury. Frequent washing, water dumping, or hydrocooling can cause surface roughness of the plywood bins and increase fruit abrasion problems. Coatings (paint or varnish-type) are available to reduce this problem. Plastic-coated plywood, while expensive, is sometimes warranted for such applications. Separate plastic liners are occasionally installed within the bin, but are difficult where side venting is required for cooling, or where water dumping is practiced.

Careful field supervision is the most critical factor in protecting fruits from injury. Physical injuries can result from improper picking procedures, excessive dropping of fruit into buckets or bags, overfilling of these containers, hitting picking containers (especially soft-sided bags) against limbs and ladders, lack of care in transferring fruits into field bins or lugs, and overfilling of these field containers. Avoiding these problems requires constant, visible, strict supervision. The importance of even short drop heights in causing impact bruising is demonstrated in the data in table 5.2 for Bartlett pears.

Transport from the field

There are many opportunities for fruit bruising to occur during field transport. Impact bruises can occur if bins or lugs are dropped or bounced. Compression bruises generally result from stacking of overfilled containers. Depth restrictions, even if bins are underfilled, may be needed to avoid compressing the bottom layers of some soft fruits.

Table 5.1. Cumulative levels of impact bruising on Bartlett pears during postharvest handling

Location	Bruised fruits (%)
Tree	0
Picker bag	14
Field bin	26
After dump	38
After size	82

Abrasion or vibration bruises may occur when fruits move or vibrate against rough surfaces or other fruits during transport.

Supervision is needed at all stages of field transport to minimize accumulation of physical injuries by the fruit. Despite good supervision to avoid injuries during loading, considerable damage may occur during truck transport. Steps that can be taken to avoid such problems include the following:

1. Avoid extended forklift movement of bins through the field from point of harvest to loading site.
2. Supervise truck or trailer loading to avoid rough handling or dropping of bins or lugs.
3. Grade farm roads to eliminate ruts, potholes, and bumps.
4. Where necessary route truck movement to avoid public roads that are in poor condition.
5. Restrict transport speeds to a level that will avoid free movement of fruits. This may require different speed limits on different roads.
6. Reduce tire air pressure on transport vehicles to reduce motion transmittal to the fruit.
7. Use suspension systems on all transport equipment. When purchasing new transport equipment, consider air suspension in place of spring suspension. Tests have shown more than 50 percent reduction in damaging motion with air suspension systems.
8. Evaluate container surfaces to determine whether cleaning, sanding, coating, or lining would reduce injury. Under difficult conditions, consider use of plastic-coated plywood bins or bin liners.

Temperature protection

The effects of delays at high temperatures on field deterioration are fully discussed elsewhere. Temperature protection in the field involves shading the harvested fruit to minimize warming (and sunscald), and prompt handling and cooling to minimize high temperature exposure. Various studies have shown that even a mild breeze will cause harvested fruit in the shade to quickly warm to near the ambient air temperature. Fruit in the sun can warm many degrees above air temperature, depending on surface color (fig. 5.1).

Temperature protection in the field should start with shading of the fruit after harvest. This may mean moving harvested fruit (bins, lugs, and so on) to the shade of trees or vines while awaiting transport. If natural shade is not available then portable shading may be needed. Inverting empty containers over the top of stacks of containers can provide some field protection. During periods of high field temperatures it may be desirable to avoid harvesting in mid-day.

During transport, handling speed is usually the only protection that is provided. Frequent scheduling of fruit transfer from the field to the cooler or packinghouse will minimize the opportunity for fruit heating and deterioration. Where transport time is extended because of distance or unavoidable delays, the covering of the load may be advantageous. Fruits exposed to the sun will warm quickly whether in the field or on top of a loaded truck. Further, air flow through the load during transport can quickly warm the fruit to the ambient air temperature. If transport of cool, early-harvested fruit is delayed until high temperatures are encountered, then the resulting warming will speed fruit deterioration and increase the cost of subsequent cooling.

If a tarpaulin is used to cover loads of fruit during transport, it should be a light color (white or silver are good choices) and should be kept clean in order to maintain good heat reflectance. It should be supported so as to maintain an air space over the load. If the tarpaulin extends down all sides of the load it will block air flow through the load and limit fruit warming during periods of high ambient air temperatures. Wetting of the tarpaulin will further reduce warming by providing an evaporative cooling surface. Care must be taken to ensure that a tarpaulin cover does not confine heat within the load and thus accelerate warming of the fruit. Under difficult transport conditions, with water tolerant fruits, top ice has been placed over fruit in field bins prior to tarping and transport.

Table 5.2. Effect of drop height on incidence and severity of impact bruising on Bartlett pears

Drop height (inch)	Fruit bruised (%)	Bruise severity score*
0	0	0
4	40	0.6
6	44	0.6
9	56	1.0
12	78	1.2
16	100	2.3

*Scored on 0-5 scale: 0 = no damage; 5 = unmarketable.

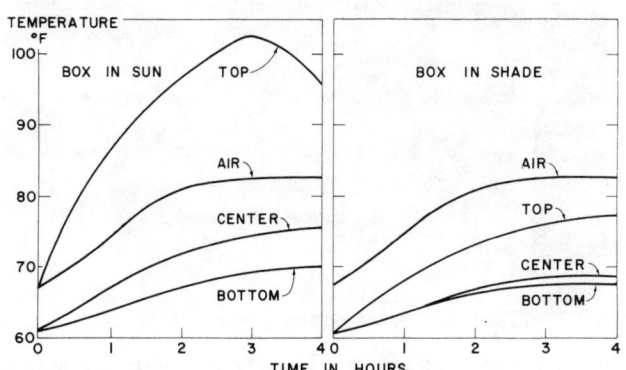

Fig. 5.1. Effect of sun exposure and position in box on field warming of sweet cherries.

Preparation for Packing

Delivery to the packer

Most fruit is dumped onto a grading and packing belt. Historically, this has been accomplished by hand dumping of small field lugs. With large scale operations this lug dumping was mechanized to provide uniform fruit flow and to reduce supervision problems. Certain commodities that are easily injured will not tolerate dumping. Table grapes are packed directly into the shipping container from field lugs.

Most fruits are now commonly dumped onto packing lines from field bins through the use of either dry dumps or water dumps. Dry dumps (fig. 5.2) are normally designed so that the bin is covered with a padded lid, then slowly inverted, and the fruit delivered through a controlled opening in the lid. Electronically-controlled delivery belts can adjust the flow of fruit to the sorting line. Properly designed dry dumps are capable of delivering a uniform product flow with minimum injury to the fruit.

Water dumps are of several types. Some dump fruits from field bins directly into water (no longer common). Others submerge the bins and float the fruits free (fig. 5.3). The most common floatation dumps submerge the entire bin as it travels along a conveyor. Pumps circulate the dump-tank water to move the free-floating fruits to an elevator. There fruits are rinsed and transferred to the sorting line. For floatation of fruits with a specific density greater than that of water, a salt (often sodium sulfate) must be added to the dump water.

Sanitation is important in water dumps. Dump water will quickly accumulate a high concentration of fungal spores, which can inoculate harvesting and handling wounds on the fruit. The dump tanks should be designed for rapid emptying and filling and for easy cleaning. Chlorine at a concentration of 50 to 200 ppm is often maintained in the dump tank water as a fungistat, although the chlorine toxicity tolerance of the fruit species and variety must be known.

Sorting line

Efficient sorting is dependent upon careful attention to a series of specific equipment and supervision requirements (fig. 5.4). Lack of adequate sorting space is probably the most limiting factor in new fruit packing facilities. This has become especially critical as volume-fill packing techniques have been widely adopted. While limited use of electronic color sorting has occurred, most separations must still be done by hand. Sorting requirements include:

Space for adequate sorting. This cannot be designated only in terms of packed containers per hour. Space requirements also depend on percent of diverted fruit (subgrade, alternate grade, and so on), number of decisions or separations required (color, shape, various defects), and relative size of fruit being sorted (number of decisions per package).

Ability to adjust flow of fruit. A supervisor must have instant control of the sorting belt speed to adjust for variabilities in fruit quality, size, and so on.

Fig. 5.3. Water submersion bin dumps are sometimes used, especially for apples and pears. Filled bins are carried under water to allow fruit to float free. The empty bin then is elevated out of the tank. For fruits with a density greater than that of water, such as pears, water density is increased by adding an appropriate salt.

Fig. 5.2. Inversion type dry bin dumps are widely used in fruit packing lines. The padded lid is clamped over the bin before inversion, then the gate in the lid is opened to obtain fruit flow during dumping.

Assignment of responsibility. Each worker must have a specific task or responsibility. This usually involves a specific area, zone, or lane of the sorting belt. When multiple decisions are made, it should also involve separating responsibilities for these decisions. It may be desirable to periodically rotate worker position on the sorting line to reduce monotony and fatigue.

Ability to view the product. Workers must have complete view of the entire surface of the product for efficient sorting. Avoid designs in which part of the belt may be hidden from the worker's view. Consider various systems to alter fruit position as the fruit progresses along the sorting belt. Slow rotation or periodic movement are desirable. Systems which both turn and rotate the fruit provide the greatest surface visibility. Adequate lighting of the product is essential.

Worker comfort requirements. Remember that workers are expected to operate at peak efficiency during long working hours. This cannot be achieved without provision for reasonable worker comfort. Platforms and stools adjusted to the proper work height for each worker are essential. Design of sorting and removal belts and chutes to eliminate unnecessary reaching and stretching are also important. Worker fatigue can also be influenced by proper lighting and noise control.

Avoidance of product injury. A sorting line that causes fruit injury is self-defeating. Thus, the delivery system, the sorting belt, and the distribution system must be designed to avoid injuries. Careful attention must be given to minimizing the height and number of drops and shears in the system. Activated sorting belts, which may aid in worker viewing of the fruit, must be carefully adjusted to avoid fruit injury. Thorough periodic cleaning of the facility to eliminate accumulated dirt will aid in reducing fruit injury. Fruits should flow along the belts only one layer deep.

Worker training and supervision. No sorting system will be effective if workers are not well-trained and supervised. Workers must have clearly specified responsibilities and must be familiar with defects and with segregation categories and limits. Posting of visual aids may help in worker training. Supervisors must be familiar with worker performance limits and be able to identify both "undersorting" and "oversorting." Supervisors must also be able to adjust product flow to stay within the limits of worker performance.

Fruit sizing

Fruit sizers are available in a wide range of designs, but they all segregate size by either weight or dimension (fig. 5.5). Dimension sizers may measure the fruit by either 2, 3, or 4 point contact. Sizers can work efficiently only if the delivery system is properly designed and adjusted to deliver a uniform flow of fruit across the full width of the sizer.

There are three important requirements to consider in selecting a sizer:

Capacity. The sizer must be capable of meeting the volume requirements of the packing operation. The practical commercial capacity of a sizer is estimated to be about two-thirds of the theoretical rated capacity. Since sizer capacity is related to fruit numbers, determine the capacity requirement based upon the smallest anticipated average fruit size to be handled.

Accuracy. The sizer must be capable of segregating fruits within the required accuracy. This may be to meet uniformity requirements for a place pack, for size segregation to meet marketing requirements, or simply to meet legal sizing requirements.

Injury. Sizers must be able to meet the performance requirement without causing injury to the fruits, including all varieties and maturities that could be commercially encountered.

Fig. 5.4. Adequate facilities for fruit sorting are essential. This sorting table limits worker reach. The primary separation is at sorting table height; fruits are rotated horizontally as they pass down the belt allowing good sorter viewing of all fruits.

Other special features may be desirable in selecting a sizer, provided the three basic requirements are met. These special features may include: (1) ease of adjustment as incoming fruit sizes change, (2) ability to adjust fruit diversion patterns as peak sizes change, and (3) ease of equipment cleaning and maintenance.

Special treatments

Depending upon the commodity to be packed, there may be other special treatments required. These may include:

Pre-sizing. Pre-sizers are usually located immediately after the dump and designed to eliminate all fruits below a minimum acceptable size. This reduces product flow over the balance of the packing line, and thus increases overall equipment capacities. Pre-sizers are of many styles and have the same basic requirements as sizers except that they are required to make only a single segregation.

Cleaning and washing. Fruits may need cleaning and washing to remove soil or contamination, or sometimes to remove natural waxes in preparation for wax applications. Detergent washes are sometimes used with soft brushes or sponges followed by clear water rinsing. Many peaches receive wet brushing to remove the trichomes (peach fuzz).

Waxing. Some fruits are waxed (fig. 5.6) as part of the packing operation. Waxes may be used to reduce water loss, to replace natural waxes removed during washing, to cover injuries such as those caused by peach defuzzing, to act as carriers for fungicides, or simply to improve the fruit's cosmetic appearance. Waxes must be approved "food grade" materials. Past studies indicate that they should not reduce the rate of water loss by more than about one-third because of the danger of interference with the gas exchange needed to maintain normal aerobic respiratory activity of the fruit.

Disease control. Some postharvest disease treatments may be applied during packing. Heat treatment, especially hot water treatment, has been studied for many types of fruits. It is widely used for papayas, typically prior to or

Fig. 5.5. (a) Weight sizers will segregate fruits according to their mass. Fruits are distributed into individual cups, which carry them forward until fruit mass exceeds the mass adjustment for the cup position. At that point fruits drop onto a cross-flow belt for transport to the packing station.

(b) Dimension sizers are of many designs—this one is a 'volumetric sizer'. Fruits are in contact with active rolls at four points. As fruits move forward, the rolls separate to provide a continuously expanding area for the turning fruits to pass through and drop onto the cross-flow belt.

at the start of the packing operation. Fungicide applications when needed are commonly applied while the fruit is spread on the conveyor belts, often immediately following washing. Fungicides are often incorporated into fruit waxes to aid in achieving a uniform surface application.

Packing the Fruit

Purposes of packing

Packing can be viewed as simply a convenience in achieving orderly marketing of fresh fruits. The shipping container provides a convenient unit to facilitate the transfer of the product from the point of production to the point of final sale or consumption. If it is to function well, the container must be designed and used in a manner that will protect individual fruits.

There are three important requirements in packing fresh fruits so as to protect them from deterioration during subsequent handling and distribution.

Fruits must be immobilized within the container. Without immobilization, fruits can become injured as a result of movement during transit. Immobilization can be accomplished by wrapping and place packing of carefully sized fruits, by placing in various types of trays, or by certain volume-fill techniques (described later). Proper design of containers and padding are essential to achieving immobilization of the fruits.

Fruits must be cushioned against impacts. Impact bruises can occur during packing as a result of unnecessary drops of the fruits. After packing, impact bruises can occur as individual packages are dropped during handling. The shift to unitized handling (usually pallet-size units) has greatly reduced rehandling of individual containers, and thus helped to reduce impact bruising. Various cushion pad materials, used as bottom pads and/or between layers of fruits, have been effective in absorbing impacts and reducing fruit bruising.

Fruits must be protected from compression. Compression bruising can result from compression of overfilled containers, either during lidding or after stacking. The packer may also cause compression bruising by pattern packing oversized fruit. Much compression bruising results from container failure after packing, a situation in which the fruit, rather than the container, assumes the stacking stresses. Thus, proper container design and specifications, selected for the particular fruit packing application, are vital in preventing fruit compression bruising (see chap. 6, Packages for Horticultural Crops).

Packing line

Fruit protection must continue as the product moves across the packing line. Thus, the packing line must be designed to minimize opportunities for fruit injuries. This includes the elimination of shears and drops whenever possible, and the incorporation of live shears (often moving belts or air jets) and padding where needed. Careful control of fruit flow, whether to packing tubs or stations or on return-flow packing belts, is needed to avoid unnecessary fruit accumulation.

For certain packing procedures, it may be desirable to design the packing line to accommodate limited resorting of fruit prior to packing. This can be especially important in volume-fill packing systems, where a large volume of fruit delivered to machine fillers taxes the system's sorting capacity. Because the fruit has been previously sized and graded, this final sorting might be more properly considered a quality control procedure—assuring that the desired fruit quality standards are uniformly met.

Proper delivery of empty containers and packaging materials to the packing stations, and adequate facilities for removal of packed containers are as important to the packer (or the operation of filling machines) as is adequate fruit delivery. Various mechanical and hand supply systems are possible, depending on the type and volume of the operation. Careful attention must be given to efficient design of this facility, however, to avoid bottlenecks in the packing operation.

Normally, fruit leaving the packing stations will be inspected prior to final padding and lidding. This inspection is a final quality control procedure prior to completion of the pack and is to assure that both fruit and packing procedures meet the packinghouse grade and quality requirements. In some larger operations, continuous third-party grading is used, with the inspector also sampling at this point to assure compliance with legal grades and standards. At this location containers are usually stamped with necessary information on variety, size, grade, and so on.

Hand-packing operations

Fruits are hand packed to create an attractive pack, often to pack a fixed count, sometimes to select for size, and always to immobilize the fruit within the shipping container

Fig. 5.6. Some fruits are washed and waxed during the preparation for market operation. Here a cold-water-emulsion wax, often containing fungicides, is applied to the fruit after washing, and allowed to dry without added heat.

(fig. 5.7). This requires fairly precise sizing of fruits, at least within a single layer of the container. Immobilization usually requires packing to lateral tightness within the container. In tray packs, the presence of an oversized fruit may prevent top pads or trays from contacting surrounding smaller fruits, which may then be subject to motion injury during transport. Similarly, an undersized fruit may prevent lateral tightness and allow surrounding fruits to turn and be injured.

Packaging materials, serving to isolate and/or immobilize the fruit, are often as important as pattern packing in preventing damage to fruits within the container. Depending on the particular packing procedure, such packaging materials may include trays, cups, wraps, shims, liners, or pads. They often add significantly to the total packaging material costs.

Hand-packing facilities vary from *return-flow* belts to a variety of tubs, bins, and chutes. The return-flow belt is normally used without mechanical sizing, with packers selecting fruits of the desired sizes from the range of sizes on the belt. Tubs, bins, and so on, are filled with pre-sized fruit from which packers can draw at random.

Older hand-pack systems simply positioned the packer and packing station within easy reach of the fruit. A modification for nonwrapped packs is the *rapid-pack* system where the work area is designed so the worker faces the fruit for easy two-hand access. A more recent innovation for tray packing is an *automated tray-pack* where the tray passes on a belt just below the fruit delivery chute. By careful regulation of the speed of the tray belt, the operator can cause fruits to fill most cups in the tray. Workers are needed to orient fruits, fill voids, remove excessive fruits, and place the trays into shipping containers.

Mechanical packs

Mechanical packing systems (volume-fill/tight-fill) are increasing in use for fruits in California. In these systems carefully sorted and sized fruits are delivered, along with empty containers, to automatic fillers. After filling, the containers pass through standard inspection, marking and closing operations, and may receive special top padding, vibration settling, and lid fastening (tight-fill). A mechanical place-packing system is available for citrus packing.

Mechanical packing systems normally handle large volumes of fruit at high speed, and afford no opportunity for further grading at the filling chute. It is therefore mandatory for the arriving fruit to have been adequately sorted and sized to assure that it meets the desired grade.

Mechanical packers should be adjusted to properly fill the volume of the shipping container. Most fillers are designed to use weight as an estimate of volume (fig. 5.8). Many fillers will fill to within a few fruits of the desired weight, but final adjustment must be done by hand as the containers pass over scales. Some fillers are designed to adjust the fill weight to the nearest fruit, so that only check weighing is then necessary.

Fig. 5.7. The face and fill pack being used for apricots is one type of hand packing for fruit. Fruits are selected for size uniformity and tightly packed in rows in the inverted lug. The remainder of the package is volume-filled with similar sized fruit before the lug bottom is nailed in place.

Fig. 5.8 Automatic volume-filling is widely used for fruit packing. Here the empty container is tilted up to the filling belt and is settled to horizontal position during filling with pre-sized fruits. The scale is adjusted to the desired fill weight.

The most frequent problem in new volume-filling installations is that the flow of fruits of certain peak sizes will exceed filler capacity, while some other sizes will flow at very low volumes. To compensate for this, one or two extra fillers may be installed so that filling capacity of peak size fruits may be doubled. Because peak sizes will vary with the fruit, variety, and orchard, the delivery system to the fillers must be easily adjustable.

A major problem with most mechanical fillers is the height of drop of fruit into the package. This can be minimized by proper design of the filling equipment. Modifications such as the tilting of containers during filling to reduce fruit drop heights are common. Padding of filler chutes and use of decelerator curtains may reduce impacts. Cushion pads placed in the bottom of the containers during filling can further reduce fruit impact injury during filling.

Tight-fill packing (fig. 5.9) is a special modification of volume-filling that is designed to assure fruit immobilization during transport. Tight-fill packing involves volume-filling the fruit into the container to an exact weight, vibration settling (with carefully controlled vibration) to eliminate voids that were left during filling, top padding with a special pad that will nest around the top fruits, and tightly fastening the lid to the body of the container. When all of these steps are properly performed, and the container is designed for the packing system, the fruits will be held tightly in place without compression bruising. There are specific requirements for container and pad design, fill density, and vibration characteristics that might be met for successful tight-fill packing. The method is described in University of California Circular 548 (6).

Fig. 5.9. Tight-fill packing unit. Machine vibrates fruit into place, seals container flaps, and fastens lid to body of container.

References

1. Gentry, J. P., F. G. Mitchell, and N. F. Sommer. 1965. Engineering and quality aspects of deciduous fruit packed by volume filling and hand placing methods. *Trans. Am. Soc. Agric. Eng.* 8:584-89.

2. Grierson, W., W. M. Miller, and W. F. Wardowski. 1978. *Packing-line machinery for Florida citrus packinghouses*. Univ. Fla. Bull. 803, 30 pp.

3. Hardenburg, R.E. 1967. *Wax and related coatings for horticultural products*. A bibliography. USDA, ARS 51-15.

4. LaRue, J. H., and F. G. Mitchell. 1964. Bulk handling of shipping fruits. *Calif. Agric.* 18(6): 6-7.

5. Mitchell, F. G., J. H. LaRue, J. P. Gentry, and M. H. Gerdts. 1963. Packing nectarines to reduce shrivel. *Calif. Agric.* 17(5): 10-11.

6. Mitchell, F. G., N. F. Sommer, J. P. Gentry, R. Guillou, and G. Mayer. 1968. *Tight-fill fruit packing*. Univ. of Calif. Agric. Exp. St. Ext. Ser. Circ. 548. 24 pp.

7. O'Brien, M., B. F. Cargill, and R. B. Fridley, eds. 1983. *Principles and practices for harvesting and handling fruits and nuts*. Westport, CT: AVI Publ. Co. 636 pp.

8. Smith, R. J. 1963. *The rapid pack method of packing fruit*. Univ. of Calif. Agric. Exp. St. Circ. 521. 20 pp.

5

Preparation for Fresh Market—*Continued*

II. Vegetables

Robert F. Kasmire

Operations

Preparation for marketing begins with harvesting (maturity selection and some quality control) and, depending upon the commodity, may include some or all of the following subsequent operations: product assembly, receiving, cleaning, trimming, sorting, grading, sizing, waxing, packaging, packing, cooling, ripening initiation, curing, storage, unitizing, and shipping. Preparation may be done mostly in the field (lettuce, celery) or in packing sheds (most vegetables), or both (broccoli, cauliflower). Vegetables may be prepared entirely at shipping point, partly at shipping point (most commodities) and partly at destination market (tomatoes, potatoes), or mostly at destination market (potatoes and other commodities shipped to prepackers in destination markets). Modern equipment and facilities used are expensive; therefore, an attempt is made to extend the shipping season for as long as possible. Equipment for some commodities (lettuce, celery) is portable, or mobile, and is used in several shipping districts each year either by the same grower-shipper or leased to others. This amortizes the costs of equipment over a longer period each year.

Harvesting

Maturity selection (at harvest). Select products of desired maturity for the intended markets. Harvest maturity influences susceptibility to handling damage, ripening required (if any), shelf-life, size, length of cooling cycles required, and market availability.

Product assembly. Assemble for field packing or for hauling to packinghouses. May include limited sorting and grading (for products to be hauled to packinghouses) or final grading (for field packed vegetables). Also includes assembling loads of field containers (bulk bins, gondolas, field boxes) at packinghouses for subsequent operations.

Packinghouse operations

Receiving. Unload products from loaded field containers onto a conveyor or into a water dump for conveying to packinghouse. May include a small-size eliminator and some sorting for removing decaying products. Dry receiving (dumping) operations often cause considerable product damage.

Cleaning. Remove soil and other foreign material from product surfaces by washing, brushing, or (sometimes) by both. The wash water may or may not be chlorinated. Recycled water should always be chlorinated.

Trimming. Remove unwanted leaves, stems, or roots prior to grading, packaging, and packing (lettuce, celery, cauliflower, asparagus, dry onions).

Grading. Separate products by market quality (grades) either before or after sorting.

Sorting. Select product by maturity, shape, color, or some other physical parameter. Some commodities are machine sorted. Culling is part of both sorting and grading.

Curing. Some products are cured (garlic, sweet potatoes, dry onions, and new crop Irish potatoes) after harvesting and prior to storage or marketing. Onions and garlic are cured to dry the necks and outer scales. Sweet potatoes and Irish potatoes are cured to develop wound periderms over cut, broken, or skinned surfaces. Curing helps heal harvesting injuries, reduces water loss, and prevents entry of decay-causing organisms during storage. Curing may be done in the field (garlic, onions), in curing rooms (sweet potatoes), or during transit (new crop Irish potatoes).

Sizing. Separate product units into physical sizes (weight, volume, length, diameter, or other parameters). Most commodities are sized by mechanical or electronic sizers, but many products are still manually sized.

Waxing. Cover surfaces of product units with a protective coating of food-grade wax to reduce water loss through epidermal openings. Waxes generally are applied only to fruit-type vegetables but may also be applied to roots (rutabagas). A fungicide may be incorporated into the wax.

Packaging. Enclose consumer units of a product in individual packages (wraps, bags, sleeves, trays, or other units) that are subsequently packed in master containers. Most materials used for consumer unit packaging are flexible laminates (plastic films comprised of two or more types of film material combined into a single film). They may be used separately or as wraps over molded polystyrene or pulp trays. Paper bags are also used. Product units in a consumer unit package should be of comparable weight, size, maturity, and grade. Some packaging involves enclosing a single product unit (a head of lettuce or cauliflower), while in other packaging several product units are enclosed in a single consumer unit (Irish potatoes, radishes, brussels sprouts, carrots). Packaging is done both automatically and manually, at both shipping point and/or destination market.

Packing. Assemble a given quantity (count or weight) of comparably sized product units or consumer unit in shipping containers. When counts are used, products are often packed in specific arrangements within containers. Counts, arrangements, and weights are often specified in, and regulated by, various government and industry codes or tariffs. Shipping containers may be bags, cartons, crates, lugs, or bulk bins. Some products are shipped unpacked to markets in bulk trucks or railroad cars.

Unitizing. Assemble packed shipping containers of products into larger units (pallet loads or slip-sheet loads) for better, less expensive handling, for load stabilization, and improved transit temperature management.

Ripening initiation. Apply ethylene or ethylene-producing materials to stimulate ripening of tomatoes and honeydew melons.

Cooling. Remove product sensible heat prior to shipping or long-term storage by cooling. Lowering product temperatures (cooling) extends product storage and shelf-life.

Loading. Load into transit vehicles for shipment.

Shipment

Transportation of products. Transport to markets in commercial or privately owned conveyances (trucks, rail cars, container vans, ships, and airplanes).

Storage

Products are held for long periods between harvesting and marketing so as to fulfill the market demand. Primarily Irish potatoes, garlic, dry onions, winter squashes, and sweet potatoes are stored in California. Carrots, celery, and cabbage and most root crops are also stored in other producing regions. Storage is generally in specially constructed storage facilities. However, storing may also be in the ground (undisturbed plant beds) prior to harvesting, but after the crop is matured (e.g., Irish potatoes in some southern California districts and dry onions). Products may be stored in bulk (usually) and prior to packing-shed operations, or in packed shipping containers. Some fall-season muskmelons are stored for a few weeks to enable shippers to capitalize on higher-priced markets later during that season. Storage involves economic risks but also provides a means of postponing shipping products to already glutted markets.

Problems

Most problems involved in preparing fresh vegetables for market involve communications among and with workers or are related to equipment or materials used. The more important types of problems confronted are the following: (1) lack of adequate training and supervision of workers; (2) lack of understanding by handlers of their roles in total product quality maintenance, and of their decisions on subsequent product quality loss; (3) lack of adequate communication among supervisors and among handlers in marketing and distribution; (4) rough handling when operations are conducted too fast in order to reduce labor costs; and (5) unsatisfactory, improper, or inadequate equipment or materials.

Reference

1. Ryall, A. L., and W. J. Lipton. 1979. *Handling, transportation, and storage of fruits and vegetables*. Vol. 1, *Vegetables and melons*, 2nd ed. 43-117. Westport, CT: AVI Publ. Co.

5

Preparation for Fresh Market—*Continued*

III. Automation Trends in Packinghouse Operations

P. Chen

In recent years there has been a marked increase towards automating packinghouse operations. Some reasons and factors that caused the change are hereby listed.

Packinghouse Automation

Reasons for automation and advantages

1. **Increase productivity**

    Higher product flow rate

    Better coordination of different operations to permit a system to operate at nearly full capacity most of the time

2. **Improve quality of products**

    More accurate sizing and quality grading

    Product units can be packed into more categories

    Less handling of products, reducing mechanical damage

3. **Reduce cost**

    More efficient use of hand labor

    Fewer conveying lines, resulting in a more compact system

    Less culls as a result of more accurate sizing and/or sorting

4. **Facilitate accountability**

    Accountability of incoming fruits by producer's identification

    Accountability of packouts—numbers of fruits, weight, size, maturity, grade, number, and type of packages

    Accountability of stored fruits

    Accountability of buying and selling prices, hours and wages of workers, and packing efficiency

5. **Facilitate easy control of operations**

    Control of packing rate

    Control of packout categories—size, grade, container

    Control of operations—receiving, fruit flow, water temperature, chemical concentration, and palletizing

    Instant feedback and instant change in operations to optimize operations and to match demand-supply conditions

Factors facilitating automation

Advancement of computer technology. In the past 10 years computer technology has advanced at an incredible rate. The size and price of computer components have decreased to less than 1 percent of what they were 10 years ago while the capacity and reliability have increased tremendously. As a result a large selection of small, low-cost, high-capacity computers are readily available for use in packinghouses.

Improvement of electronic and sensing components. Electronic components are also getting better, cheaper, smaller, and more reliable. The availability of low-cost, high-quality sensing elements, such as tranducers and photo-electric cells (electric eyes), makes it possible to measure and detect such parameters as temperature, weight, color, and surface blemishes at an extremely high speed and to transform the parameter values into electric signals that can be fed into the computer for further processing.

Increased knowledge of physical properties of agricultural products. The responses of agricultural products to different kinds of inputs are important to the design of automated systems. The following are examples of some physical properties and their applications to packinghouse operations.

1. **Physical**

    Parameters—size, shape, volume, weight, density, and surface area

    Application—singulation, orientation, sizing, packing, quality evaluation, waxing, and coloring

2. **Mechanical**

    Properties—force-deformation, firmness, response to impact, vibration, and stress-strain

Application—fruit handling, cause and prevention of damage, quality evaluation

3. **Optical and radiation**

   Related properties—light reflectance, transmittance, absorption and emission, X-ray and gamma-ray absorption

   Application—color sorting, blemish detection, maturity evaluation, composition analysis, and other quality evaluations, sizing, counting

4. **Vibrational and sonic**

   Properties—vibration of individual fruit, vibration of fruits in containers, natural frequency, sonic and ultrasonic transmission

   Application—quality evaluation, material sorting, and prevention of damage

Present automated systems

There are several automated fruit packing systems presently available. Basically each system consists of an electronic sizing unit, a quality evaluating unit, and a computer, which analyzes the size and quality information (data) and determines the destination of fruits as dictated by a program, which can be easily changed at any instant. The computer also keeps track of each fruit and accounts for its size, weight, grade, and destination, and finally prints out the packout sheet. In addition to the electronic-computer system, some systems also have mechanical devices for automated operations, such as bin dumping, chemical mixing, tray packing, bagging, box filling, and palletizing.

Future Trends

1. **Broader application of computer and electronic devices**

   Integration of all operations via the computer

   Control of each operation and optimization of the entire system

   Analysis of supply-demand conditions and optimization of packout categories

2. **Increase in efficiency**

   Higher speed—increased production

   Use of scanning and multi-spots viewing equipment

   Elimination of singulation requirement—fruit can go through grading devices in batch instead of one at a time

3. **Improved techniques for quality evaluation**

   Improved accuracy in evaluation of quality factors

   Inclusion of more quality factors—bruise, external and internal defects, sugar-acid ratio, and overall quality

4. **Improved fruit handling system to reduce damage**

References

1. Birth, G. S. 1979. Radiometric measurement of food quality—a review. *J. Food Sci.* 44:949-53, 957.
2. Chen, P. 1978. Use of optical properties of food materials in quality evaluation and materials sorting. *J. Food Process Eng.* 2(4):307-22.
3. Finney, E. E., Jr. 1973. *Measurement techniques for quality control of agricultural products*. Am. Soc. Agric. Eng., St. Joseph, MI. 53 pp.
4. ———. 1978. Engineering techniques for nondestructive quality evaluation of agricultural products. *J. Food Protection* 41(1): 57-62.
5. Foraker, J. D. 1980. Computer age apple packing. *Fruit Grower*, March.
6. Gaffney, J. J., comp. 1976. *Quality detection in food*. ASAE Publ. 1-76, Am. Soc. Agric. Eng., St. Joseph, MI. 241 pp.
7. McClure, W. F., and R. P. Rohrbach. 1978. Asynchronous sensing for sorting small fruit. *Agric. Eng.* 59(6): 13-14.

Personal correspondence

1. Brown, Ian. Decco Tiltbelt. P. O. Box 98, Three Rivers, CA, 93271.
2. Johnson, Maurice. Sunkist Growers. 760 E. Sunkist St., Ontario, CA, 91761.
3. Thornton, Wayne. FMC Corp. P.O. Box 219, Lindsay, CA, 93247.

5

Preparation for Fresh Market—*Continued*

IV. Cull Utilization

JAMES F. THOMPSON

Packinghouses in California reject as culls about 10 to 15 percent of the fruit and vegetables that are delivered. These rejects are culled because of scars, split pits, deformities, mechanical injuries, sunburn or sunscald, mold or insect damage, immaturity, overripeness, softness, or small size. These defects result in a huge quantity of material that must be utilized or disposed of in some manner.

Disposition of Culls

The most obvious approach to reducing the magnitude of this problem is to reduce the quantity of culls that reach the packinghouse. The farmer should use the best cultural practices to produce a well-sized, unblemished commodity. It should be harvested and handled with care to minimize injury and harvesting personnel should be encouraged to discard poor quality fruit and vegetables in the field.

It is also possible to reduce the quantity of culls by lowering the quality standards. However, this often results in the poorer quality fruit and vegetables being culled by the retail distributer or the consumer. The added costs of handling and shipping poor quality fruit and vegetables will increase the cost of good quality produce. A packer shipping high quality fruits and vegetables has a competitive advantage.

However, even careful attention to reducing the quantity of culls will not entirely eliminate the need for management of culls at the packinghouse. Culls are a cost liability to the packinghouse operator. The cost can be minimized if culls are sold as a by-product rather than just returned to the field and dumped. In California, the largest use of cull fruit (about 20% of the total produced) is as a cattle feed.

Cattle feed

Cull fruit is very palatable and a good source of energy for animals but is low in protein and has other characteristics that make it different from other feed sources. For example, stone fruit (peaches, plums, nectarines) contain 85 percent water, 9 percent digestible dry matter, 4 percent pits, and 2 percent indigestible dry matter. High water content dilutes the real value as feed because it makes culls expensive to transport, requires large trough volumes, and allows the feed to spoil quickly. If fed in large proportions, cull fruit will cause almost continuous urination and, consequently, the animals will have a high salt requirement. The only potential advantage to the high water content is that animals in a remote, dry location will need no extra water hauled to them.

Low protein levels in cull fruit limit the quantity that can be fed. Where rapid weight gain is important, in feed lots, for example, only about 20 percent of the ration can be composed of cull fruit. As a maintenance ration up to 80 percent of the feed can be culls.

Stone fruit pits are not an animal health problem. Cattle spit out some pits while eating and many of the remaining pits will be regurgitated with the cud and spit out. The pits rarely cause internal injuries or choking. In fact, the main problem with pits is disposing of them as they tend to fill feed troughs.

Typically cull fruit is bought for $2 to $5 per ton. In terms of feed value, this is equivalent to buying barley for $20 to $50 per ton. However, the added costs of handling and transporting culls must be added to the cost. Also, some cost must be added to account for the risk of using a feed which has not been thoroughly tested.

Cull potatoes are another good source of feed for animals. Like stone fruit they are high in water content (about 77%), high in energy value, and low in protein. Beef steers can be fed up to 50 percent potato waste in finishing rations and still have acceptable weight gain. However, the cattle must be carefully adapted to a potato ration and the ration should not be changed rapidly.

In addition to cattle feed, there are several other potential uses for culls. Fruit that is too ripe for long distance shipment can be utilized in local markets, if they are available. Some culls can be used for processing into juice or a canned, frozen, or dried product. A large quantity of cull apples, pears, oranges, and papayas are used for juice extraction. Cull avocados are made into guacamole and undersized artichokes are marinated and canned. Some culls can be used to produce dried fruit for human consumption. However, good quality dried fruit is made only from good quality fresh fruit. Only undersized or slightly over-ripe fruit should be considered for drying.

Alcohol production

Most fruit and some cull vegetables (especially roots and tubers) can be used for alcohol production. Alcohol for human consumption has a much higher value than alcohol for motor fuel. Some cull pears, kiwifruits, and apples are used for fruit wine production in California with some of the apple wine converted to vinegar.

The use of culls for fuel alcohol production is limited mainly by the low sugar content of most fruits and vegetables. The 8 percent to 12 percent sugar content of most cull fruits results in an alcohol yield of about l0 gallons per ton of fruit. Potatoes have one of the best yields of alcohol for culls at 20 to 25 gallons per ton but this is still low compared to better feedstocks, such as corn with a yield of 90 gallons per ton. The low yield prevents hauling the culls any significant distance. If fuel alcohol production from culls is to be economical, it must be produced at the packinghouse. The low sugar content also results in 4 percent to 5 percent alcohol "wines" which require considerable energy per gallon of alcohol to process and distill.

Waste Management

Stillage waste left after distillation of alcohol is a waste management problem. It has very little protein so it is not suitable as an animal feed but it has a high pollution potential as measured by B.O.D. (biological oxygen demand). The stillage waste may be useable as a feedstock for methane generation. The effluent from a methane generator would be low in pollution potential.

In general, there are several key constraints to any method of utilization. First, transportation is expensive, 80 percent to 90 percent of the weight of culls is water which usually has no value to the one using the culls. Second, culls are produced in large quantities during a short time period. A year-round operation must be able to store the culls, usually by drying, which adds a cost to the product, or by utilizing other products during the off-season. Operations that run only during the season cannot make large capital investments in equipment that will only be utilized a few months a year. Finally, there must be a market for the by-product. For example, in California dried fruit has a very specialized and limited market which cannot absorb extra amounts of product produced from culls.

Pollution potential

Unfortunately, the limitations to utilization often result in large portions of culls being discarded. Improper disposal will result in fly, odor, health, or pollution problems.

Flies and odor problems are prevented by insuring rapid drying. Fly maggots hatch into adults within 7 to 10 days and odor problems can develop before fly problems. The culls should be crushed and spread no more than one or two layers deep; sometimes this is done on orchard roads. Discing culls into the soil can be utilized although discing tends to cover the fruit with soil and slows drying. Also, insects or diseases which may have caused the fruit to be culled in the first place may infect a future crop. Disposal sites should be as far away from neighbors as possible. Flies can travel up to 5 miles from the area where they hatch.

Culls should not be dumped near stream beds. Fruit dump sites will attract the dumping of all kinds of refuse. If available, use municipal solid waste disposal sites if culls are deposited away from the point of production.

6
Packages for Horticultural Crops

F. Gordon Mitchell

Packages for horticultural crops provide convenient units for marketing and distribution of the product and they have many special requirements. Packages must protect the contents against undue damage during distribution and must maintain the shape and strength, often for long periods at a relative humidity near saturation and sometimes after water drenching. They often must facilitate rapid cooling of the contents from high field temperatures to low storage or transport temperatures, and must allow the removal of heat produced by the contents. Because of the fragile nature of many horticultural products, packages usually must assume all stacking stresses throughout distribution and be adaptable to high volume packing operations. If used for display, packages must be attractive to the consumer.

Present Horticultural Packages

Many materials, sizes, and shapes are represented in packages for horticultural products. Over 500 different containers are currently used for produce in the U.S. Past efforts at standardization have had only limited success. Major changes have been in response to economic considerations—the use of less expensive materials, package adaptability to less expensive packing and handling procedures, or ability to increase load density during transport. In the United States, major shifts have been from wood to corrugated containers (with limited use of plastic containers), from hand packing to mechanical volume-fill packing operations, and from single package handling to unitized handling on pallets. These changes have forced a general review of the requirements of packages for use with horticultural products.

Product Requirements

The successful development of a package for horticultural products is based upon emphasizing the requirements of the product being packaged. These requirements will vary widely with the commodity, marketing program, packing method, and so on. However, there are many generalities that will transcend most commodities.

Protection from injuries

All physical injuries to the product must be avoided wherever possible during handling and distribution. Some more obvious open wounds (e.g., cuts, punctures) often occur prior to packaging and can be eliminated by good supervision and sorting. Certain bruises, however, may accumulate throughout all stages of handling, including packaging and distribution.

Impact bruises (fig. 6.1). Impact bruising results from dropping the product onto a hard surface, either individually or in packages. Impact injury may not be visible from the surface, so careful quality control is needed to protect against it. Dropping the product into the package is a common source of impact injury during packing. Careful padding at drop points, use of decelerator strips at filling chutes, and designing fillers to raise empty containers to reduce drop heights during volume-filling can all reduce incidence and severity of impact bruising. Cushion pads placed in the bottom of empty containers before volume-filling may provide further protection.

Packaged products can receive impact bruising from excessive drops during palletization, loading, unloading, commodity segregation, and so on. Bottom cushion pads in the containers may reduce impact force and reduce bruising. Unit handling reduces container rehandling, and thus reduces the number of impacts. However, even rough

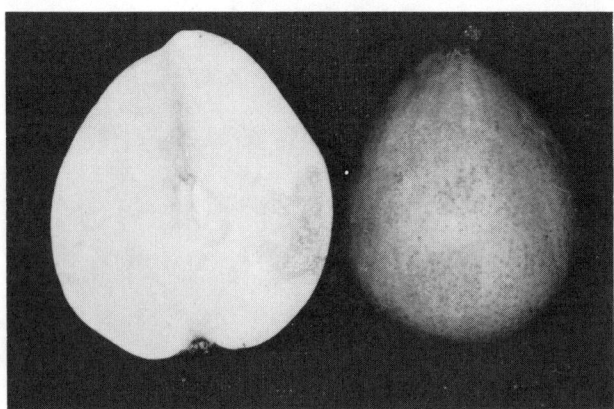

Fig. 6.1. Impact bruise on Anjou pears. Bruising extends into the flesh and may or may not be visible on the surface.

handling by forklifts can cause impact bruising. Careful supervision is essential in effectively reducing produce exposure to damaging impacts.

Compression bruises (fig. 6.2). Compression bruising results from improper packing and from inadequate package performance. The package dimensions must be carefully adjusted to accommodate the mass of product being packed. For pattern packs, product size must be carefully selected to avoid compression injuries during packing. Probably the greatest single cause of compression bruising is intentional overpacking of containers to create a bulge pack. Whatever historical significance this container undersizing may have had, a bulge pack will not convince modern buyers that they are getting something for nothing.

Compression of the product after it has been packed into the container is a major cause of bruising. The most obvious problem occurs when overfilled containers are stacked. Because of container distortion, the commodity absorbs much of the stacking force. Considerable bruising also occurs because a container is not strong enough to support containers stacked on top of it. Containers must be able to withstand the weight of additional containers.

Avoid stacking containers beyond their design limits. It is not economically feasible to design corrugated containers to withstand three- or four-pallet high stacking during storage. Even two-high pallet stacking should be considered a temporary requirement—shippers should recognize that temporary two-pallet high stacking will often occur during distribution. For multipallet stacking heights and for storage, pallet racks or "crutches" (discussed later), are less expensive than extra-strength containers.

Vibration/abrasion bruises (fig. 6.3). Less well-recognized is the damage that results when products are able to move within the container during transit. The resulting vibration bruising is usually restricted to the product surface, except on certain very soft commodities. To prevent vibration injury the product must be immobilized within the container. This is achieved by properly sizing the container for the product, and by properly adjusting the fill density.

In hand packing, the product is immobilized by achieving lateral tightness within the container. Supplemental packaging materials such as wraps, trays, cups, shims, liners, and pads may be useful. Volume-fill packs can be immobilized by proper filling, settling, padding, and lidding. A special procedure called tight-fill packing is designed specifically to immobilize the product after volume-filling.

After being filled, the containers must perform properly if the product is to remain immobilized. Any bulging of the container will loosen the contents and enable transit injury to occur. Thus, to immobilize the product the container must be sufficiently strong to resist bulging throughout the distribution period, even during high-humidity storage and transport.

Facilitate temperature management

The horticultural package must accommodate the special temperature requirements of the product. Good temperature management depends upon achieving contact between the product and the external environment. For some products, provision for air flow past container surfaces may be sufficient. Usually ventilation is needed to facilitate air flow through the container to achieve rapid heat removal. Within limits, increasing ventilation will speed heat exchange. For corrugated containers, 5 percent venting of side or end panels is sufficient for rapid cooling without overly weakening the container. A few large vents perform better than many small vents. Vertical slots, kept at least two inches from all container edges, perform well.

Certain fruits require ripening prior to retail marketing. For rapid, uniform ripening they must be uniformly warmed to ripening temperature, and often need treatment with ethylene gas. A container that is properly vented for cooling will perform well for both warming and gassing.

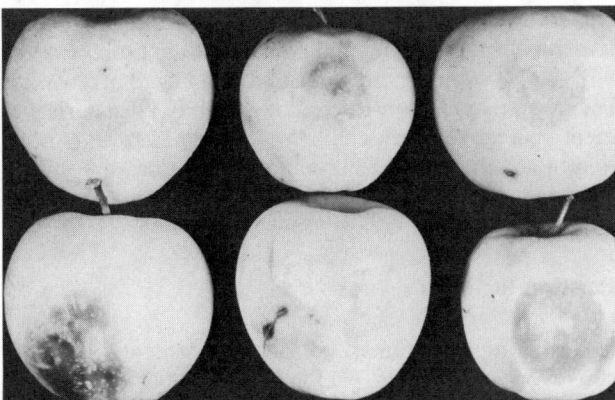

Fig. 6.2. Compression bruise on Golden Delicious apples. Bruise damage occurs on surface and into flesh of fruit.

Fig. 6.3. Vibration bruise on Bartlett pears. Bruise damage appears on fruit surface and usually does not extend into flesh.

Container vents should be relatively unrestricted by internal packaging materials. Liners, wraps, trays, shims, or pads should be designed and located to have a minimum effect on air flow through the package. If these restrictive materials are essential to successful packaging, then the air flow should be increased to compensate for their effects.

Occasionally packages are intentionally designed to restrict heat flow into the package. Some containers used for air transport are designed without ventilation to delay product warming during transit without refrigeration. Flower boxes are designed so vents can be closed after forced-air cooling. In using the package as a heat barrier, it must be recognized that heat flow restriction will be equal in both directions. Thus, the heat of respiration of the product will be retained within the container to the same degree that outside heat is excluded. Package-icing may be useful in such a system if the commodity is packaged cold and both the product and package will tolerate prolonged water contact.

Protection from water loss

Many horticultural products suffer deterioration (wilting, shriveling, drying) as a result of water loss during handling and marketing. Water loss occurs because of a water vapor pressure gradient between the product, which is normally at saturation (100% RH), and the surrounding environment (which is at a lower relative humidity). During storage it is desirable to hold most products at a high relative humidity to minimize water loss. During transport and marketing there is usually little or no environmental humidity control available. Thus the package may be designed to provide a partial barrier against movement of water vapor from the product.

Many package moisture barriers are available. Plastic liners, usually designed with small perforations to allow some gas exchange, have often been used. These maintain an essentially saturated atmosphere within the container, which can cause surface cracking problems with some commodities. Plastic curtains, which may be open at the container ends and folded around the product sides and top, provide a partial moisture barrier, and have been a successful compromise for some fruits. Corrugated containers can be treated with various coatings, which act as moisture barriers. Because most package materials are hygroscopic, surface coatings also slow water vapor uptake by the packages, and delay their deterioration. The most widely used, called "curtain-coating," is a poly-wax emulsion that is coated on the surface of the corrugated board.

Moisture barriers inside a package must not impede essential air flow through container vents. Poly liners are particularly troublesome because they completely block all vents in the container. With poly curtains, tests with tray-packed peaches have shown that increased air flow during cooling and storage could compensate for this blockage. Containers with surface coatings can be vented normally, so that the moisture barrier does not affect air movement. In most conditions, container venting will cause only a limited increase in water loss from the product. For long-term storage purposes, water loss can be reduced in such containers by partial restriction of ventilation, with compensating upgrading of air flow requirements for cooling.

Facilitate special treatments

Certain commodities have special treatments which must be considered in packaging selection and design. Examples are sulfur dioxide fumigation of grapes for disease control and methyl bromide fumigation of various commodities for insect control. These treatments require well-vented containers through which the fumigant can readily flow. Normally, venting that is sufficient for rapid cooling is adequate for fumigation.

Ethylene gas may be beneficial or deleterious to various horticultural products. The need for package venting to achieve uniform warming and ethylene treatment during fruit ripening has been mentioned. Conversely, some commodities must be protected from detrimental effects of ethylene gas. Certain in-package ethylene scrubbing procedures, which are under study, may require special container modifications to be effectively used. Other room scrubbers that circulate air from the storage room through the scrubbing unit may require good container venting to be effective.

Modified atmosphere storage has been achieved with some commodities, especially apples and pears, with the use of partially-sealed poly liners. Accumulation of an approximately 2 percent to 3 percent carbon dioxide concentration inside the liners has reportedly improved fruit storage life. Because such poly liners inhibit heat exchange and slow product cooling, the real benefits under modern handling conditions are questionable. Use of this technique has been declining in recent years.

Compatibility with Handling Systems

The package may need special design features to make it compatible with packing equipment and handling procedures. Container top flaps may be a handling advantage or detriment, depending on the packing procedure. The container must be sized to facilitate unitization and mixed load handling. Weather and contamination problems that will likely be encountered must be considered in advance. Packages may be constructed of a variety of materials to meet special requirements. Wood containers (fig. 6.4), long a standard in the horticultural industry, are still in use where long-term storage or severe moisture conditions are encountered.

Packing facilities

In designing a container, consider its compatibility with conveyors and other packing equipment. Equipment changes that must be made to accommodate a new package are expensive. In volume-fill operations, container top flaps (which pose problems with hand packing) are often an advantage in containing the product during filling. This advantage may justify the cost of redesigning the conveyor system to accommodate the increased height of the flaps.

A new container for field use must be compatible with field conditions. Some container treatments, such as wax coatings, may deteriorate at high temperatures, discoloring the container and losing much of their effectiveness. Some containers (e.g., polystyrene) are very lightweight when empty, and special precautions may be needed to stabilize them during strong winds. Other field problems such as rains, heavy dews, or container soilage may require special attention.

As a new container is introduced, inventory problems must receive consideration. New containers should simplify operations wherever possible, so elimination of existing containers should be scheduled. With careful planning, the inventory problem during container changeover can be minimized.

Unitized handling

With few exceptions, horticultural packages in the United States must be capable of unitized handling. This requires designing for secure palletization, either in register or cross-stacking. This is difficult to achieve with many bulge packs, so attention to proper container mass/volume requirements must accompany new container design. The container must be strong enough to withstand expected stresses in the stacking column. If cross-stacked, container vents must be located to facilitate air circulation through all containers on the pallet. This will require a geometric relationship between horizontal container dimensions and vent locations on both the sides and ends of containers.

Container dimensions must be compatible with the pallet dimensions that will be encountered. For some years receivers have voiced a desire to standardize on 48 inch by 40 inch (or 1200 mm by 1000 mm) pallets. The trend toward such a goal has been slow partly because of poor space utilization in transport vehicles, and partly because of many vested interests in existing containers, pallets, and so on. While standard pallet dimensions are desired, the paramount need is for the container to be compatible with the pallet size likely to be used for that commodity. In selecting container dimensions, check their legality in the production and marketing areas.

Some containers are designed to facilitate stabilization of the shipping pallet. Stacking tabs on certain styles of containers are examples (fig. 6.5). Many containers can be glue-bonded between layers on the pallet with special "break-away" palletizing glues. Normally, at least one horizontal pallet strap of nylon or steel is required to assure stability of the glue-bonded pallet unit. Where gluing is not done, two, or three vertical straps may also be needed to stabilize the pallet load.

Container standardization

Anyone who has observed produce handling in distribution warehouses is appalled at the large number of container types, shapes, and dimensions in use. These containers are not designed to be compatible in loading, and consequently unnecessary and often serious product injury occurs as commodities are loaded together in mixed load shipments or for retail distribution. Most products would fit into containers of just a few horizontal dimensions.

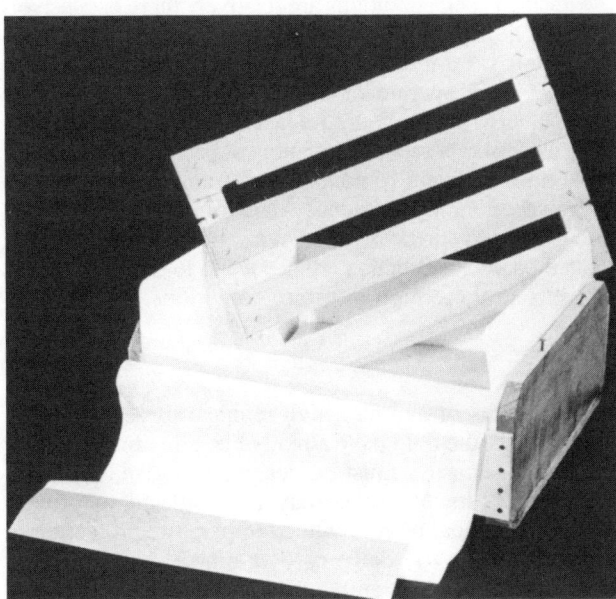

Fig. 6.4. Wood lug used as a shipping container for grapes. A similar style and design container has been widely used for many horticultural crops. These containers often have wood ends and a wood/paper veneer side and bottom wrap.

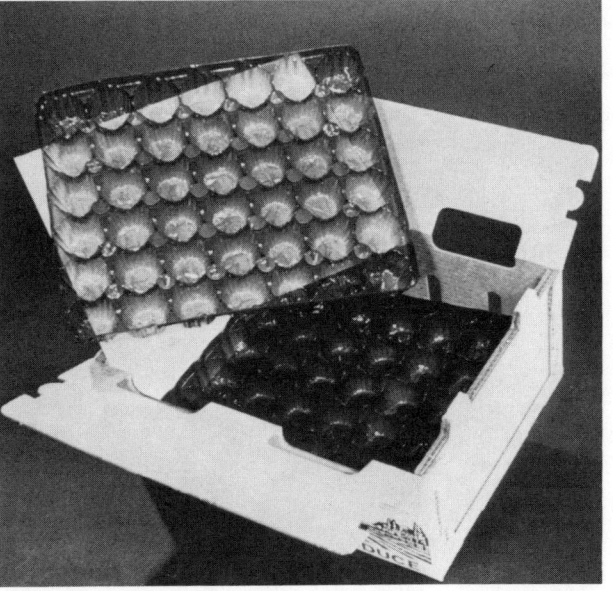

Fig. 6.5. Corrugated "Bliss type" lug used as a shipping container for many fruits. Container shown is equipped with plastic trays that are often used for hand or hand-assisted packs of fruits. Note end stacking tabs used to stabilize pallet loads of containers.

This need has triggered a serious effort to establish a list of standard-sized containers. In developing these standards there has been interest in making new dimensions meet the international metric dimensions.

There are many considerations in developing new containers for the so-called "metric pallet" (1200 mm by 1000 mm). Outside horizontal dimensions being suggested for U.S. horticultural products include 600 mm by 400 mm, 500 mm by 400 mm, 500 mm by 300 mm, and 400 mm by 300 mm. A few additional dimension combinations that afford stability and good coverage of the pallet area may be needed to accommodate some commodities. For example, a 400 mm by 333 mm container will approximate the current widely used "Los Angeles lug" dimensions and may allow conversion with a minimum effect on packing and handling equipment.

Because of side bulge, containers of full dimensions will not fit a 1200 mm by 1000 mm pallet. How much bulge to allow will depend on the package specifications, including board weight, container height and design, and relative container size. If containers are too wide or long, with bulge, to fit the pallet, then pallet dimensions will be exceeded and some container walls will not be supported by the pallet. Narrowing of container dimensions by up to 10 mm may be necessary to compensate for bulging.

There is some effort to also standardize container depths for all commodities. This poses many more serious problems in meeting packing and marketing requirements and in adequately protecting the product.

It is important that product requirements be carefully considered in selecting a new container design. If a telescope-style container must be used (a two-piece container with a lid that fits over the body) (fig. 6.6), then the inside dimensions will be less than in a single-wall container. The density of the product within the container must be carefully calculated based upon the useable inside space of the container. Fill density information for many fruits in volume-fill and tight-fill packs is available. Some of these packs have very specific container depth requirements to perform satisfactorily. Tight-fill containers must allow a filling depth of three to four times the diameter of the largest size fruit to be packed. Some soft commodities require shallow packing depths to avoid damage.

Trays for tray-packed commodities must be redesigned to fit the new container dimension. With many commodities, 8 to 10 different sizes of trays are used. Because of the expense of retooling for new trays, it is essential that the exact container inside horizontal dimensions be well-established prior to changeover. Further, the tray packs may fit poorly into containers of certain depths, resulting in poor packing density. All of these requirements must be carefully considered for each commodity before conversion to standardized containers commences.

Adaptability to Handling Requirements

The package must perform efficiently under the handling conditions that will be encountered. This includes not only product marketing and distribution, but also inventory and handling of packaging materials.

Moisture

Some commodities, especially certain vegetables, encounter water contact during and after packing. Containers used for these commodities must be able to withstand water contact, often for prolonged periods. Such water contact may occur when a packed product is hydrocooled, when ice is placed into the package (package-icing), or when top ice is placed over a load during transport. Corrugated containers especially designed and manufactured to withstand such conditions are relatively more expensive.

Most horticultural packages must tolerate exposure to high relative humidity after packing. While the storage facilities may operate at a moderate relative humidity (often 85% to 95%), water released from the product will create near 100 percent RH within the package. In terminal markets, containers removed from transport vehicles or storage rooms may condense water from the atmosphere onto their cold surfaces. Special treatments or heavy corrugated board construction are needed to tolerate prolonged exposure to these high moisture conditions.

High temperature

The problems of sun and high temperature exposure of containers in the field have already been mentioned. Containers for such use must be designed to avoid such deterioration. Flower boxes may be insulated to protect flowers from heat or cold when temperature protection is not available (e.g., delivery of a single container at a rural bus stop). High ambient temperatures encountered by carry-over inventories of corrugated containers can reportedly speed delamination of the corrugated board, and thus pose potential problems during the subsequent season.

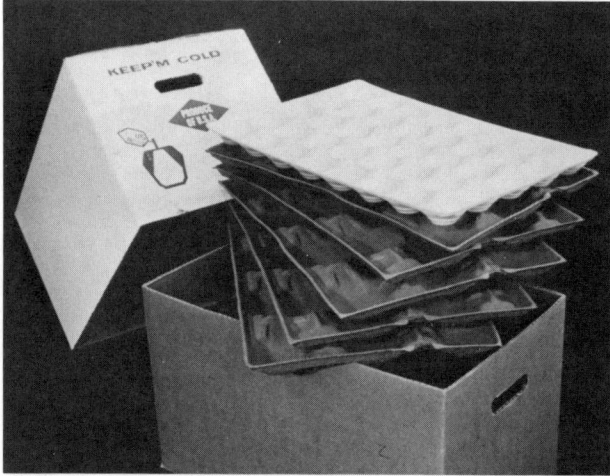

Fig. 6.6. Full telescope corrugated container used as a shipping container for apples. Similar containers are used for many other horticultural commodities. Apples are layer-packed into trays shown.

Product storage

The package must withstand the necessary environmental conditions during the entire storage and marketing season of the commodity. To minimize problems during storage, stacking aids such as pallet frames, racks and "crutches" (rigid steel or wood corners fitted around the loaded pallet) are used. Even with these aids, the container must be capable of withstanding handling abuses at the end of the storage and distribution period.

Inspection

The package should facilitate easy inspection of the contents. Telescope-style corrugated containers are easily inspected by removal of the lid. Snap-on lids and inspection ports designed into other styles of containers serve the same function. Whatever the procedure, the container must be capable of reclosing and protecting the product during the balance of the distribution period.

Economic Considerations

In considering the cost of any new container, all of the costs of adapting it into the marketing system must be considered. These will involve costs of packaging material, labor, modifications in packing and handling operations, and potential changes in product condition.

1. **Packaging costs**

   Container component costs

   Container make-up costs—labor and materials

   Needed internal packaging materials—liners, shims, pads, trays, wraps

   Storage costs of container components

2. **Packing costs**

   Adaptability to mechanized package distribution

   Effect on packing operation

   Effect on packing labor efficiency

   Number of packing steps required

   Costs of modifying or converting packing facilities

3. **Palletizing and handling costs**

   Effect on pallet stacking efficiency

   Effect on costs of strapping labor and materials

   Adaptability to pallet gluing

   Adaptability to various pallet materials and substitutes

4. **Marketing costs**

   Effect on load density in storage and in transport vehicles

   Special labor or equipment needed for handling

   Adaptability of container as a display unit

5. **Product value costs**

   Effect of package in modifying product deterioration

   Value of adjustments due to container failures

   Value of brand "reputation" related to container performance

Simulated Transit Testing

Selection of good testing procedures are an important aspect of any container development program. Objective laboratory testing is the first step in evaluating new containers and packing procedures. Such testing can reduce costly and time-consuming trial shipments and allow more precise evaluation of a large number of variables. Following such laboratory testing, a promising package or treatment can be compared with the standard in a trial shipment. Use of laboratory tests involves a minimum time expenditure and provides a high degree of confidence that major problems will not be encountered.

Basis for test procedures

To be effective, laboratory test procedures must meet certain conditions.

- Duplicate the types of insults to products that might be encountered in actual transport and handling
- Laboratory treatment should equal a severe transport and/or handling condition
- Laboratory procedure should allow rapid testing of a large number of variables
- Tests should emphasize transport and handling effects on the product as well as the containers and packaging materials

Test procedures

The actual test procedure used in the Pomology Postharvest Laboratory at U.C. Davis includes a series of impact and vibration treatments.

Vertical impact test. Number and height are adjustable, but 30 impacts from a 5 cm (2-inch) height are standard.

Horizontal impact test. Number and intensity are also adjustable but 30 impacts at 3.2 km (2 miles) per hour are standard.

Transit vibration test. Stroke, frequency, and time are adjustable, but 30 minutes vibration at 1.1 g acceleration (using a .6 cm [1/4-inch] stroke at a frequency of 550 cycles per minute) is standard.

Actual transit test. Normally the best components of all variables tested in the laboratory are combined in designing the new container or packing method. This is first compared with the industry standard in a laboratory test. If promising, it is then tested in a well-replicated trial shipment in comparison with the industry standard.

References

1. Gentry, J. P., F. G. Mitchell, and N. F. Sommer. 1965. Engineering and quality aspects of deciduous fruit packed by volume filling and hand placing methods. *Trans. ASAE* 8:584-89.

2. Guillou, R. 1964. *Orderly development of produce containers.* Proc. Fruit & Veg. Perish. Handling Conf., Univ. of Calif., Davis.

3. Guillou, R., N. F. Sommer, and F. G. Mitchell. 1962. Simulated transit testing for produce containers. *TAPPI* 45(1): 176-79A.

4. Hardenburg, R. E. 1966. Packaging and protection. In *Protecting our food supply.* USDA Agric. Yearb. (1966): 102-117.

5. Heiss, R., ed. 1970. *Principles of food packaging, an international guide.* P. Keppler Verlag KG, 332 pp.

6. Hochart, B. 1972. *Wood as a packaging material in developing countries.* United Nations Publication #E.72. II. B. 12. 111 pp.

7. Mitchell, F. G., N. F. Sommer, J. P. Gentry, R. Guillou, and G. Mayer. 1968. *Tight-fill fruit packing.* Univ. Calif Agric. Ext. St. Circ. 548. 24 pp.

8. O'Brien, M., J. E. Gentry, and R. C. Gibson. 1965. Vibrating characteristics of fruits as related to in-transit injury. *Trans. ASAE* 8:241-43.

9. O'Brien, M., and R. Guillou. 1969. An in-transit vibration simulator for fruit handling studies. *Trans. ASAE* 12(1): 94-97.

10. Smith, R. J. 1963. *The rapid pack method of packing fruit.* Univ. Calif. Agric. Exp. St. Circ. 521. 20 pp.

11. Sommer, N. F., and D. A. Luvisi. 1960. *Choosing the right package for fresh fruit.* Packaging Engineering 5:37-43.

12. Stokes, D. R., and G. W. Woodley. 1974. *Standardization of shipping containers for fruits and vegetables.* USDA Mark. Res. Rpt. 991. 118 pp.

7
Cooling Horticultural Commodities

F. GORDON MITCHELL

Many changes have occurred in packing, handling, transporting, marketing, and distributing horticultural crops that directly influence cooling requirements and results. New packages and packing materials, packing methods and palletization procedures, often make produce cooling more difficult. The current desire to reach more distant markets, prolong storage life, and market a product that better satisfies consumers, often requires more uniform or faster cooling. Discussed here are some of these changes and their implications for product cooling requirements, and some possible responses that might improve the cooling system.

The Need for Cooling

An understanding of the cooling requirements of horticultural commodities begins with adequate knowledge of their biological responses. All fresh horticultural crops are living organisms, carrying on the many biological processes that are essential to the maintenance of life. They must remain alive and healthy until processed or consumed. The energy that is needed for these life processes comes from the food reserves that accumulated while the commodities were still attached to the plant (with some exceptions among flowers).

The process by which these food reserves are converted into energy is called respiration. In a complex series of steps, the stored food reserves (starches and sugars) are converted first to organic acids, then to more simple carbon compounds. Oxygen from the surrounding air is utilized in the process, and carbon dioxide is released. If oxygen is severely limited, anaerobic respiration will occur, aldehydes, alcohols, and other undesirable materials will be produced, and the tissue will ultimately die.

Horticultural commodities are not the tight, dense, closed tissues that they appear to be. Rather, they have an amazingly open structure, with a complete network of interconnecting air spaces (called intercellular spaces) throughout. This can be illustrated with many commodities by injecting air while the intact tissue is under water. Air bubbles appear almost instantly and simultaneously over the entire surface. The oxygen concentration in the intercellular spaces in the center of many commodities is almost as high as in the surrounding atmosphere as a result of this structure.

Some of the energy that is produced through respiratory activity is utilized in maintaining life processes. Excess energy is released in the form of heat, called "vital heat." The amount of vital heat varies with the type of product, variety, maturity or stage of ripeness, injuries, temperature, and other stress related factors. This heat must be considered in any temperature management program.

Product temperature is a major determinant of the rate of respiratory activity. Since the final result of this respiratory activity is product deterioration and senescence, it is normally desirable to achieve as low a respiratory rate as possible without danger of tissue injury or death. An exception would be during controlled ripening of fruits. Each 10°C (18°F) temperature reduction reduces respiratory activity by a factor of 2 to 4. For example, the respiratory rate of a product at 5°C (41°F) would be only 1/4 to 1/16 of what it would be at 25°C (77°F). Good cooling and temperature management practices are, therefore, critical to lowering the rate of physiological deterioration.

Ethylene effects

Temperature affects both the rate of ethylene production and the sensitivity of products to ethylene. Ethylene gas is a naturally-produced material in most if not all plant tissue. This simple chemical compound is generally recognized as a fruit-ripening hormone. It can have important beneficial or detrimental effects on fresh commodities, depending on management needs. For these ethylene effects to occur, a minimum concentration must accumulate within the internal atmosphere of the product, and the temperature must be above a minimum level. These threshold concentration and temperature levels are not well-defined. However, since the rates of both production and action of ethylene are temperature-dependent, rapid cooling and good temperature management are vital if fruit ripening and other deterioration processes are to be delayed.

In terms of respiratory activity, fruits—including fruit-type vegetables—have been grouped into two classifications, *climacteric* and *nonclimacteric*. Climacteric fruits are those which normally ripen after harvesting, during which time sugars normally increase and volatile constituents (flavors and odors) develop. Flesh softening accompanies ripening; but if such fruits are picked while immature they may soften without development of sweetness and flavors. Ethylene gas will initiate ripening, which is accompanied by a rapid rise in the respiratory activity in the fruit (called the climacteric rise). Included are apples, peaches, papayas, cantaloupes, tomatoes, and many fruits.

The nonclimacteric fruits, including citrus, grapes, strawberries, and others, do not ripen after harvest and exhibit no rise in respiratory activity. Even under optimum ripening conditions, dessert quality does not improve noticeably. Ethylene gas may affect color changes among these fruits; for example, the breakdown of the green chlorophyll pigment causes oranges to "color." However, the sugar, acid, and flavor of the fruit are not influenced by the treatment.

Ethylene gas is also implicated in a number of product injury and deterioration problems. Among these are russet spotting of lettuce, sleepiness in carnations, leaf abscissions, rind pitting in citrus, bitter flavor in carrots, yellowing of cucumbers, softening of kiwifruits, and others.

Moisture loss

Fresh commodities constantly lose water to the surrounding environment. After harvest this lost water cannot be replaced by the plant (with the exception of flowers), and weight loss will occur. Many products show visual shriveling or wilting after losing 3 percent to 5 percent of their initial weight. They lose water as a result of a water vapor gradient between their essentially saturated internal atmosphere (within the intercellular spaces) and the less-saturated external atmosphere. Water vapor migrates in the direction of lower concentration, primarily through natural openings on the fruit surface, but also through surface injuries. The rate of migration is controlled by the vapor-pressure difference between the product and its environment, which is governed by temperature and relative humidity. Warm air can hold much more water vapor than cold air. Relative humidity is a measure of the amount of water vapor in the air as a percent of the amount air can hold at that temperature. At 25°C (77°F) and 30 percent relative humidity, the product will lose water 36 times faster than it would at 0°C (32°F) and 90 percent relative humidity. Thus, maintenance of low product temperature is essential in reducing water loss and subsequent product shriveling and wilting.

Rot organisms

Postharvest management of fresh horticultural commodities actually involves two living systems—the product and the microorganisms that attack it. Of thousands of potential microbial pathogens, only a few cause problems. These few organisms, however, cause extensive direct loss of fresh fruits, vegetables, flowers, and ornamentals.

Temperature affects the rate of growth and spread of these microorganisms in the same way that it affects the commodity—the lower the temperature the slower their life processes progress. Certain organisms that can cause severe losses will not grow at low storage temperatures. For example, Rhizopus rot ceases growth at about 5°C (41°F), and germinating spores have been found to be chilling-sensitive, and to be killed after 2 days exposure at 0°C (32°F). Other organisms will continue growth, but at a very slow rate at safe storage temperatures near 0°C. In studies with Botrytis rot of strawberries during a 7-day marketing period, at 2°C (36°F) or below, germinating spores would not penetrate into the fruit, and at 0°C mycelium (the vegetative fungal growth) would not penetrate a healthy fruit from an adjacent invaded fruit. Good temperature management, thus, plays a vital role in reducing microbial loss problems.

Injuries

Physical injuries can result from abuses to fresh commodities at any temperature, but temperature affects the severity of the product response to those injuries. Bruises and other wounds cause increased ethylene production, which may accelerate respiration, cause deterioration problems, or initiate fruit ripening. Bruising usually damages the natural barriers on the product surface, thus increasing the opportunity for water loss and for entry for rot organisms. Prompt cooling and maintenance of low temperatures reduces the results of injuries by affecting all of these processes.

Low temperatures

Good temperature management is the single most important factor in delaying product deterioration. Prompt cooling and maintenance of proper temperatures are both essential parts of the temperature management system. For many products, this means maintaining as low a temperature as possible without danger of freezing. The freezing point will vary with soluble solids content; for some products freezing point guides are available. How closely the freezing point can be approached will depend on the accuracy and sensitivity of the temperature controls.

Many horticultural crops are subject to chilling injury at temperatures considerably above their freezing point. Often these crops must be held at modified temperatures. Chilling injury symptoms include surface or internal browning, surface pitting, failure to degreen, failure to ripen, increased susceptibility to microorganisms, texture changes (mealy or woolly texture) and loss of flavor. Most crops of tropical and subtropical origin and even some deciduous fruits are subject to chilling injury. Guides to the lowest safe temperature for various commodities are available.

Speed of cooling

Many products benefit from prompt, thorough cooling. For example, strawberries experience increasing deterioration losses as delays between harvesting and cooling exceed 1 hour. A similar pattern exists for sweet cherries when delays exceed about 4 hours. These fruits benefit from cooling even when rewarming will occur during subsequent handling; deterioration is proportional to the total exposure time to warm temperature, and not to the pattern of cooling and warming.

Some Bartlett pears should be cooled to −0.5°C (31°F) within 24 hours of harvest to reduce losses from an enzymatic breakdown problem called "watery breakdown." This problem is especially serious among fruits harvested in mid- and late-season. This rapid cooling to low temperature also limits decay and ensures maximum storage life, and thus is generally recommended for all Bartlett pears that are to be stored.

There are exceptions to the need for prompt cooling after harvest. For many years freshly harvested freestone peaches in South Africa have been held at ambient temperature for about 36 hours prior to cooling. This fruit is subject to chilling injury (dry and/or brown tissue), and the delayed cooling treatment was found to delay onset of the problem. A recent report suggests a similar treatment may be useful for honeydew melons if they are gassed with ethylene to initiate ripening prior to shipment at chilling temperatures.

While most fruits would not be expected to be harmed by moisture condensation during warming, some table grapes have reportedly developed berry cracking (and subsequent rot) where condensation moisture remained in the cluster for prolonged periods. For that reason, prompt cooling might be undesirable unless low temperatures could be maintained. Thus, it is important to know the product and its temperature and marketing requirements, before selecting a cooling program.

Changes Affecting Cooling

Cooling is affected by many different factors throughout the handling system—from the field to the consumer. Some handling changes have increased the need for more rapid, thorough cooling. Included here are the desire to market more mature (or even ripe) products, to extend the storage period, to supply more distant markets, and to meet new consumer quality standards. Some other changes have made fast, thorough cooling more difficult to attain, and have caused the development of new cooling procedures. Included here are the changes from field lugs to large bins, new package designs, unitized handling, and changing transport equipment. The effects of these changes must be understood if improved product performance is to be achieved.

Extended market life

Handlers want to extend the postharvest life of fresh horticultural commodities for a variety of reasons. Processors may store to accumulate product for processing, to protect quality when deliveries exceed processing capacity, to hold the product over weekend periods in order to avoid labor complications, or simply to extend the processing season. Storage may be for a few days to several months.

Packers may want to store the product in order to extend their marketing season, to accumulate a supply for holiday periods, to facilitate orderly marketing, to attempt avoiding price declines during periods of oversupply, or to develop export markets. Thus, longer shelf-life may be required. Since deterioration is a function of time and temperature, faster cooling can significantly extend shelf-life.

Consumer demands

There is increasing evidence that consumer satisfaction with fresh commodities is related to certain "quality" aspects, especially appearance, flavor, and maturity or ripeness, or ability to quickly ripen after purchase. With consumers purchasing larger quantities of fresh commodities, handlers are anxious to make the changes needed to sustain consumer demand. To meet these requirements fresh produce shippers are shipping slightly more mature products that have developed more of the characteristic aromas and flavors. Fast cooling and good temperature management are vital to protecting such commodities.

Transportation

Refrigerated transport equipment has changed during the past generation from the use of ice in bunkers to mechanical refrigeration systems. The correct use of ice bunker equipment, which included proper loading patterns, frequent re-icing and adequate air circulation capacity, generally allowed reasonable cooling of most products during transport. Most mechanical refrigeration equipment in current use is designed to maintain temperature, but lacks the air flow and refrigeration capacity needed for rapid cooling. Further, packages and loading patterns have changed. High-density loads are used to minimize transportation costs. These factors all inhibit cooling during transit. Therefore, thorough product cooling before loading and protection against warming are very important.

Pallet bins

In harvest operations, small field lugs have generally been replaced by large pallet bins, often holding about one-half ton of product. Those field lugs were separated by cleats, which facilitated air flow when cooling was needed. In contrast, the larger dimensions of pallet bins result in much of the product being remote from the top or side air flow surfaces. This allows little opportunity for cold air to penetrate into the bin in normal room cooling operations.

Shipping containers

Corrugated fiberboard has essentially replaced wood in the construction of shipping containers. Corrugated containers normally stack more tightly than wooden containers and lack cleats, which on wooden containers facilitate air circulation during storage and transport. Further, they generally have much less ventilation than the wood containers they replaced. Internal packaging materials, such as wraps, trays, liners, pads, and plastic bags further restrict heat removal. Despite these adverse properties, proper stacking and adequate spacing of corrugated shipping containers on storage pallets, together with provision for high velocity air flow, can provide reasonable cooling speed.

Unitization

Many horticultural products are now handled and shipped unitized on pallets. When corrugated containers are tightly stacked on a pallet for unitized handling, cooling problems become more severe. Container vents, even if adequately sized, are too often blocked when containers are cross-stacked on pallets, and much of the product has little access to cold air. Cooling of such units in conventional cold rooms is slow and irregular, and often inadequate without some modification in container design and venting, pallet-stacking patterns, fan operation, and/or cooling systems.

Meeting Changing Needs

Conflict between the need to achieve more rapid cooling and the increasing difficulty of cooling appears almost impossible to resolve through the use of conventional room cooling procedures. However, rapid cooling systems are available that can achieve good results if properly used.

Cooling: Aspect of handling system

The entire product handling system must be considered when cooling facilities are planned because any change in the system may affect cooling rate, uniformity, and requirements. The packing method will dictate how, and sometimes when, the product is presented for cooling. Packaging materials and design will affect access of the coolant to the product, and pallet stacking patterns will influence coolant flow through and around containers. Loading patterns, transport equipment, and marketing procedures all greatly affect cooling requirements. The maximum market life of many commodities may be approached when they are shipped for long distances by refrigerated ocean transport for export marketing.

Cooling and storage: Two separate operations

Refrigeration capacity for fast cooling and for cold storage are quite different. For example, it takes about 100 times more refrigeration to cool pears in 24 hours than to cold store them for 24 hours. Even when fruit is cooled over a 6-day period, the daily refrigeration capacity during cooling is almost 25 times that for cold storage. This ignores the refrigeration required to remove heat entering the facility through walls, doors, fans, and forklifts.

Other differences between cooling and cold storage must be considered. During fast cooling, high air velocity will not increase total water loss from the product if it occurs only while the product is being cooled. However, during subsequent cold storage, high air velocity will desiccate horticultural commodities because of the long period of exposure to rapidly moving air. High relative humidity is essential to prevent excessive water loss during cold storage, but is not as important during the short cooling period.

Thus, cooling and storage are two separate operations that have vastly different requirements. The specific requirements for achieving fast, uniform cooling must be considered independently of the cold storage requirements.

Cooling Methods

Several cooling methods are available for use with horticultural commodities—room cooling, forced-air cooling, hydrocooling, package icing, and vacuum cooling are used before storage or loading for shipment. Top-icing, channel-icing, and mechanical refrigeration in transport vehicles are used for transit cooling. A few cooling methods (e.g., room cooling and hydrocooling) are used with a wide range of commodities, but most commodities respond best to one or two of the cooling methods. The table includes cooling methods and the types of commodities for which they may be suited.

Economic considerations may dictate which cooling method is used. Some, such as hydrocooling, forced-air cooling, and vacuum cooling, require considerable initial capital investment. Other cooling methods require less capital investment, but may require expensive water-tolerant shipping containers and may have higher product loss and higher operating costs.

Room Cooling

This widely used cooling method involves placing field or shipping containers of produce into a cold room. Typically cold air is discharged into the room near the ceiling, moves horizontally across the ceiling, and then sweeps past the produce containers to return to the heat exchanges (fig. 7.1). For best results containers should be stacked so the moving cold air can contact all container surfaces. An air flow of at least 61 to 122 m (200 to 400 feet) per minute is needed to provide the turbulence to achieve heat removal. Well-vented containers can greatly speed room cooling by achieving some air exchange.

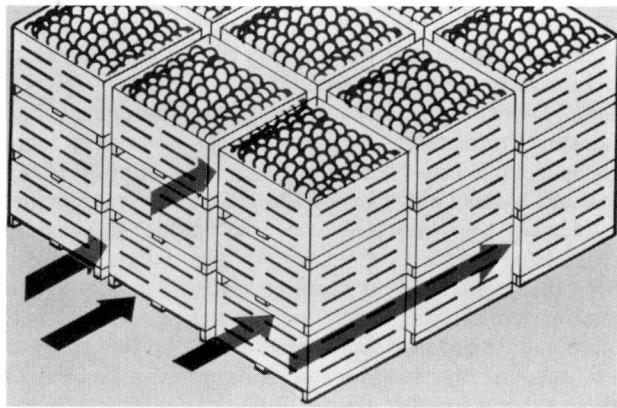

Fig. 7.1. Diagrammatic view of air path during room cooling of produce in bins. Air circulating through the room passes over surfaces and through forklift openings in returning to the cooling coils. In this system the air takes the path of least resistance in moving past the product. Cooling from surface to the center of bins is largely by conduction.

Table. Cooling methods and suitable commodities

Cooling method*	Commodities cooled	Comments
Room cooling	All commodities	Too slow for many perishable commodities. Cooling rates vary extensively within loads, pallets, and containers.
Forced-air cooling (Pressure cooling)	Fruits, berries, fruit-type vegetables, tubers, cut flowers, cauliflower	Much faster than room cooling; cooling rates very uniform if properly used. Container venting and stacking requirements are critical to effective cooling.
Hydrocooling	Stems, leafy vegetables, some fruits and fruit-type vegetables	Very fast cooling; uniform cooling in bulk if properly used, but may vary extensively in packed shipping containers; daily cleaning and sanitation measures essential; product must tolerate wetting; water-tolerant shipping containers may be needed.
Package-icing	Roots, stems, some flower-type vegetables, green onions, brussels sprouts	Fast cooling; limited to commodities that can tolerate water-ice contact; water-tolerant shipping containers are essential.
Vacuum cooling	Leafy vegetables; some stem and flower-type vegetables	Commodities must have a favorable surface-to-mass ratio for effective cooling. Causes about 1% weight loss for each 6° C cooled. A procedure that adds water during cooling prevents this weight loss but equipment is more expensive, and water-tolerant shipping containers are needed.
Transit Cooling:		
Mechanical refrigeration	All commodities	Cooling in most available equipment is too slow and variable; generally not effective.
Top-icing and channel-icing	Some roots, stems, leafy vegetables, cantaloupes	Slow and irregular, top-ice weight reduces net pay load; water-tolerant shipping containers needed.

*For these methods to be effective, cold-storage rooms are needed to hold the commodity after cooling.

Room cooling allows produce to be cooled in the same location where it will be stored, thus requiring less rehandling. Best room cooling, however, requires more spacing than is needed for good storage management, thus rehandling may be needed to achieve good storage space utilization.

Products for room cooling must be tolerant of slow heat removal, for much of the cooling is by heat conduction to the container surface. For such commodities there is a potential economy from room cooling, because the refrigeration load is distributed over a longer period, and because air flow requirements are often less than for fast cooling. Because the air velocity needed for cooling is greater than that needed for storage, products stored in the cooling room will lose water faster than when under more ideal storage conditions.

Room cooling has become increasingly difficult as more commodities are being handled in large field bins or in containers that are tightly unitized on pallets. The increasing difficulty results from the longer path for conduction cooling and the greater difficulty of effecting air movement within the large product units.

Ceiling jets

This modification of room cooling directs air flow past container surfaces in an attempt to improve heat removal. A ceiling is constructed in the room, with small inverted open cones installed to direct air into the space above the ceiling. This creates a slightly positive pressure which results in the air flowing from the cones. The floor of the cooling room is carefully marked for spaced stacking of pallets or bins so that air from the cones will sweep down the corners of the stacked units and spread into the channels that were created during stacking. The air return plenum then draws the air in one direction.

Cooling bays

For both cooling and storage, a single large room is divided into bays by installing partitions part way into the room from each side. Air supply channels are constructed to direct the air into each bay. When a single bay is filled with warm product, supply ducts can be opened to direct a large volume of cold air behind the product. Air return generally occurs down the center forklift aisle of the room. When cooling is completed the air supply can be reduced in that one bay to create desired storage conditions. With this system a cold product located in one bay is not warmed by warm product in other bays.

Forced-air Cooling (Pressure Cooling)

Forced-air cooling, which has been described in detail in various publications, can solve many difficult cooling problems because it provides for cold air movement through, rather than around, containers. The system, which creates a slight pressure gradient to cause air to flow through container vents, achieves rapid cooling as a result of the intimate contact between cold air and warm product. With proper design, fast, uniform cooling can be achieved through stacks of pallet bins or unitized pallet loads of containers. Various cooler designs can be used, depending on specific needs. Converting existing cooling facilities to forced-air cooling is often simple and inexpensive, provided sufficient refrigeration capacity and cooling surfaces (e.g., evaporator coil surfaces) are available. Some variations in forced-air cooler design are described here.

Forced-air tunnel

This is the more traditional forced-air cooling system. Essentially, two rows of palletized containers or bins are placed on either side of an exhaust fan leaving an aisle between rows. The aisle and the open end are then covered to create an air plenum tunnel (fig. 7.2). With the exhaust fan operating, a slight negative air pressure is created within the plenum tunnel. Cold air from the room then moves through any openings in or between containers toward the low pressure zone, sweeping warm air from around the product as it moves. The exhaust fan can be a portable unit that is placed to direct the warm exhaust air toward the air return of the cold room, or it can be a permanent unit which also circulates the air over the cooling surface and returns it to the cold room (fig. 7.3).

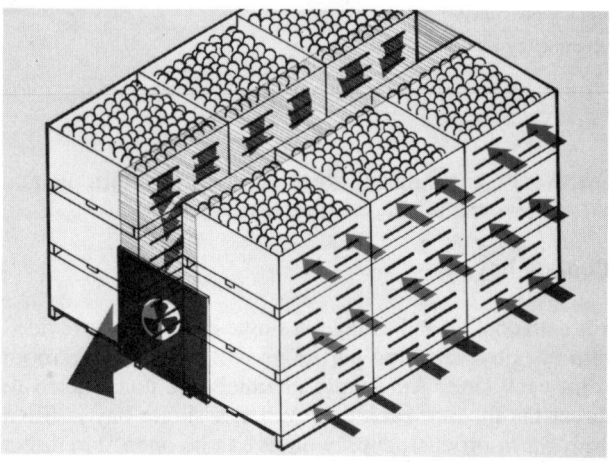

Fig. 7.2. Diagrammatic view of a forced-air cooling tunnel. Either bins or palletized containers can be placed to form a tunnel from which air is exhausted. The negative pressure then causes cold air from the room to pass through ventilation slots to directly contact the warm product.

Fig. 7.3. Forced-air cooling tunnel in serpentine cooling of packaged produce on unitized pallets. Air circulating fan is built into wall and circulates air through fruit and over cooling coils. Canvas plenum cover is designed to fit varying cooling loads.

Cold wall

This is a permanently constructed air plenum equipped with an exhaust fan. It is often located at one end or side of a cold room, with the exhaust fan designed to move air over the cooling surface. Openings are located along the room side of the plenum against which stacks on pallet loads of containers can be placed (fig. 7.4). Various damper designs have been developed so that air flow is blocked except when a pallet is in place. Each pallet will start cooling as soon as it is in place; thus there is no need to await deliveries to complete a tunnel. Shelves are often built so that multilayers of pallets can be cooled with this system. Different packages, and even partial pallets, can be accommodated by proper design of the damper system. This is a benefit in some operations where a range of commodities or varieties is handled. Each pallet must be promptly moved from the cooler as soon as it is cold in order to avoid unnecessary desiccation from continued rapid air flow over the product.

Fig. 7.4. Cold-wall type forced-air cooler for use with stacks of flower containers. Open end-vents allow air to be pulled through the containers for rapid cooling, but are designed to allow closing during shipment. Courtesy of Mel Gagnon.

Serpentine cooling

This system is used for forced-air cooling produce in pallet bins that are constructed with bottom ventilation slots, with or without side ventilation. It requires modification of the cold-wall design to allow the forklift openings between bins to be used as air supply and return plenums. The cold air moves vertically through the product within each bin in response to the slight pressure difference between plenums (fig. 7.5). Bins may be stacked several rows deep against the cold wall, depending on the cooling speed desired and the available air flow. The air flow capacity of the small forklift opening plenums is the primary limitation. To achieve the desired air flow pattern, openings are provided in the cold wall to match alternate forklift openings, starting one bin up from the floor. On the room side of the bins, these same openings are then blocked (fig. 7.6). Thus, air flowing into an open slot between

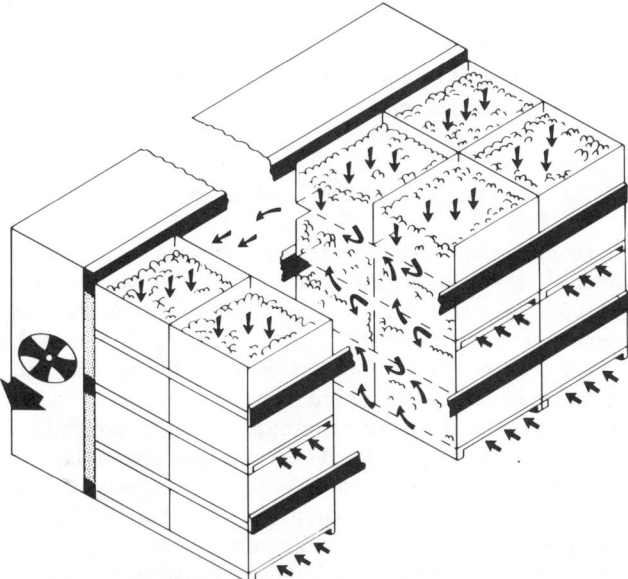

Fig. 7.5. Serpentine forced-air cooling diagram showing pattern of air flow. This system is specific for cooling fruit in field bins. By blocking alternate forklift openings on cold wall and room sides, with fans operating, air is forced to pass vertically through bins to cool fruit.

bins must pass up or down through one bin of product to reach the cold wall. This system requires no space between rows of bins and is not limited in height. Rapid cooling is achieved because air flows vertically through a relatively shallow layer of product.

Container venting

Effective container venting is essential for forced-air cooling to work efficiently. Cold air must be able to pass through all parts of a container. For this to happen, container vents must remain open after stacking. Thus, venting patterns are important. If containers are palletized in-register, container side or end vents will suffice, provided they are properly located in relation to trays, pads, and so on. If cross-stacking is used, then matching side and end vents are very desirable. For the 400 mm by 300 mm container cross-stacked on the 1200 mm by 1000 mm pallet, vertical vent slots on 100 mm centers around the container perimeter should be considered, because they will remain matched when cross-stacked.

Too little venting will restrict air flow; too much venting will weaken the container. A reasonable compromise appears to be about 5 percent side and/or end area venting. A few large vents are more effective than many small vents for speeding the cooling rate. Locating vents midway from top to bottom is adequate unless trays or other packing materials will isolate some of the product. Any type of bag, liner, or vertical divider inside the package will eliminate much of the effect of vents. Vertical slots, at least 12 mm (1/2-inch) in width are better than round vents.

Hydrocooling

The use of cold water to quickly cool produce is an old and effective cooling method used for cooling a wide range of fruits and vegetables in bins or in bulk before packing (fig. 7.7). Its use is limited for packed commodities because of the difficulty of achieving sufficient water flow

Fig. 7.6. Serpentine forced-air cooler in operation. Plastic straps are placed over every other forklift opening from bottom to top. These close off the openings in the room side of the cooler. Air entering the open channels then must move up and down through the product to return to the cold wall. Note that bins can be tightly stacked in rows since no center air flow plenum is needed.

Fig. 7.7. A conveyor type bin hydrocooler in operation. Ice water is pumped into top pan where it "rains" down through the product in the bin. Dwell time in the cooler is controlled by conveyor speed.

through the containers, and because the containers must be water-tolerant. Guides for effective design and operation of hydrocoolers are available. The product must be tolerant of wetting, not susceptible to water beating damage, and tolerant of chemicals that are used to prevent the spread of disease organisms via the hydrocooling water. Sanitation, including frequent cleaning and the maintenance of a low chlorine concentration, is important. Product tolerance to chlorine must be known. Shower-pan holes must be cleaned daily to avoid plugging, which causes uneven water flow over the product.

Potential limitations of hydrocooling must also be considered. When the hydrocooler is operating at capacity, arriving warm produce must remain at ambient temperatures to await cooling. Further, when cooling is completed, the product must be moved to a cold room, or rapid rewarming will occur. Hydrocooling operations can require rehandling of the pallet bins prior to packing or storage. Cooling efficiency may be low, unless the hydrocooler is operated continuously at maximum capacity, or is inside a cold room or an insulated enclosure.

Package-Icing

Some commodities are cooled by filling packed containers with predetermined quantities of ice, depending upon the initial product temperature. Initially the direct contact between product and ice causes fast cooling. However, as soon as the ice in contact with the product has melted, the cooling rate slows considerably. Package ice may be finely crushed ice, flake-ice, or a slurry of ice and water called liquid-ice. Liquid-icing distributes the ice throughout the container, achieving better ice/product contact (figs. 7.8 and 7.9). Package-icing requires use of more expensive, water-tolerant shipping containers. In small operations package-ice is hand raked or shoveled into containers; in large volume operations it is done automatically with mechanical icers. Liquid-icers for automatically icing pallet loads of packed cartons are now used for cooling some field-packed vegetables.

Vacuum Cooling

Vegetables that have a favorable surface-to-mass ratio, such as leafy vegetables, can be rapidly cooled by this method. It is also used to cool cauliflower, celery, and some sweet corn, carrots, and bell peppers. Its use with carrots and peppers is primarily to dry the surface and stems, respectively, to inhibit postharvest decay. Cooling is achieved by reducing the atmospheric pressure inside a large, strongly constructed steel chamber containing the product. Reducing the atmospheric pressure also reduces the pressure of water vapor in the chamber. When the water vapor pressure in the chamber is reduced below that in the product's intercellular spaces, water will evaporate from the product resulting in cooling. Vacuum cooling causes about 1 percent product weight loss (mostly water) for each 6°C of product cooling. This amount of weight loss can be objectionable for celery and some leaf lettuce. In one patented process now in use, especially for celery, weight loss is greatly reduced by addition of water in the form of a fine spray during the vacuum cooling cycle (fig. 7.10).

Some vacuum coolers are permanently located. However, most vacuum cooling equipment is now portable, and used in two or more districts each year. This allows amortization of the high costs of vacuum coolers over a longer period each year. While early vacuum coolers were primarily of the stem-jet type, most presently used coolers use mechanical, rotary vacuum pumps.

Cooling before Packing

Cooling problems with products in unitized pallets, or poly-packed products, can be avoided by cooling the

Fig. 7.8. A pallet liquid-icing machine in operation. High volume flow of the ice-water mix is pumped into chamber and flows through container vents to deposit the ice throughout the package.

Fig. 7.9. A package of broccoli that has been liquid-iced is opened to show the penetration of ice throughout the package.

product before packing. This increases the basic cooling cost because products that are subsequently removed (culls or diverted products) are also cooled. With cantaloupes, this problem is avoided by removal of most cull fruit before hydrocooling. If 20 percent cullage occurs after cooling, the cooling cost increases 25 percent. If 50 percent is removed (diverting pears to a processor, for example), then the cooling cost per ton of packed product doubles. The cost may become prohibitive unless the cooling adds value to the diverted product.

Some rewarming will occur if produce is packed after cooling. A mild breeze can rewarm products to near ambient temperatures within 30 minutes. Some packers minimize this rewarming by only partially cooling the product before packing, followed by complete cooling after packing. One packer solves this problem another way: fruit is forced-air cooled in bins on arrival from the field and the bin dump is located in the forced-air bin cooling room. The cold, dumped fruit moves from cold room to packing area, where it is sorted, sized, and tight-fill packed into containers within 3 or 4 minutes. Packed containers are then conveyed into a cold room for palletizing within 6 to 7 minutes after leaving the bin cooler. Fruit temperature rise in that system is minimal.

Summary

Effective cooling and temperature management in the changing horticultural crops handling system requires a complete understanding of product and market requirements, and of cooling methods now available.

- Rapid cooling and good product temperature management are increasingly recognized as essential for successful marketing. This recognition results from increased knowledge of product temperature requirements together with changing handling and marketing requirements.

- Cooling is part of the total perishables handling system. Effects on cooling rate must be considered whenever a change is made in any part of the system.

- Requirements for cooling and cold storage are quite different and should be considered as two separate operations.

- Several cooling methods and variations are available to achieve reasonably fast cooling, even under the conditions imposed by newer handling procedures. Effective use of these options requires a complete understanding of the principles involved.

- Introduction of field bins and unitized handling of packed containers has made rapid cooling more difficult to achieve. Contemporary cooling operations often involve large product masses with restricted access. Modifications to cooling systems can provide effective cooling even under such conditions.

- Cooling efficiency can often be improved by attention to details of air management, package design, or pallet stacking patterns.

- Increased costs involved in achieving faster cooling may be relatively small when total cost of the cooling system is considered.

- Fast cooling methods can often be accomplished through minor modifications of existing cooling facilities. Requirements should be determined by a qualified refrigeration engineer after evaluating the complete refrigeration system.

References

1. Isenberg, F. M. R., R. F. Kasmire, and J. E. Parson. 1982. *Vacuum cooling vegetables*. Cornell Univ. Coop. Ext. Bull. No. 186. 10 pp.

2. Jeffrey, J. J. 1977. Engineering principles related to the design of systems for air cooling of fruits and vegetables in shipping containers. *Proc. 29th Int'l. Conf. on Handling Perishables Agricultural Commodities*, 151-64. Mich. State Univ.

3. Kasmire, R. F., and F. G. Mitchell. 1974. *Perishables Handling Issue 36*. Univ. Calif. Coop. Ext. 14 pp.

4. Kasmire, R. F., and J. E. Parson. 1979. *Operator's guide to effective vacuum cooling*. Univ. Calif. Coop. Ext. (unnumbered).

5. Lutz, J. M., and R. E. Hardenburg. 1968. *The commercial storage of fruits, vegetables, and florist and nursery stocks*. USDA Agric. Handb. 66. 94 pp.

6. Mitchell, F. G., R. Guillou, and R. A. Parsons. 1972. *Commercial cooling of fruits and vegetables*. Univ. Calif. Agric. Exp. St. Ext. Serv. Manual 43. 44 pp.

7. Mitchell, F. G., and R. F. Kasmire. 1978. *Perishables Handling Issue 39*. Univ. Calif. Coop. Ext. 12 pp.

8. Rij, R. E., J. F. Thompson, and D. S. Farnham. 1979. *Handling, precooling, and temperature management of cut flower crops for truck transportation*. Univ. Calif. Coop. Ext. Leaf. 21058.

9. Thompson, J. F., and R. F. Kasmire. 1981. An evaporative cooler for vegetable crops. *Calif. Agric.* 35(3&4): 20-21.

Fig. 7.10. Vacuum cooler being loaded. Batches of product are filled into the chamber which is then closed and the vacuum drawn. The unit shown uses a patented process to introduce water during the cooling cycle in order to reduce water evaporation from the product. Courtesy of Mel Gagnon.

8

Cooling Operations: Evaluation of Efficiency

ROBERT F. KASMIRE

Cooling efficiency might be considered as the amount of product cooling achieved per unit of time in proportion to the total costs expended to obtain that cooling. Cooling costs that must be considered include capital investment and amortization for facilities and equipment, materials, labor, power, operations, and maintenance. To perform efficiently, cooling systems must be properly planned, designed, operated, and maintained. *Energy efficiency* in cooling is the amount of product cooling achieved in proportion to the amount of energy consumed to produce that amount of cooling. *Cooling rate* is the amount of product cooling obtained per unit of time.

Energy for cooling is a small percentage of the total energy input (TEI) needed to produce, harvest, package, and cool fresh fruits and vegetables. For example, only 4.5 percent of the TEI was needed to cool summer cantaloupes an average of 16.7°C (30°F) (Chancellor et al. 1978). Hydrocooling Florida sweet corn required only 0.8 percent of the TEI from production through cooling, according to Gaffney (1980). Even within these limited energy use estimates, there are opportunities for reducing energy use in cooling operations. However, the economics of any potential reductions need to be carefully evaluated.

Planning

Product(s) to be cooled

1. Types of cooling methods used must be compatible with types of products to be cooled. Fruits are best cooled by forced-air cooling, roots and stems by hydrocooling, and leafy vegetables by vacuum or hydro-vac cooling. With products that can be effectively cooled by more than one method, economic considerations may be more important in planning the method to use.

2. Size and density affect rate of heat transfer within products. Small sizes cool faster than larger sizes of a given commodity.

3. Final product temperature desired will affect the length of cooling cycle needed. Longer cooling cycles are needed to achieve final product temperatures.

4. Quantity to be cooled per unit of time will influence the size of cooler and refrigeration costs. Cooler size should be estimated by the average daily maximum amount of product to be cooled during a season.

5. Future expansion of shipping (and cooling) operations should be considered. It is less expensive to plan ahead for expansion than to completely remove and rebuild cooling equipment and facilities.

Packaging affects cooling

Packages (shipping containers) and packaging materials to be used will affect cooling rate. They all inhibit cooling somewhat and thereby reduce the amount of product that can be thoroughly cooled per day.

1. Types of shipping containers used (bulk bins, crates, cartons, bags) will affect cooling rate and uniformity, and will, therefore, affect the efficiency of a cooler. Operators should determine cooling rates for products in each type of container used.

2. Packaging materials (consumer packages, container liners, and so on): Cooling efficiency is reduced with increasing tightness and thickness of packaging materials. Ventilating or perforating packaging materials improves cooling efficiency somewhat but such openings in the materials must match those in shipping containers.

3. Container venting, stacking, and loading patterns during cooling: Vents in adjacent containers must be accurately aligned for rapid cooling to occur. Container venting can improve cooling efficiency if the containers are properly designed, constructed, and stacked. Stacking and loading patterns in coolers should be such that cooling rates (and efficiency) are minimally impaired.

Cooling method selection factors

1. Familiarity of management and operations personnel with possible cooling methods: For the most efficient use of a cooler, management and operations personnel should be thoroughly familiar with the principles, requirements, and individual characteristics of the specific cooler used.

2. Susceptibility of products to damage by a cooling system, e.g., damage to cherries and leafy vegetables by hydrocooling.

3. Length of cooling season.

☐ Long season—fixed installations and equipment. More capital investment may be justified to reduce operation costs (energy, labor, management, and maintenance), e.g., better insulated cold rooms. Greater capacity refrigeration units can help cool more products faster, reducing labor costs (fig. 8.1).

☐ Short season. Portable cooling facilities (vacuum coolers, hydrocoolers, refrigeration units for cold rooms and forced-air coolers, and liquid-icing units) can be used in two or more shipping districts each year (fig. 8.2). Combinations of fixed and portable facilities with permanent cold rooms with basic refrigeration capacity and capability of adding refrigeration from mobile units can be used during peak season.

☐ Harvesting products during the coolest part of the day and then moving them promptly to the cooler minimizes product temperature increase. Keeping products under shades prevents product heating prior to cooling (fig. 8.3). These practices lower the amount of refrigeration capacity needed for cooling, can require shorter cooling cycles, and provide for more product cooling per day.

☐ Cooler protection from heat is desirable. Coolers exposed to direct sunlight absorb more heat, use refrigeration capacity needed for product cooling, and operate less efficiently. Less heat is conducted across cooler surfaces and motors operate more efficiently in coolers located under shades or in cold rooms.

Fig. 8.1. A permanently-located vacuum cooler. Courtesy of Mel Gagnon.

Fig. 8.2. A portable vacuum cooler.

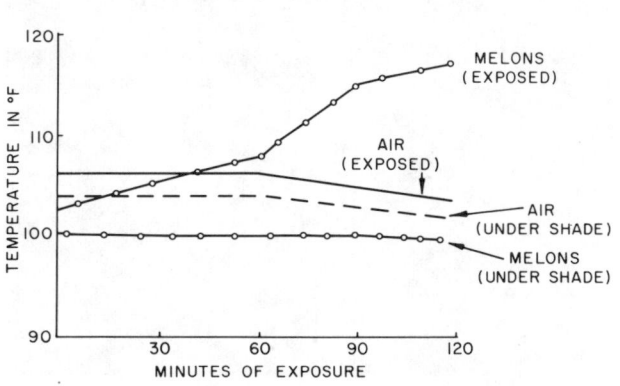

Fig. 8.3. Comparison of fruit temperatures of cantaloupes kept in the shade vs those exposed to the sun.

Cooler Design

Cooler design involves engineering the cooler for the specific requirements determined. Design factors that must be considered include the following:

1. Type of refrigeration system to use: Choices include mechanical refrigeration with an evaporatively cooled coil; mechanical refrigeration with a packed cooling tower; purchased ice for use in room, forced-air, hydrocooling, or package-icing; or evaporative cooling. Combinations of systems are also used, e.g., mechanical refrigeration is used to make flake-ice that is then added directly or with water as a liquid-ice slurry to packed containers of produce for cooling.

2. Total cost of refrigeration sources in the location of cooler: Factors that must be considered are initial capital investment (first cost), operating costs (energy, labor, maintenance, ability to obtain minimum temperature desired), costs of shipping containers, and additional product handling costs.

3. Product exposure to cooling medium (cold air, water, ice): Amount of exposure influences the amount of cooling achieved and, thus, the efficiency of a cooler. Efficiency is increased when the cooling medium makes thorough contact with the product, e.g., products that are hydrocooled in bulk have much greater exposure to the cooling water than products in packed containers. Containers prevent the water from having equal access to all of the product in the package (fig. 8.4).

4. Cooling rate: This rate increases when the cooling medium circulation rate (cfm, gpm) is adequate to maintain a minimum temperature difference between the refrigeration source and the cooling medium.

5. Exposure of cooling medium to refrigeration source (coils, ice): Increasing the amount of contact between the cooling medium and the refrigeration source by use of larger coil surface or more ice surface provides greater heat transfer from the cooling medium to the coil or ice, which increases cooler efficiency at completion of cooling.

6. Final product temperatures desired or needed: For each degree of product cooling achieved, both operating and capital costs increase when commodities are cooled to very low final temperatures, e.g., within 1°C (2°F) of the cooling medium's temperature. This applies especially to hydrocooling and vacuum cooling but also to forced-air cooling. This is because for each additional one-half cooling period used, the amount of cooling achieved (heat removed from product) is less in quantity (fig. 8.5).

7. Temperature measurement: Proper measurement of product temperature (fig. 8.6) and cooling medium temperature is essential to understand the amount of cooling being achieved and to obtain maximum performance from a cooler. This requires knowledge of the cooling pattern within shipping containers, pallet bins, or a cooling chamber.

8. Heat removal: Average maximum product heat load to be removed will influence the amount and cost of refrigeration needed and the potential cooler efficiency.

9. Design features: Operations heat load features that must be considered in the design of a cooler include lights; loading and unloading lift-truck traffic; types of lift-trucks used; fans; pumps; leaks around doors; cold room outside surface type and color; type, amount, and condition of insulation; and cold room slab construction. Use of double flap-doors can help reduce heat leakage into cold storage rooms during loading and unloading. Use of insulated "bumper seals" around truck loading doors of cold rooms also helps to reduce heat gain during loading. These subjects are discussed in detail in chapter 9. Cooling systems should be designed for minimum utilities (energy) costs. This includes minimum energy use per unit of commodity cooled and minimum energy use during peak electrical rate periods.

10. Labor requirements: Coolers with high labor inputs require optimal management for good cooler efficiency. During periods of maximum cooling need, e.g., in summer,

Fig. 8.4. A hydrocooler with heavy water shower splashing over and around cartons of produce being cooled.

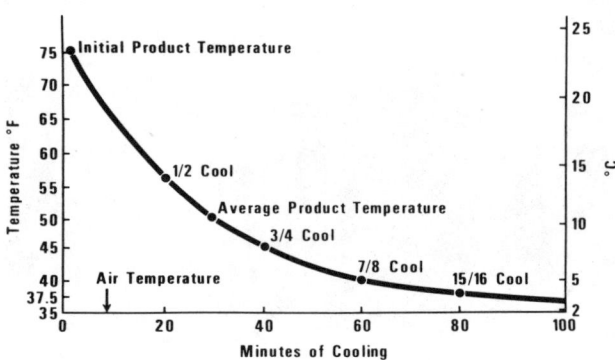

Fig. 8.5. Changes in cooling rate with time of cooling.

this requirement is often neglected, resulting in inefficient and costly cooling.

11. Maintenance requirements: Coolers designed for easiest maintenance tend to be better maintained and more efficient.

12. Size of cold room needed with cooler: More lift-truck activity is needed to load and unload large vs small cold rooms. This adds considerably more heat to large rooms, especially if lift-trucks are LP gas powered. Thus, less heat from lift-trucks would be added to a series of small cold rooms having a combined area equal to that of one large room.

13. Volatile levels: Air changes needed to prevent accumulation of harmful levels of volatiles (e.g., ethylene). Provisions must be made to maintain safe levels of ethylene and possibly other harmful volatiles. Cold room temperatures are better maintained when the number of air changes is reduced. Use of heat exchangers in the port where the air exchange occurs increases cooler efficiency but requires greater initial investment.

Cooler Operation

The key to efficiency in cooler costs and energy inputs can be found in cooler operation. Knowledgeable, skilled operators are essential. Factors involved include:

1. Temperature measurement: Measuring product temperatures before and after cooling, if done properly, provides a measure of the amount of cooling achieved per unit of time, information that is essential to obtaining maximum cooler efficiency. Thermometers used for temperature monitoring should be regularly calibrated for accuracy.

2. Cooling cycle length: Determining length of cooling cycle needed can be achieved through use of cooling schedules. This is determined with the before-and-after cooling temperatures recorded. Adjusting length of cooling cycle according to incoming product temperatures and final temperatures desired (after cooling) increases cooler efficiency.

3. Variable cooling rates: Determining variable cooling rates for different areas of the cooler (room and forced-air) is necessary because all areas of most coolers do not cool at the same rate. Operators need to know the variability in cooling rates for different areas of a cooler. This applies mainly to room coolers and forced-air coolers, but can also apply to hydrocoolers.

4. Labor costs: The cost of labor to operate coolers can be considerable if not properly planned for. Forced-air coolers and vacuum coolers require lift-truck loading and unloading. Loads need to be located so that only the minimum amount of lift-truck activity is needed to load and unload coolers. Excessive, unnecessary activity is costly and adds heat to cold rooms. This requires more energy use to provide the refrigeration needed to absorb the heat from lift-trucks. Cooler operators generally perform cooler maintenance tasks in addition to operating the cooler.

5. Power expense: Power costs provide a clue to cooler operating efficiency. As power costs increase, it is essential to keep energy use as low as possible to achieve the needed results. Utilize the lowest rates whenever possible where variable rate schedules apply. Initiate cooling of each load soon after it arrives at the cooler, instead of waiting for several loads to arrive before beginning to cool them.

6. Loading of cold rooms: Careful management of loading and unloading cold rooms helps to limit the amount of heat entering through open doors and from excessive lift-truck activity. Reducing the amount of lift-truck operations in cold rooms can increase cooler efficiency. One way to accomplish this is by conveying pallets of packed containers of produce on an electric-powered conveyor through flap doors into cold rooms and then using only electric-powered lift trucks in the cold rooms.

7. Experienced personnel: Cooler operators must be knowledgeable about cooling systems and methods used. Knowledgeable operators would more likely be able to operate coolers more efficiently and effectively.

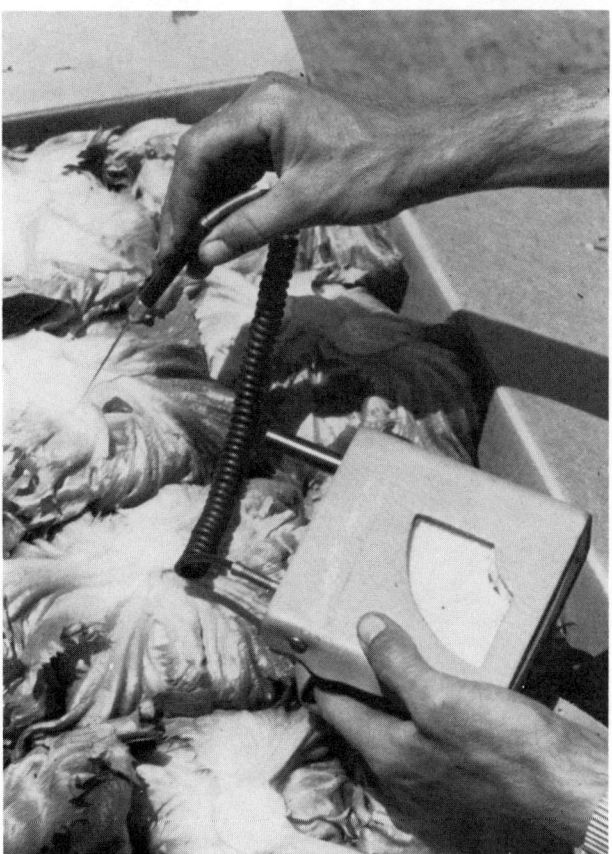

Fig. 8.6. Measuring lettuce temperature after vacuum cooling.

Maintenance

Proper maintenance is essential to efficient cooler operation. This includes (1) conducting operations checks before, during, and after cooling; (2) cooler cleaning; (3) monitoring safety procedures and equipment; (4) periodic and annual maintenance; and (5) use of maintenance manuals. All the physical features of a cooler must be properly maintained for efficient operation. These include refrigeration system components, cooling medium circulation components, conveyors, lubrication points, electrical systems, walls, doors and door seals, and flap doors. Unfortunately, some of these features are ignored or inadequately maintained, resulting in lower cooler efficiency and excessive power costs.

1. Pre-operation checks before starting the cooler each season and before each day's operation. These are especially important with vacuum coolers.

2. Constant checking for defective operation of any mechanical parts and refrigerant levels. Replace damaged flaps on flap doors to reduce heat leakage into cold rooms.

3. Regular cleaning and refilling of cooling medium reservoirs (hydrocoolers, icing systems). Regularly cleaning debris from hydrocooler shower pans or nozzles to provide uniform application (flow) of cold water in the cooler is essential.

4. Safety: Safe operations are less costly than unsafe operations. Any down time caused by accidents can greatly reduce cooler efficiency. Efficiency is also low if workers have to spend extra time avoiding hazards. Be careful of unnecessarily hazardous working conditions. In addition to the pain and cost of injuries, delays can occur when replacement workers must be trained.

Comparing energy costs of various cooling methods would be futile without first knowing the energy efficiency of the individual facilities being compared, including effects of their design, operation, and maintenance.

References

1. ASHRAE. 1982. Methods of precooling food. In *ASHRAE guide and data book, applications volume*, chap. 27. Am. Soc. Heating, Refrigeration, and Air Conditioning Engineers. Atlanta, GA.

2. Barger, W. R. 1963. *Vacuum precooling—a comparison of the cooling of different vegetables*. USDA Mark. Res. Rept. 600. 12 pp.

3. Gaffney, J. J. 1977. Engineering principles related to the design of systems for air cooling of fruits and vegetables in shipping containers. In *Proceedings 29th Int'l. Conf. on Handling Perish. Agr. Commodities*, 151-64. Mich. State Univ.

4. Isenberg, F. M. R., R. F. Kasmire, and J. Parson. 1982. *Vacuum cooling vegetables*. Cornell Univ. Coop. Ext. Info. Bull. 186. 10 pp.

5. Kasmire, R. F., and J. Parson. 1979. *Operator's guide to effective vacuum cooling*. Univ. Calif. Coop. Ext. Service, Misc. Publ.

6. Kasmire, R. F., and R. A. Parsons. 1971. *Precooling cantaloupes—a guide for shippers*. Univ. Calif. Agric. Ext. Serv. Misc. Publ.

7. Mitchell, F. G., R. Guillou, and R. A. Parsons. 1972. *Commercial cooling of fruits and vegetables*. Univ. of Calif. Agric. Sci. Publ. Manual 43. 42 pp.

8. Stewart, J. K., and M. H. Couey. 1963. *Hydrocooling vegetables; a practical guide to predicting final temperatures and cooling times*. USDA Mark. Res. Rpt. 637. 32 pp.

9
Storage Systems

JAMES F. THOMPSON

Introduction

Orderly marketing of perishable commodities often requires some storage to balance day-to-day fluctuations between product harvest and sales or for long-term storage to extend marketing beyond the end of harvest season. The goals of storage are:

1. Slow biological activity of product by maintaining the lowest temperature that will not cause freezing or chilling injury and by controlling atmospheric composition
2. Slow growth of microorganisms by maintaining low temperatures and minimizing surface moisture
3. Reduce product drying by reducing the difference between product and air temperatures and maintaining high humidity in the storage room

With some commodities, the storage facility may also be used to apply special treatments. For example, potatoes and sweet potatoes are held at high temperature and high relative humidity to cure wounds sustained during harvest, oranges may be degreened prior to shipment, and pears treated to ripen more quickly and uniformly. This chapter will describe the equipment and techniques commonly used to control temperature, relative humidity, and atmospheric composition in a storage facility.

Storage Considerations

Temperature

The temperature in a storage facility normally should be kept within about 1.1°C (2°F) of the desired temperature for the commodities being stored. For storage very close to the freezing point, a more narrow range may be needed. Temperatures below the optimum range for a given commodity will cause freezing or chilling injury; temperatures above it will result in reduced storage life. In addition, wide temperature fluctuations can result in both water condensing on the stored products and more rapid water loss. USDA Agricultural Handbook 66 lists recommended storage temperatures and humidities for horticultural products.

Maintaining proper storage temperatures within the prescribed range is a result of several important design factors. The refrigeration system must be sized to handle the maximum expected heat load. Obviously, an undersized system will allow the air temperature to rise during peak heat load conditions. An oversized system is unnecessarily expensive. The system should also be designed so that the temperature of the air leaving the refrigeration coils is close to the desired temperature in the room. This will prevent large temperature fluctuations as the refrigeration system cycles on and off. Large refrigeration coils will allow the use of a small temperature difference between the air leaving them and the air in the room and still maintain adequate refrigeration capacity.

Temperature variation within the room is minimized by incorporating adequate levels of insulation in the walls and by maintaining adequate air circulation in the room. The circulation system should be designed to provide 0.25 to 0.35 meters per second (50 feet to 75 feet per minute) air flow around the stacked containers. As a rule of thumb, this can be obtained with a circulation system which provides an air volume of at least 7.5 air changes per hour based on the volume of the empty storage room. When the room is filled, the containers must be stacked to allow air passage past at least one side of each container.

Thermostats are usually placed at a height of 1.5 m (5 feet) from the floor for ease of checking and placed in a representative location in the room. They should not be placed next to sources of heat such as a door or on walls with an exterior surface. Similarly, they should not be placed in a typically cold area such as near the air discharge of the refrigeration unit. A calibrated thermometer should be used to periodically check the thermostat. Remember, errors of only a few degrees can affect product quality.

Humidity

For most perishable commodities, the relative humidity in a storage facility should be kept in the range of 90 percent to 95 percent. Humidity levels below this range will result in unacceptable moisture loss. Humidities close to 100 percent may cause excessive growth of microorganisms and may cause surface cracking, although it is unusual to find a storage facility with relative humidities that are too high. Refrigeration equipment must be especially designed to maintain high relative humidity. In normal systems the evaporator coils (the units which produce the cold air) operate at a temperature about 6°C (11°F) lower than the desired air temperature in the room. This causes an excessive amount of moisture to condense on the coils resulting in relative humidities of 70 percent to 80 percent in the storage room. The large coils achieve the same refrigeration capacity as smaller coils, but can operate at a higher temperature, thus reducing the moisture removed from the air. The coils should be large enough to allow no more than a 3°C (5°F) difference between the coil temperature and the room air temperature. Mechanical humidifiers, fog spray nozzles, or steam systems are sometimes used to add moisture to the storage room and reduce the drying effect of the condenser coils. However, this added moisture may necessitate more frequent defrosting of coils and increase operating problems.

Some refrigeration systems utilize a wet coil to maintain a high humidity. In this system, water is cooled to 0°C (32°F) or a higher temperature if higher room temperatures are desired. The water is sprayed down through a coil and the storage area air is cooled and humidified as it moves upward through the coil. However, as this air moves through the storage area it will pick up heat and the increase in temperature will reduce the relative humidity. Supplemental humidification may also be needed with this system.

Atmospheric composition

The atmospheric composition in a storage facility is controlled by adding gases, allowing the commodity to produce or consume gases, or by physically or chemically removing undesirable gases from the storage area. Gases such as carbon monoxide (CO), carbon dioxide (CO_2), ethylene (C_2H_4), and nitrogen (N_2) can be added to a facility from a bottled supply (or dry ice in the case of CO_2) or produced by on-site generators. As perishable commodities undergo respiration, they consume oxygen (O_2) and release CO_2. This effect can be successfully used to control the concentration of these gases in storage. High concentrations of undesirable gases can be removed by scrubbing devices. For example, CO_2 can be absorbed in water or lime; C_2H_4 and other volatiles can be removed by potassium permanganate, catalytic oxidation or UV light. Oxygen (O_2) can be removed by using it in a combustion process or by a molecular sieve. In some cases, outside concentrations of gases are desirable and the build-up of unwanted gases can be controlled by ventilation.

Refrigeration

Mechanical refrigeration

Most storage facilities use mechanical refrigeration to control storage temperature. This system utilizes the fact that a liquid absorbs heat as it changes to a gas. The simplest method for doing this is to release liquid nitrogen in the storage environment but this requires a constant outside supply of refrigerant. This system is used only to a limited extent with highway vans where high nitrogen concentrations and a low O_2 level may also be of value. The more common mechanical refrigeration systems use a refrigerant such as ammonia or a variety of halide fluids (sometimes referred to by the trade name "freon") where the vapor can be easily recaptured by a compressor and heat exchanger.

Figure 9.1 illustrates the components of a typical vapor recompression (or mechanical) refrigeration system. The refrigerant fluid passes through the expansion valve where the pressure drops and the liquid will evaporate at temperatures low enough to be effective in removing heat from the storage area. Heat needed to cause evaporation comes from the material to be cooled. Heat is transferred to the air in the storage room and it is forced past the evaporator (cooling coil). It is usually a finned tube heat exchanger which transfers the heat in the air to the liquid refrigerant causing it to evaporate. The evaporator is located in the storage area. After fully changing to a gas, the gas is repressurized by the compressor and then passes through a condenser where it is cooled to a liquid. The condenser is located outside the storage area and rejects heat. Liquid is stored in the receiver and is metered out of the receiver as needed to produce the needed cooling.

Expansion valves

A mechanical refrigeration system is controlled primarily by the expansion valve. The valve controls the low side pressure. Low pressures cause the liquid refrigerant to evaporate at low temperatures. The valve also controls the flow of refrigerant, which affects the amount of refrigeration capacity available. Capillary tubes and thermostatic expansion valves are the most common types of expansion valves.

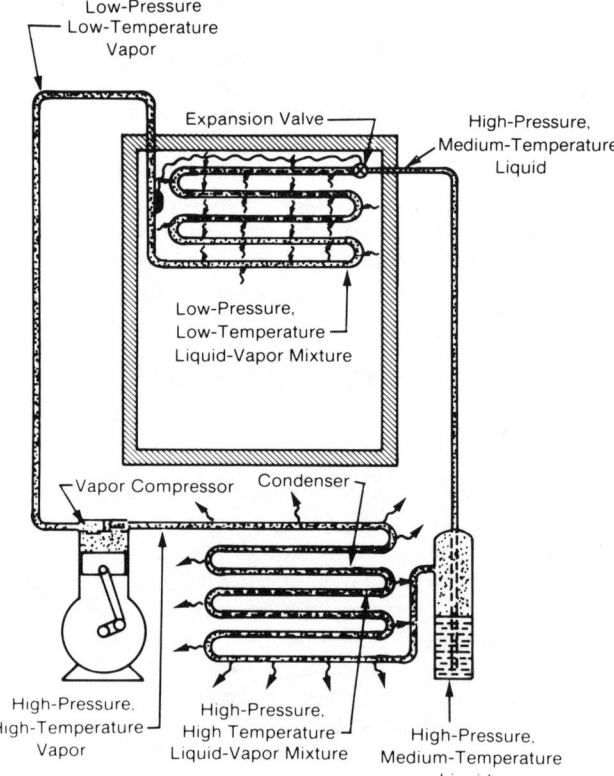

Fig. 9.1. Schematic of a typical vapor recompression or mechanical refrigeration system.

The capillary tube is used with small refrigeration systems (less than one horsepower). It is a 0.6 to 6 m (2 feet to 20 feet) long tube with a very small inside diameter of 0.6 to 2.3 mm (0.025 inch to 0.090 inch). The resistance of the liquid flowing through such a small tube causes the needed pressure drop between the low pressure side and high pressure side of the system. A capillary is inexpensive and has no moving parts to maintain, but it cannot be adjusted, is subject to clogging, and requires a relatively constant weight of refrigerant in the system.

The thermostatic expansion valve is the most widely used expansion valve. It regulates the flow of refrigerant to maintain a constant temperature difference between the evaporator inlet or evaporating temperature and the temperature at the coil outlet (maintains a constant super heat). This type will allow the low side pressure to vary, so when high refrigeration loads are required, the temperature of the evaporator coil will increase.

Large refrigeration systems may use an evaporator coil which is designed to always have liquid refrigerant in it (flooded coil). A flooded coil has a greater heat transfer efficiency than an equivalently-sized nonflooded coil. Refrigerant flow is controlled primarily with a float control. It insures a constant level of refrigerant in the coil. The float control may operate in parallel with a thermostatic expansion valve.

Other controls such as suction pressure regulators may be used in conjunction with float controls. These are especially useful in maintaining a minimum evaporator coil temperature to prevent excess dehumidification.

Evaporators

Modern cold storages usually use finned tube evaporators. Air from the storage is forced past the tubes by fans which are a part of a complete evaporator unit. Evaporators operating near 0°C (32°F) build up frost and it must be removed to maintain good heat transfer efficiency. Defrosting may be done by periodically flooding the coils with water, by electric heaters, by directing hot refrigerant gas to the evaporators, or by continuously defrosting with a brine or glycol solution.

Compressors

The most common types of refrigeration compressors are reciprocating (piston), centrifugal, rotary, and rotary screw (fig. 9.2). The type of compressor used in an operation depends on size needed, refrigerant type, nature of the installation, and availability.

Condensers

Condensers are categorized primarily on the basis of the method used to cool them. Small systems usually use an air-cooled unit. Home refrigerators have a coiled tube in the back which allows a natural draft of air to flow past. Larger systems use a fan to guarantee a good flow of air past the condenser. Large condensers also can be water cooled. Water is a better heat conductor than air, allowing water cooled condensers to be smaller than equivalently sized forced-air units. However, water cooled units may require large quantities of water, which can be expensive to obtain and dispose of. Evaporative condensers reduce water consumption by evaporatively cooling the heated condenser water and recycling it; they require close attention to water quality to prevent damage to the heat exchanger.

Refrigerants

The choice of which refrigerant to use in a vapor recompression system is based on the following factors:

1. Cost of refrigerant—halide refrigerants are more expensive than ammonia
2. Compatibility—ammonia cannot be used with metals that contain copper; halide refrigerants cannot be used with magnesium and may damage plastic materials
3. Toxicity—ammonia at very low concentrations can cause injury to perishable commodities (including humans)

Control systems

Successful operation of a large refrigeration system requires a good control system and a system for displaying the system's operating condition. At a minimum, panel lights should be installed to indicate whether fans and compressors are operating and critical fluid levels in surge and receiver tanks. Controls should be set up to allow manual operation of motors.

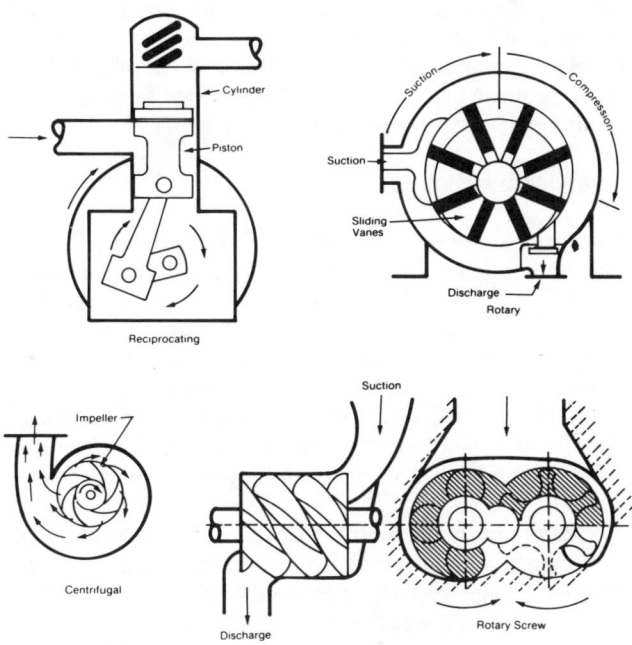

Fig. 9.2. Common types of refrigeration compressors.

Microcomputers and programmable controllers are being used to allow even more precise control of large refrigeration systems. They are especially valuable in reducing electricity use during peak rate periods. Defrost cycles can be programmed to activate only at night and unnecessary fans and compressor motors turned off during peak rate periods.

Absorption refrigeration is used in a few cold storage operations. It differs from mechanical refrigeration in that the vapor recovery is accomplished primarily through use of heat rather than mechanical power. However, absorption refrigeration is less energy efficient than mechanical refrigeration and is usually used when an inexpensive source of heat is available.

The capacity of a refrigeration system is based on adding all the heat inputs to a storage area. Heat inputs include:

1. Heat conducted through walls, floor, and ceiling
2. Field and respiration heat from the product
3. Heat from air infiltration
4. Heat from equipment such as lights, fans, forklifts, and personnel

Refrigeration equipment for storage facilities generally is not designed to remove much product field heat because of the large capacity required to do this. A separate cooling facility is used for this purpose.

Alternate Refrigeration Sources

In many developing countries, mechanical refrigeration is prohibitively expensive to install and maintain. There are a number of other techniques which can be used to produce refrigeration. In some cases these will provide nearly recommended storage conditions. In others, it is a compromise between proper storage conditions and low equipment, capital, and operating costs.

Evaporative cooling

Evaporative cooling techniques are very energy efficient and economical. A well-designed evaporative cooler will produce air with a relative humidity greater than 90 percent. The main limitation to evaporative cooling is that it will cool air only to the wet bulb temperature. During the harvest season in the U.S., wet bulb temperatures vary from 10° to 25°C (50° to 75°F) depending on location, time of day, and weather conditions. This temperature range is acceptable for many chilling-sensitive commodities.

Minimum temperatures from an evaporative cooling system can be reduced by using a multiple-stage system. In a two-stage system, an evaporative cooler is used to cool water to the wet bulb temperature which is, in turn, used in a water-to-air heat exchanger to cool outside air. The cooled air has a reduced wet bulb temperature (a 10°C [18°F] drop in dry bulb temperature produces a 3° to 4°C [5.4° to 7.2°F] drop in the wet bulb) and is passed through an evaporative cooler for final cooling. This system can reduce the temperature produced by an evaporative cooler by as much as 6°C (10°F) compared to a single-stage system. A system can be set up to use more than two stages and the theoretical minimum temperature which a multistage system can produce is the dew point temperature of the air.

The water for cooling in the systems mentioned above comes from domestic sources. It is also practical to cool by evaporating water from the commodity. Snap beans have been cooled in transit by erecting an air scoop above the cab of the truck which forces outside air through a bulk load of beans. This system prevented heat buildup and kept the beans at or below the outside air temperature. It is not advisable to use this system for any great length of time to avoid excessive weight loss.

Nighttime cooling

In some parts of the world there are significant differences between night and day temperatures allowing nighttime ventilation to be a good source of cold. In parts of California, the difference between maximum and minimum temperatures can be as high as 22°C (40°F) during the summer. Nighttime cooling is commonly used for unrefrigerated storage of potatoes, onions, sweet potatoes, hard rind squashes, and pumpkins. As a rule, night ventilation is effective in maintaining a given product temperature when the outside air temperature is less than product temperature during 20 percent to 30 percent of the time.

Low nighttime temperatures can be used to reduce field heat simply by harvesting produce during early morning hours. Several cantaloupe growers in California are experimenting with using artificial lighting to allow nighttime harvest.

It is theoretically possible to produce air temperatures below nighttime minimums by radiating heat to a clear, nighttime sky. A clear night sky is very cold and a good radiating surface such as a black metal roof will lose heat and cool below the air temperature. Computer aided simulation studies have indicated this method could cool air about 4°C (7.2°F) below night air temperature. This concept has not been used commercially in California.

Well water

In some areas, well water can be an effective source of cold. The temperature of the ground at depths greater than about 2 m (6 feet) below the surface is equal to the average annual air temperature. Well water temperatures are usually very near this.

Naturally-formed ice

Before the development of mechanical refrigeration, refrigeration was provided by ice formed naturally in shallow ponds during the winter. The ice was insulated from spring and summer heat with straw and hauled to cities as needed.

Unfortunately, energy costs for transportation have, in most cases, increased faster than electrical costs and it is not feasible to transport ice any significant distances. However, cooling facilities in climates where winter ice making is feasible can store ice nearby for summer use. In some cases, it may be feasible to transport perishable commodities to the ice for storage. This is especially practical when the ice location is between the location of production and consumption.

High altitude cooling

High altitude can also be a source of cold. As a rule of thumb, air temperatures decrease by 10°C (18°F) every 1 km (0.62 miles) increase in altitude. It is not possible to bring this air down to ground level because it naturally heats up as it drops in altitude (adiabatic compression heating). However, in some cases it may be possible to store the commodity at high elevations in mountainous areas. For example, in California most perishable commodities are produced in the valley floors near sea level. However, much of the production is shipped east across the Sierra Mountains where the passes are about 1,828 m (6,000 feet) high. Air temperature has the potential of being 18°C (32°F) cooler and it may reduce energy costs to store perishables there rather than on the valley floor.

Underground storage

Cellars, abandoned mines, and other underground storages have been used for years for storage of fruits and vegetables. As mentioned earlier, their temperature is approximately equal to the average annual air temperature. Cool underground spaces work well for storing already cooled produce but not for removing field heat. The soil has a poor ability to transfer heat. Once the cold is depleted from an area it will not regenerate rapidly.

The Storage Building

The storage must be sized to handle peak amounts of product. The floor area can be calculated knowing the volume of the produce and maximum storage height and allowing for aisle ways, room for forklift maneuvering, and staging areas. Maximum storage height can be increased by use of shelves or racks and forklifts with suitable masts. Multistory structures are generally not used because of the difficulty of moving the product from one level to another.

The building should have a floor perimeter generally in the shape of a square. A rectangular configuration has more wall area per square foot of floor area resulting in higher construction cost and higher heat loss than a square configuration. Entrances, exits, and storage areas should be arranged so that product generally moves in one direction through the facility. This is especially important when the storage facility is used in conjunction with a cooler to remove field heat.

Plan for future expansion

Many facilities are enlarged as business grows and the building should be located to allow expansion. The building should be within easy access of utilities. Extending roads and energy utilities to a facility can be very expensive.

Refrigerated facilities have been constructed out of a wide variety of materials. The floor and foundation is usually a concrete slab. A vapor barrier is installed to prevent moisture movement through the slab and rigid insulation is usually placed above the barrier and below the concrete. Walls can be made of concrete block, tilt-up concrete, metal, or wood frame construction. Batt, rigid board, or foam insulation is always installed and protected with a vapor barrier. Wood frame and concrete block are losing favor to metal and tilt-up concrete construction in the U.S. Ceilings can be constructed with rigid board or foam materials, or built separately from the roof and loose fill or batt insulation used.

If modified atmosphere techniques are used in the storage facility, the vapor barrier may also serve as a gas barrier and special precautions must be taken to insure a gas tight seal.

References

Specific cold storage construction information:

1. Hallowell, E. R. 1980. *Cold and Freezer Storage Manual*. Westport, CT: AVI Publ. Co. 356 pp.

Product storage recommendations:

2. Lutz, J. M., and R. E. Hardenburg. 1968. *The commercial storage of fruits, vegetables, and florist and nursery stocks*. USDA Agric. Handb. 66. 94 pp.

See also:

3. Bartsch, J. A., and G. D. Blanpied. 1984. *Refrigerated storage for horticultural crops*. Agric. Eng. Ext. Bull. 448. Cornell Univ. Ithaca, NY.

4. Davis, D. C. 1980. Moisture control and storage systems for vegetable crops. In *Drying and storage of agricultural crops*. C. W. Hall, 310-59. Westport, CT: AVI Publ. Co.

10
Psychrometrics and Perishable Commodities

JAMES F. THOMPSON

Introduction

Psychrometrics is the measurement of the heat and water vapor properties of air. Commonly used psychrometric variables are temperature, relative humidity, dew point temperature, and wet bulb temperature. While these may be familiar, they are often not well understood.

Psychrometric Chart

The psychrometric chart describes the relationships between these variables. Figures 10.1 and 10.2 are psychrometric charts in English and metric units, respectively, which will help to illustrate the meaning of various terms.

Temperature, sometimes called *dry bulb temperature* after the nonwetted thermometer in a psychrometer, is the horizontal axis of the chart. The vertical axis is the moisture content of the air called *humidity ratio* (or mixing ratio or absolute humidity). The units of humidity ratio are mass of water vapor per mass of dry air. Under typical California conditions, the humidity ratio of outside air will vary from .004 to .015 kg/kg. Even though water vapor represents only .4 to 1.5 percent of the weight of the air, water vapor plays a very significant role in the effect of air conditions on the postharvest life of perishable commodities.

The maximum amount of water vapor that air can hold at a specific temperature is given by the left most, upward-curved line in figure 10.1 or 10.2. Notice that air holds increasingly more water vapor at increasing temperatures. As a rule of thumb, the maximum amount of water that the air can hold doubles for every 11°C (20°F) increase in temperature. This line in figure 10.1 or 10.2 is also called the *100 percent relative humidity line*. A corresponding 50 percent relative humidity line is approximated by the points which represent the humidity ratio when the air contains one-half of its maximum water vapor content. The other relative humidity lines are formed in a similar manner.

Notice that relative humidity without some other psychrometric variable does not determine a specific air condition on the chart and is not very meaningful. As will be shown, 80 percent relative humidity at 0°C (32°F) is

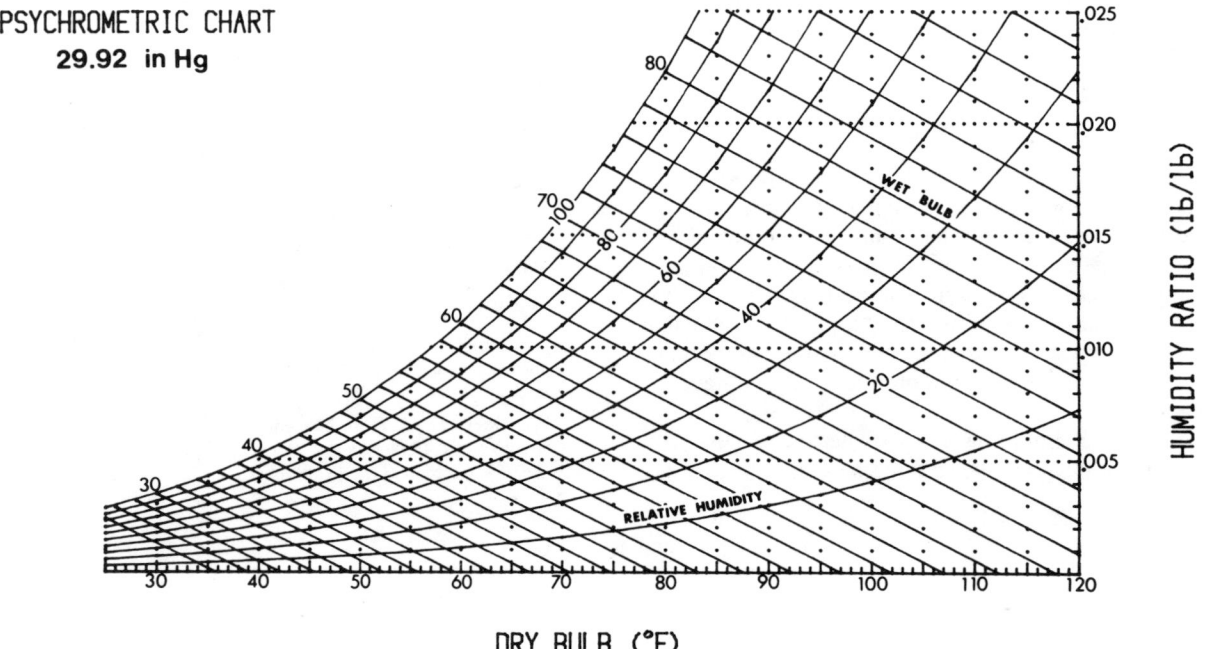

Fig. 10.1. A psychrometric chart in English units.

a much different air condition than 80 percent relative humidity at 20°C (68°F).

If air is cooled without changing its moisture content, it will lose capacity to hold moisture. If cooled enough, it will become saturated (have 100 percent relative humidity) and if cooled further, will lose water in the form of dew or frost. The temperature that causes condensation to form is called the *dew point temperature* if it is above 0°C (32°F) or the *frost point temperature* if it is below 0°C (32°F).

Another commonly used psychrometric variable is *wet bulb temperature*. On the chart this is represented by lines that slope diagonally upward from right to left. Basically, these lines approximate the temperature and water vapor conditions of a thermometer covered with water-soaked gauze. In practice, wet bulb lines are used to determine the exact point on the psychrometric chart which represents the air conditions in a given location as measured by a psychrometer. The intersection of the diagonal wet bulb temperature line (equal to the temperature of a wet bulb thermometer) and the vertical dry bulb temperature line defines the temperature and humidity conditions of air.

Vapor pressure is not shown on most psychrometric charts, but is an important concept in handling perishables. It is a function of the humidity ratio and temperature of air. As humidity ratio and temperature increase, vapor pressure increases.

Psychrometric charts and calculators are based on a specific atmospheric pressure, usually a typical sea level condition. Precise calculations of psychrometric variables will require adjustment for barometric pressures different from those listed on a particular chart you may be using. Consult the ASHRAE Handbook listed in the references for more information on this. Most field measurements will not require adjustment for pressure.

Effect of Psychrometric Variables on Perishable Commodities

Temperature

Air temperature is the most important variable because it tends to control the flesh temperature of perishable commodities. All perishables have an optimum range of storage temperatures. Above the optimum, they respire at unacceptably high rates and are more susceptible to ethylene and disease damage. In fact, horticultural commodities respire at rates which double, triple, or even quadruple for every 10°C (18°F) increase in temperature. Temperatures below the optimum will result in freezing or chilling damage. Accurate control of temperature is vitally important in maintaining maximum shelf-life.

Vapor pressure

The rate of moisture loss from a perishable is primarily controlled by the difference in vapor pressure between the air in the intercellular spaces of plant material and the air surrounding it. Remember, vapor pressure increases as the air moisture content and air temperature increase. The air in fresh plant material is nearly saturated or, in other words, is close to 100 percent relative humidity. Therefore, the humidity ratio of this air is determined solely by the temperature of the plant material. From the psychrometric chart it is apparent that low temperatures

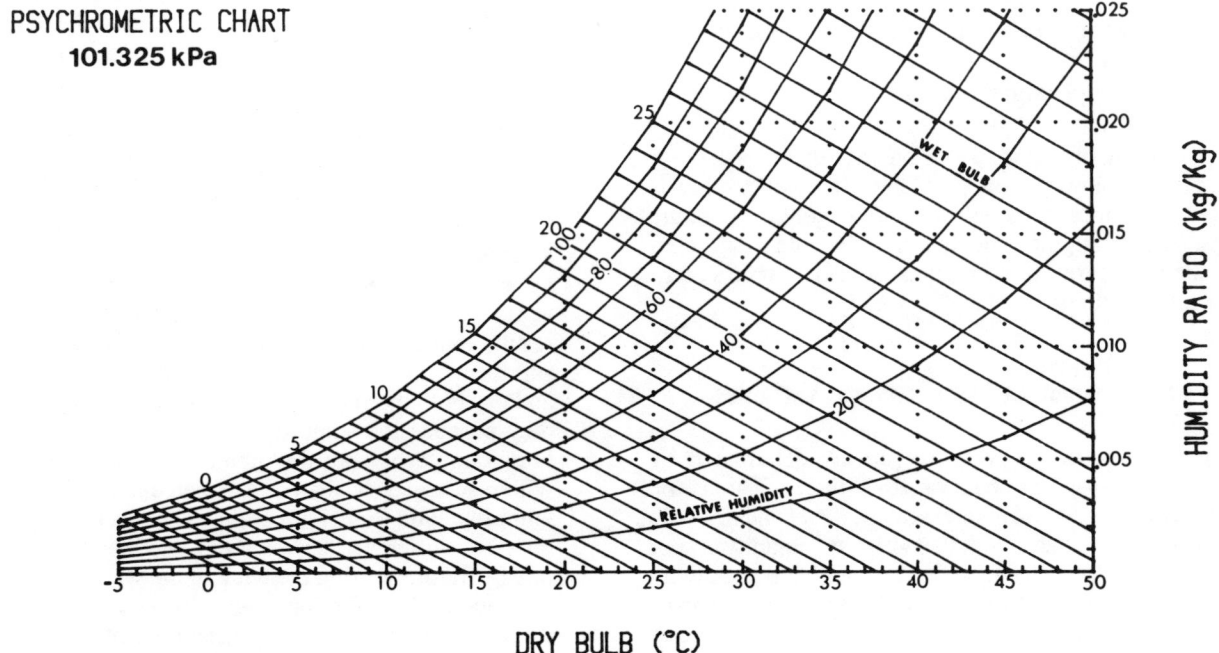

Fig. 10.2. A psychrometric chart in SI (metric) units.

result in low humidity ratios and high temperatures cause high humidity ratios.

Consider several examples of how the drying of perishables is influenced by vapor pressure differences. If an apple were precooled to 0°C (32°F) and placed in a refrigerated room with saturated air at 0°C (32°F), the apple would not lose moisture because the humidity ratio and temperature of the air in the apple and the surrounding air are the same. However, if the apple were at 20°C (68°F) because it was not precooled before being placed in the refrigerator, the air in the apple would have a high vapor pressure (high temperature and humidity ratio) compared to the refrigerated air, causing the apple to dry. If the apple were precooled to 0°C (32°F) but the refrigerated air were at 70 percent relative humidity, drying would also occur because the refrigerated air is at a lower humidity ratio than the saturated air in the apple. However, the rate of moisture loss is much greater when the apple is not precooled than when the apple is at storage temperature but the storage room air is not saturated. The difference in humidity ratio between the air in the apple and the storage air is over nine times more when the apple is not precooled than when it is cooled and put in unsaturated storage air.

Drying of perishables in refrigerated storage is reduced by decreasing the difference in humidity ratio between air in the perishable commodity and air surrounding it. Both temperature of the commodity and humidity ratio in the surrounding air must be controlled.

Relative humidity

Relative humidity is a commonly used term for describing the humidity of the air but is not particularly meaningful without knowing the dry bulb temperature of the air. These two variables allow the determination of humidity ratio which is a better index of the potential of desiccation.

Dew point temperature

Condensation of liquid water on perishables can be a factor in causing disease problems. If a commodity is cooled to a temperature below the dew point temperature of the outside air and brought out of the cold room, condensation will form. Condensation can also occur in storage if air temperatures fluctuate too greatly.

Measurement of Psychrometric Variables

All psychrometric properties of air are determined by measuring two psychrometric variables (three, if barometric pressure is considered). For example, if wet and dry bulb temperatures are measured, then relative humidity, vapor pressure, dew point, and so on, can be determined with the aid of a psychrometric chart. While many variables can be measured to determine the psychrometric state of air, dry bulb temperature, wet bulb temperature, dew point temperature, and relative humidity are most commonly measured.

Dry bulb temperature

Dry bulb temperature can be simply and inexpensively measured by a mercury-in-glass thermometer. The thermometer should be marked in divisions of at most 0.2°C or 0.5°F divisions if the thermometer is used in conjunction with a wet bulb thermometer for determining cold storage air conditions. The thermometer should be shielded from radiant heat sources such as motors, lights, external walls, and people. The shielding can be accomplished by placing the thermometer so it cannot "see" the warm object or protecting it with a radiant heat shield assembly.

Hand-held thermistor, resistance bulb, or thermocouple thermometers can also be used. They are more expensive than a mercury-in-glass thermometer but are not necessarily more accurate. Often these instruments can be purchased with a sharp probe allowing them to be used for fruit pulp temperature measurement. Inexpensive alcohol-in-glass and bi-metallic dial thermometers are not recommended unless they have been calibrated.

Wet bulb temperature

The use of a wet bulb thermometer in conjunction with a dry bulb thermometer is a very common method of determining the state point on the psychrometric chart. The wet bulb thermometer is basically an ordinary glass thermometer (although electronic temperature sensing elements can also be used) with a wetted, cotton wick secured around the reservoir. Air is forced over the wick causing it to cool to the wet bulb temperature. The wet and dry bulb temperatures together determine the state point of the air on the psychrometric chart allowing all other variables to be determined.

An accurate wet bulb temperature reading is dependent on: (1) sensitivity and accuracy of the thermometer, (2) maintaining an adequate air speed past the wick, (3) shielding the thermometer from radiation, (4) use of distilled or deionized water to wet the wick, and (5) use of a cotton wick.

The thermometer sensitivity required to determine an accurate humidity varies according to the temperature range of the air. At low temperatures more sensitivity is needed than at high temperatures. For example, at 65°C a 0.5°C error in wet bulb temperature reading results in a 2.6 percent error in relative humidity determination but at 0°C a 0.5°C error in wet bulb temperature reading results in a 10.5 percent error in relative humidity measurements. In most cases, absolute calibration of the wet and dry bulb thermometer is not as important as ensuring they produce the same reading at a given temperature. For example, if both thermometers read 0.5°C low this will result in less than a 1.3 percent error in relative humidity at dry bulb temperatures between 65°C and 0°C (at a 5°C difference between dry and wet bulb temperatures). Before wetting the wick of the wet bulb thermometer, operate both thermometers long enough to determine if there is any difference between their readings. If there is a difference and you must use the thermometers, assume that one is correct and adjust the reading of the other accordingly when determining relative humidity.

The rate of evaporation from the wick is a function of air velocity past it. A minimum air velocity of about 3 meters per second (500 feet per minute) is required for accurate readings. An air velocity much below this will result in an erroneously high wet bulb reading. Wet bulb devices that do not provide a guaranteed air flow, such as those that sit on a desk, cannot be relied on to give an accurate reading.

As with the dry bulb thermometer, sources of radiant heat such as motors, lights, and so on, will affect the wet bulb thermometer. The reading must be taken in an area protected from these sources of radiation or thermometers must be shielded from radiant energy.

A buildup of salts from impure water or contaminants in the air will affect the rate of water evaporation from the wick and result in erroneous data. Distilled or deionized water should be used to moisten the wick and the wick should be replaced if there is any sign of contamination. Care should be taken to ensure that the wick material has not been treated with chemicals such as sizing compounds that would affect the water evaporation rate.

Special care must be taken when using a wet bulb thermometer when the wet bulb temperature is near freezing. Most humidity tables and calculators are based on a frozen wick at wet bulb temperatures below 0°C (32°F). At temperatures below 0°C, touch the wick with a piece of clean ice or another cold object to induce freezing. (Distilled water can be cooled below 0°C without freezing.) Your psychrometric chart or calculator must use frost bulb temperatures, not wet bulb temperatures, below 0°C to be accurate with this method.

Under most conditions wet bulb temperature data are not reliable when the relative humidity is below 20 percent or the wet bulb temperature is above 100°C. At low humidities, the wet bulb temperature is much lower than the dry bulb temperature and it is difficult for the wet bulb thermometer to be cooled completely because of heat transferred by the glass stem. Under normal circumstances, liquid water does not reach temperatures above 100°C so wet bulb temperatures above the boiling temperature of water cannot be measured with a wet bulb thermometer.

In general, properly designed and operated wet and dry bulb psychrometers can operate with an accuracy of less than 2 percent of the actual relative humidity. Improper operation will greatly increase the error.

Relative humidity

Direct relative humidity measurement usually employs an electric sensing element or a mechanical system. Electric hygrometers operate using substances whose electrical properties change as a function of their moisture content. As the humidity of the air surrounding the sensor increases, its moisture increases proportionally affecting the sensor's electrical properties. These devices are more expensive than wet and dry bulb psychrometers, but their accuracy is not as severely affected by incorrect operation. An accuracy of less than 2 percent of the actual humidity is often obtainable. Sensors will lose their calibration if allowed to become contaminated and some lose calibration if water condenses on them. Most sensors have a limited life. Mechanical hygrometers usually employ human hairs as a relative humidity sensing element. Hair changes in length in proportion to the humidity of the air. The hair element responds slowly to changes in relative humidity and is not dependable at very high relative humidities. These devices are acceptable as an indicator of a general range of humidity but are not especially dependable for accurate relative humidity measurement.

Dew point indictors

Two types of dew point sensors are in common use today: a saturated salt system and a condensation dew point method. The saturated salt system will operate at dew points between −12° to 37°C (10° to 100°F) with an accuracy of less than 1°C (2°F). The system is lower in cost than the condensation system, is not significantly affected by contaminating ions, and has a response time of about 4 minutes. The condensation type is very accurate over a wide range of dew point temperatures (less than .5°C from −73° to 100°C or less than 1°F from −100° to 212°F). A condensation dew point hygrometer can be expensive.

There are a variety of other methods for measuring psychrometric variables. Some are extremely accurate and have some characteristics which make them suited to particular sampling conditions. However, most are not commercially available and are used primarily as laboratory instruments.

References

1. ASHRAE. 1981. *ASHRAE handbook—fundamentals volume*. Am. Soc. Heating, Refrigeration and Air Conditioning Engineers. Atlanta, GA.

2. Gaffney, J. J. 1978. Humidity: Basic principles and measurement techniques. *HortScience* 13(5): 551-55.

3. Wexler, A., and W. G. Brombacher. 1951. *Methods of measuring humidity and testing hygrometers*. National Bureau of Standards, Circ. 512.

11

Modified Atmospheres and Low-pressure Systems during Transport and Storage

ADEL A. KADER

Introduction

Modified atmospheres (MA) or controlled atmospheres (CA) mean removal or addition of gases resulting in an atmospheric composition surrounding the commodity that is different from that of air (78.08% N_2, 20.95% O_2, 0.03% CO_2). Usually this involves reduction of oxygen (O_2) and/or elevation of carbon dioxide (CO_2) concentrations. The MA and CA differ only in the degree of control; CA is more exact.

The use of modified or controlled atmospheres should be considered as a supplement to proper temperature and relative humidity management procedures. The potential for benefit or hazard from using MA is dependent upon the commodity, variety, physiological age, atmospheric composition, and temperature and duration of storage. This helps explain the wide variability in results among published reports for a given commodity.

The atmospheres, MA or CA, can be used during transport, temporary storage, or long-term storage of horticultural commodities destined for fresh market or processing. A prestorage treatment with elevated CO_2 can also be used for some fruits. Recently, carbon monoxide (CO) has been tested as an added component to MA for slowing down brown discoloration and controlling decay.

Effects of Controlled Atmospheres

Potential beneficial effects

Used properly, MA or CA can supplement proper temperature management and can result in one or more of the following benefits which translate into overall reduction in quantitative and qualitative losses during postharvest handling and storage of some horticultural commodities.

1. Retardation of senescence (ripening) and associated biochemical and physiological changes, i.e., slowing down respiration and ethylene production rates, softening, and compositional changes.

2. Reduction of fruit sensitivity to ethylene action at O_2 levels below about 8 percent and/or CO_2 levels above 1 percent.

3. Alleviation of certain physiological disorders such as chilling injury of various commodities, russet spotting in lettuce, and some storage disorders of apples.

4. Modified atmosphere can have a direct or indirect effect on postharvest pathogens and consequently decay incidence and severity. For example, elevated CO_2 levels (10% to 15%) significantly inhibit development of Botrytis rot on strawberries, cherries, and other fruits.

5. Modified atmosphere can be a useful tool for insect control in some commodities.

Potential harmful effects

In most cases, the difference between beneficial and harmful MA combinations is relatively small. Also, MA combinations that are necessary to control decay and/or insects, for example, cannot always be tolerated by the commodity and may result in faster deterioration. Potential hazards of MA to the commodity include the following:

1. Initiation and/or aggravation of certain physiological disorders such as blackheart in potatoes, brown stain on lettuce, and brown heart in apples and pears.

2. Irregular ripening of fruits such as banana, pear, and tomato can result from O_2 levels below 2 percent and/or CO_2 levels above 5 percent.

3. Development of off-flavors and off-odors at very low O_2 concentrations as a result of anaerobic respiration.

4. Increased susceptibility to decay when the commodity is physiologically injured by too-low O_2 or too-high CO_2 concentrations.

5. Stimulation of sprouting and retardation of periderm development in some root and tuber vegetables such as potatoes.

Modified Atmospheres: Requirements and Recommendations

After 50 years of research and development, large-scale use of MA and CA is still limited largely to apples and pears for long-term storage. Use of MA during transport includes several other horticultural crops. Based on published reports and some unpublished results, a summary of current MA recommendations is included in table 11.1 (fruits) and table 11.2 (vegetables). Also included is an estimate of potential benefits and extent of current commercial use. There is no doubt that some of these MA combinations will be changed as more research is completed. The possibility of adding carbon monoxide to MA for some commodities may change its potential for benefit. Hypobaric or low-pressure systems may also provide new opportunities for making CA a more useful treatment for some commodities.

Carbon Monoxide as a Supplement

For several years, carbon monoxide (CO) at 2 percent to 3 percent has been used as a supplement to MA during transit of lettuce as a discoloration inhibitor. Recently, some additional benefits of CO have been demonstrated. These benefits as well as possible hazards of CO are summarized below.

Characteristics of carbon monoxide

Carbon monoxide is a colorless, tasteless, odorless, and flammable gas with explosive limits between 12.5 percent and 74.2 percent (volume) in air. It is a product of incomplete combustion and is generally associated with automobile exhaust. Carbon monoxide is extremely toxic to living systems in which hemoglobin respiratory pigments in the blood serve a major respiratory function as O_2 carriers.

Beneficial effects

1. Carbon monoxide (1% to 5%) added to reduced (2% to 5%) O_2 atmospheres inhibits discoloration of lettuce butts and mechanically damaged tissue. Similar effects

Table 11.1. Summary of recommended CA or MA conditions during transport and/or storage of selected fruits

Commodity	Temp. range (°C)*	CA† %O_2	%CO_2	Potential for benefit‡	Remarks§
Deciduous tree fruits					
Apple	0-5	2-3	1-2	A	About 40% of production is stored under CA
Apricot	0-5	2-3	2-3	C	No commercial use
Cherry (sweet)	0-5	3-10	10-12	B	Some commercial use
Fig	0-5	5	15	B	Limited commercial use
Grape	0-5	None	None	D	Incompatible with SO_2 fumigation
Kiwifruit	0-5	2	5	A	Some commercial use
Nectarine	0-5	1-2	5	B	Limited commercial use
Peach	0-5	1-2	5	B	Limited commercial use
Pear	0-5	2-3	0-1	A	Some commercial use
Persimmon	0-5	3-5	5-8	C	No commercial use
Plum and prune	0-5	1-2	0-5	B	No commercial use
Strawberry	0-5	10	15-20	A	Increasing use during transit
Nuts and dried fruits	0-25	0-1	0-100	A	Effective insect control method
Subtropical and tropical fruits					
Avocado	5-13	2-5	3-10	B	Limited commercial use
Banana	12-15	2-5	2-5	A	Some commercial use
Grapefruit	10-15	3-10	5-10	C	No commercial use
Lemon	10-15	5	0-5	B	No commercial use
Lime	10-15	5	0-10	B	No commercial use
Olive	8-12	2-5	5-10	C	No commercial use
Orange	5-10	10	5	C	No commercial use
Mango	10-15	5	5	C	No commercial use
Papaya	10-15	5	10	C	No commercial use
Pineapple	10-15	5	10	C	No commercial use

*Usual and/or recommended range. A relative humidity of 85% to 95% is recommended.
†Best CA combination may vary among varieties and according to storage temperature and duration.
‡A = excellent, B = good, C = fair, D = slight or none.
§Comments about use refer to domestic marketing only; many of these commodities are shipped under MA for export marketing.

have been observed on other commodities. This inhibition of discoloration is lost upon removal of the commodity from MA to air during destination marketing.

2. Carbon monoxide (5% to 10%) added to MA has been found to inhibit growth of several important postharvest pathogens and to prevent decay development on several fruits and vegetables. The fungistatic effects of CO are maximized at O_2 levels below 5 percent.

3. Although CO alone was not found to be an effective fumigant for insect control in harvested lettuce, its possible use in addition to other MA combinations merits further investigation.

Possible hazards

1. Carbon monoxide may aggravate certain physiological disorders. For example, if CO is used in a situation where CO_2 accumulated above 2 percent during transit of lettuce, it will increase the severity of brown stain (a carbon dioxide-induced disorder).

2. Carbon monoxide is known to mimic ethylene (C_2H_4) effects such as enhancing fruit ripening and induction of certain physiological disorders. However, when CO is used in combination with reduced O_2 and/or elevated CO_2, such effects are minimized and become insignificant except for those commodities which are extremely C_2H_4-sensitive, such as kiwifruits.

Table 11.2. Summary of recommended CA or MA conditions during transport and/or storage of selected vegetables

Commodity	Temp. range (°C)*	CA† %O_2	%CO_2	Potential for benefit‡	Remarks§
Artichokes	0-5	2-3	3-5	B	No commercial use
Asparagus	0-5	air	5-10	B	Limited commercial use
Beans, snap	5-10	2-3	5-10	C	Potential for use by processors
Beets	0-5	None	None	D	98-100% rh is best
Broccoli	0-5	1-2	5-10	B	Limited commercial use
Brussels sprouts	0-5	1-2	5-7	B	No commercial use
Cabbage	0-5	3-5	5-7	B	Some commercial use for long-term storage of certain cultivars
Cantaloupes	3-7	3-5	10-15	B	Limited commercial use
Carrots	0-5	None	None	D	98-100% rh is best
Cauliflower	0-5	2-5	2-5	C	No commercial use
Celery	0-5	2-4	0	C	Limited commercial use in mixed loads with lettuce
Corn, sweet	0-5	2-4	10-20	B	Limited commercial use
Cucumbers	8-12	3-5	0	C	No commercial use
Honeydews	10-12	3-5	0	C	No commercial use
Leeks	0-5	1-2	3-5	B	No commercial use
Lettuce	0-5	2-5	0	B	Some commercial use with 2-3% CO added
Mushrooms	0-5	air	10-15	C	Limited commercial use
Okra	8-12	3-5	0	C	No commercial use; 5-10% CO_2 is beneficial at 5-8°C
Onions, dry	0-5	1-2	0	B	No commercial use; 75% rh
Onions, green	0-5	1-2	10-20	C	Limited commercial use
Peppers, bell	8-12	3-5	0	C	Limited commercial use
Peppers, chili	8-12	3-5	0	C	No commercial use; 10-15% CO_2 is beneficial at 5-8°C
Potatoes	4-12	None	None	D	No commercial use
Radish	0-5	None	None	D	98-100% rh is best
Spinach	0-5	air	10-20	C	No commercial use
Tomatoes, mature-green	12-20	3-5	0	B	Limited commercial use
partially-ripe	8-12	3-5	0	B	Limited commercial use

*,†,‡,§ See Table 11.1 for footnotes. A relative humidity of 90% to 95% is recommended unless otherwise indicated under Remarks.

3. Because of its extreme toxicity to humans and flammability at concentrations between 12.5 percent and 74.2 percent in air, adequate safety measures should be followed if and when CO is used.

Prestorage Treatments with Elevated Carbon Dioxide

Tests conducted at several Experiment Stations indicated that treating apples for 2 weeks or pears for 2 to 4 weeks with 12 percent CO_2 (at 0° to 5°C) before CA storage delayed fruit softening. However, this treatment resulted in varying amounts of internal and external CO_2 injury depending on variety, season, and production area. Its commercial application is currently limited to some Golden Delicious apples in the Northwest of the U.S.

Elevated CO_2 treatments have also been shown to alleviate chilling injury symptoms on some subtropical and tropical fruits, but this treatment is not recommended yet for commercial application.

Ethylene Removal in Modified Atmosphere Storage

Most research workers have assumed that ethylene removal from MA storage rooms is not necessary because ethylene effects on fruit ripening at 0° to 5°C (32° to 41°F) and under MA conditions are negligible. However, some recent studies indicate that the presence of ethylene, at concentrations likely to occur in MA and CA rooms, can enhance fruit softening during long-term storage. Thus, ethylene (C_2H_4) removal is recommended in long-term CA storage of apples and pears. It is particularly important for storage of kiwifruits, carnations, or other commodities which are extremely sensitive to C_2H_4 action. Further research is needed to evaluate the possible effects of ethylene and nonethylenic volatiles on various quality attributes of other commodities under MA conditions. Also, there is a need for new methods that are effective and economical for the removal of ethylene and other volatiles from MA storage rooms.

Methods of Atmospheric Modification

Atmosphere generators

Oxygen control. This can be accomplished by a recirculating system where air is recirculated from the CA room into the generator and back into the room or by a purge system where fresh air is introduced into the generator, then the generated low-O_2 air is fed into the room. Open-flame burners, which were used in the past, have been largely replaced by catalytic burners or converters. Many types of catalytic burners or converters are available in various sizes to fit CA rooms of different capacities; these include Gen-O-Fresh units (E.R. Krueger, Inc., California), COB units (Pacific Columbia Co., Washington), and Sulzer burners (Switzerland). The Smit Oxydrain system (Holland; Controlled Atmosphere Instrumentation, Washington) utilizes ammonia (NH_3) that is thermally cracked into H_2 and N_2, H_2 reacts with O_2 to form H_2O and the N_2 is used to purge O_2 from the CA room while most of the water vapor condenses. A promising new development is nitrogen separators or generators that take compressed air and circulate it through molecular sieve beds, which separate N_2 from other components. These generators operate in a cyclic process on the principle of pressure swing adsorption. The produced N_2 can then be used to flush the CA room and reduce O_2 level. This system is currently under evaluation.

Carbon dioxide control. Addition of CO_2 is usually from pressurized gas cylinders. Dry ice is sometimes used as a source of CO_2 during transport. Removal of CO_2 is accomplished by one of the following scrubbing methods: sodium hydroxide, water, activated charcoal, hydrated lime ($Ca[OH]_2$), or molecular sieve. The most common CO_2 scrubber utilizes brine water which is pumped over the evaporator coil where it absorbs CO_2. This may be combined with placement of a small quantity of lime inside the CA room. A lime box located adjacent to the CA room with a circulating system to pass the room atmospheres through the box is also commonly used. The box is usually sized to hold about 12 kg lime/ton of fruit and the spent lime is replaced with fresh lime. The amount of lime needed to absorb CO_2 may be placed inside the CA room. Several types of activated carbon scrubbers, such as Sulzer and Charcasorb units, are currently used. During the past few years, there has been an increase in use of molecular sieve scrubbers.

Carbon monoxide addition. Carbon monoxide can be added from pressurized gas cylinders using a blending system with nitrogen to avoid exceeding 10 percent CO.

Ethylene removal. A few methods can be used to remove ethylene from cold storage facilities. Adequate ventilation (one air exchange per hour) to reduce C_2H_4 concentration in cold storage facilities cannot be used in CA storage. Also, the use of ozone to oxidize C_2H_4 requires O_2 levels above those available in CA storage. Use of ethylene absorbers, such as potassium permanganate alone or in combination with activated and brominated charcoal, can be effective in CA storage facilities provided that the air is circulated through the filters containing these materials, which must be replaced when spent. Use of catalytic burners for ethylene removal from CA storage facilities is currently being evaluated and appears promising.

Hypobaric or low pressure system (LPS)

Reducing the total pressure (under partial vacuum conditions) results in reducing the partial pressures of individual gases of air. This can be an effective method for reducing O_2 tension, and for accelerating the escape of C_2H_4 and other volatiles. LPS has some advantages over other methods of atmospheric modification, i.e.: (1) more exact control of O_2 concentrations that permit the use of lower O_2 tensions than is possible with CA, and (2) removal of ethylene and other volatiles.

However, LPS has some limitations when CO_2 or CO addition is important for a given commodity. Transit vehicles with LPS were tested to a limited extent on a commercial scale for transporting some animal and plant products. Stationary storage structures with LPS were being developed by Grumman Dormavac before it terminated its efforts related to hypobaric storage.

Commodity-generated modified atmosphere

In this case, the commodity, through respiration, is used to reduce O_2 and increase CO_2 under restricted air-exchange conditions using one or more type of barrier as shown in figure 11.1. If elevated CO_2 is not desirable, CO_2 scrubbers mentioned above are used. Restriction of air exchange may be achieved by one or more of the following methods: use of air-tight cold storage rooms, packaging in film wraps or bags, use of polyethylene liners in shipping containers, use of pallet shrouds (plastic covers), manipulation of shipping container vents, application of waxes and other surface coatings, or use of plastic covers with diffusion windows (polymeric membranes). Atmospheric modification by these methods is usually slow and much of the benefit of MA may be lost.

Atmospheres during transit

Modified atmospheres in rail cars, trucks, and seavans. Gas tightness of the transit vehicle is essential to the maintenance of MA during transit. This is a limiting factor to expanded use of MA. The Tectrol system (TransFresh Corp.), which is used to modify the atmosphere in rail cars and marine containers, utilizes the following procedures: 1) reduced O_2 is achieved by N_2 flushing, 2) elevated CO_2 and/or CO is accomplished by adding them using gas blending manifolds, 3) CO_2 removal by placing bags of fresh hydrated lime in the transit vehicle (amount of lime depends upon commodity), and 4) breather bags are used to compensate for barometric pressure fluctuations.

The Nitrol system utilizes a liquid N_2 tank, which is carried along with the container van. The van is equipped with an O_2 sensor for controlling N_2 release as needed to maintain the desired O_2 concentration.

The Nhytemp system uses liquid N_2 as a refrigerant and as a means of maintaining reduced O_2 levels during transport. The quantity of liquid N_2 carried in a special compartment in the van determines the duration during which temperature and O_2 concentration are maintained at the desired level.

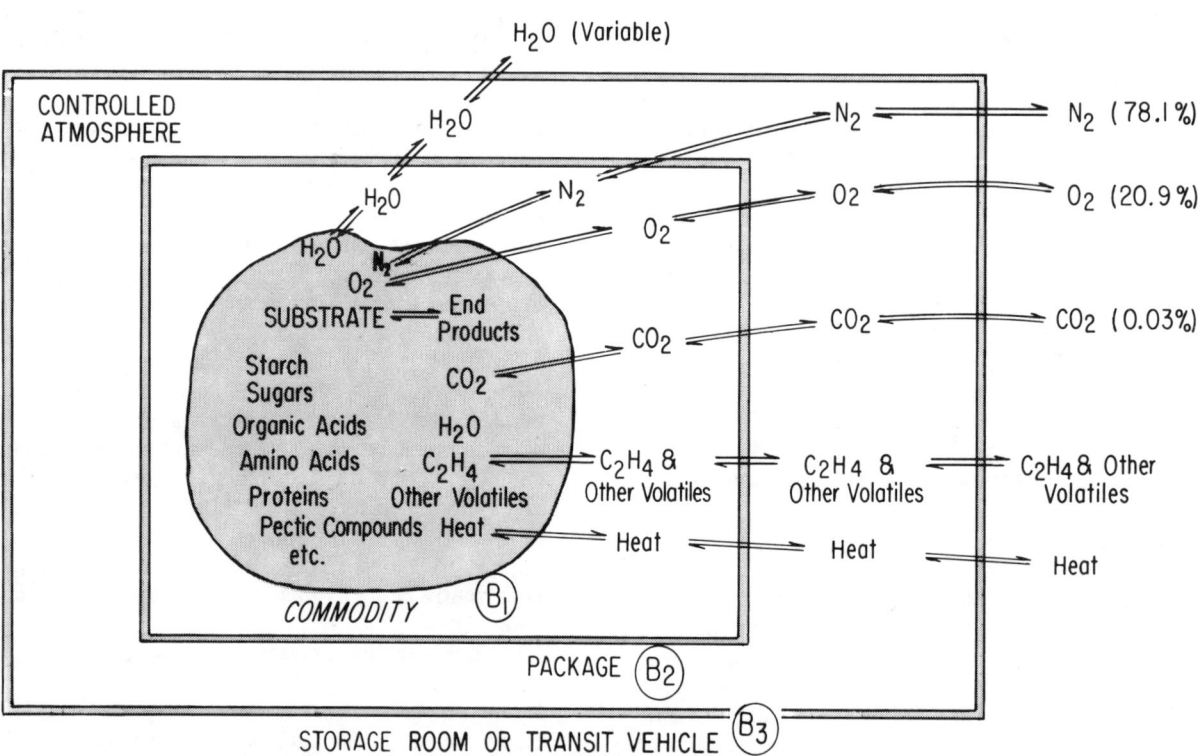

Fig. 11.1. Types of barriers which can be used to establish a modified atmosphere.

B_1—Natural epidermis, skin, peel, or rind. Wax coating, film wrap.

B_2—Package—Wood, paperboard, plastic (may include additional liner in package).

B_3—Storage room wall or vehicle wall, may be sealed against gas exchange.

Additional barriers may include—consumer packages inside the master package and pallet covers over several packages.

Modified atmosphere in a pallet cover. Polyethylene pallet covers (or shrouds) are used to cover all shipping containers on a pallet and are sealed by various means (tape, heat seal, and so on) onto a plastic sheet placed on the wooden pallet base. Then, a partial vacuum is established within the pallet cover and the desired gas mixture is introduced. This method is in common use on strawberries and limited use on bush berries, cherries, and other commodities. It can facilitate mixing of commodities which require different MA conditions during transit at the same temperature. Potential problems are primarily related to loss of seal due to tearing of the pallet cover or imperfect seal of the cover to the base.

Modified atmospheres in individual shipping containers. Some examples of commercial use of commodity-generated MA during transit include: polyethylene liners in cherry boxes, polyethylene bags for bananas destined for distant markets (Banavac system), and for cut lettuce and other vegetables. Cut (shredded or chopped) lettuce may be packaged into 5-mil plastic bags, then partial vacuum is established and a gas mixture (30% to 50% O_2 + 4% to 6% CO) is introduced into the bag which is then sealed. This procedure is currently in limited commercial use.

Monitoring of atmospheric composition

Accurate monitoring of O_2 and CO_2 concentrations is essential to successful CA or MA storage. It is required for certification of CA storage in some states and by some insurance companies. The various methods of gas sampling and analysis will be discussed in chapter 12 of this book.

Controlled Atmosphere Storage Structures

Considerations in construction

Building materials. Most new CA storage facilities utilize tilt-up concrete buildings.

Size. Usually one large facility is composed of several smaller CA units of a capacity that depends on commodity, amount of fruit of each variety, and marketing requirements.

Refrigeration system. A separate refrigeration unit for each room is preferred over a larger capacity unit to refrigerate all units of CA facility.

Type of refrigerant. Freon has less danger to commodities than ammonia. If ammonia is used, careful checking of the refrigeration system before each season, and checking the use of ammonia monitoring and alarm systems, are essential.

Insulation. Maximum resistance to heat and moisture penetration is required.

Gas tightness. Proper sealing of walls, ceiling, and floor is essential to achieve gas tightness.

Accessibility during storage. Inspection windows are useful for observing product without opening the CA room. Sampling ports are needed for collecting gas samples for analysis from various locations in the room.

Plastic tents

A less expensive structure for CA storage involves use of a single sheet of 75-μm polyethylene film supported by a framework (constructed of wood or PVC piping) to create a plastic tent within a cold storage room. The size of the CA tent depends on the quantity of the commodity to be stored. Floor sealing is accomplished by sealing the polyethylene walls into troughs using gas rubber tubing. Air circulation within the tent is facilitated by using a fan and CO_2 control is maintained by placing bags of fresh hydrated lime in front of the fan. Inlet and outlet air exchange pipes, installed into the front panel of the tent, are used to control O_2 and tent volume by adjusting leak rates through them. Initial reduction of O_2 levels can be achieved by using the commodity through its respiration (which would take 2 to 3 weeks) or CA generators. Commercial use of CA tents is still very limited and their future application will depend on evaluation of their advantages and disadvantages as compared to more permanent structures.

Facilities and equipment

1. Selection of appropriate CA generating systems from among those discussed above should be based on relative effectiveness in attaining and maintaining desired atmospheric composition, cost of operation and maintenance, and initial capital costs of the system.

2. Breather bags or other methods may be needed to compensate for fluctuations in barometric pressure.

3. Air purification systems to absorb objectionable odors and/or to remove ethylene may be needed for some commodities.

Monitoring of environmental conditions

1. Thermocouples and temperature recorders are used to monitor commodity temperature during CA storage.

2. Use of automatic and sensitive gas sampling and analysis systems to monitor O_2 and CO_2 levels is increasing. Special alarm systems may also be used to alert the storage operator when O_2 levels drop below a certain point and when CO_2 levels exceed the desired concentration.

Safety considerations

The atmosphere in CA rooms will not support human life and people have died of asphyxia while working inside CA rooms without a breathing apparatus. A danger sign should be posted on the door. The access hatch in the door should be large enough to accommodate a person with breathing equipment strapped to his or her back. At least two persons with breathing equipment should work together at all times with one inside and one outside the room watching the first person.

References

1. Bartsch, J. A., and G. D. Blanpied. 1984. *Refrigerated storage for horticultural crops*. Agric. Eng. Ext. Bull. 448. Cornell Univ. Ithaca, NY.

2. Brecht, P. E. 1980. Use of controlled atmospheres to retard deterioration of produce. *Food Technol.* 34(3): 45-50.

3. Burton, W. G. 1974. Some biophysical principles underlying the controlled atmosphere storage of plant material. *Ann. Appl. Biol.* 78: 149-68.

4. Dalrymple, D. G. 1967. *The development of controlled atmosphere storage of fruits*. USDA, Div. Mark. Utilz. Sci. 56 pp.

5. Dewey, D. H., ed. 1977. *Controlled atmospheres for the storage and transport of perishable agricultural commodities* (Proc. 2nd Natl. CA Res. Conf., April, 1977). Hortic. Rept. No. 28, Dept. Hortic., Mich. State Univ. 301 pp.

6. Dewey, D. H., R. C. Herner, and D. R. Dilley, eds. *Controlled atmospheres for the storage and transport of horticultural crops* (Proc. 2nd Natl. CA Res. Conf., January, 1969). Hortic. Rept. No. 9, Dept. Hortic., Mich. State Univ. 155 pp.

7. El-Goorani, M. A., and N. F. Sommer. 1981. Effects of modified atmospheres on postharvest pathogens of fruits and vegetables. *Hort. Rev.* 3:412-61.

8. Hardenburg, R. E. 1971. Effect of in-package environment on keeping quality of fruits and vegetables. *HortScience* 6:198-201.

9. Isenberg, F. M. R. 1979. Controlled atmosphere storage of vegetables. *Hortic. Rev.* 1:337-94.

10. Jamison, W. 1980. Use of hypobaric conditions for refrigerated storage of meats, fruits, and vegetables. *Food Technol.* 34(3): 64-71.

11. Kader, A. A. 1980. Prevention of ripening in fruits by use of controlled atmospheres. *Food Technol.* 34(3): 51-54.

12. Kader, A. A., and L. L. Morris. 1977a. *Modified atmospheres*. An indexed reference list with emphasis on horticultural commodities, Supplement #2 (May 1, 1974 to February 28, 1977). Vegetable Crops Series 187, Univ. of Calif., Davis. 28 pp. (386 titles).

13. _____. 1977b. Relative tolerance of fruits and vegetables to elevated CO_2 and reduced O_2 levels. Mich. State Univ. *Hortic. Rept.* 28: 260-65.

14. _____. 1981. *Modified atmospheres*. An indexed reference list with emphasis on horticultural commodities, Supplement #3 (March 1, 1977 to December 31, 1980). Vegetable Crops Series 213, Univ. of Calif., Davis. 36 pp. (467 titles).

15. Lipton, W. J. 1975. Controlled atmospheres for fresh vegetables and fruits. Why and When. In *Postharvest biology and handling of fruits and vegetables*, eds. N. F. Haard and D. K. Salunkhe, 130-43. Westport, CT: AVI Publ. Co.

16. Lougheed, E. C., D. P. Murr, and L. Berard. 1978. Low pressure storage for horticultural crops. *HortScience* 13(1): 21-27.

17. Morris, L. L., L. L. Claypool, and D. P. Murr. 1971. *Modified atmospheres*. An indexed reference list through 1969, with emphasis on horticultural commodities. Univ. of Calif., Div. of Agric. Sci. 115 pp. (2326 titles).

18. Murr, D. P., A. A. Kader, and L. L. Morris. 1974. *Modified atmospheres*. An indexed reference list with emphasis on horticultural commodities, Supplement #1 (January 1, 1970 to April 30, 1974). Vegetable Crops Series 168, Univ. of Calif., Davis. 39 pp. (395 titles).

19. Pantastico, Er. B., ed. 1975. *Postharvest physiology, handling and utilization of tropical and subtropical fruits and vegetables*, 175-218. Westport, CT: AVI Publ. Co.

20. Richardson, D. G., and M. Meheriuk, eds. 1982. *Controlled atmospheres for storage and transport of perishable agricultural commodities* (Proc. 3rd Natl. CA Res. Conf., July, 1981). Beaverton, OR: Timber Press. 390 pp.

21. Smock, R. M. 1979. Controlled atmosphere storage of fruits. *Hortic. Rev.* 1:301-36.

12
Methods of Gas Mixing, Sampling, and Analysis

ADEL A. KADER

Gas Mixing

Principles

In postharvest research and technology we are usually concerned with monitoring atmospheric composition and with mixing two or more of the following gases: air, nitrogen, oxygen, carbon dioxide, ethylene, and carbon monoxide. Procedures for gas mixing are based on mass, volume, or pressure relationships. Remember the following laws and definitions for gases:

1. Boyle's Law: $V = K(\frac{1}{P})$; $P_1V_1 = P_2V_2$

   Where: V = volume
   K = proportionality constant
   P = pressure

2. Charles Law: $PV = KT$

   $$\frac{P_1V_1}{T_1} = \frac{P_2V_2}{T_2}$$

   Where: P = pressure
   V = volume
   T = temperature
   K = constant

3. Ideal gas law: If K is proportional to the number of moles of gas (n), then: $PV = nRT$

   Where: R = molar gas constant
   $P, V, \& T$ = same as above

4. Density $= \frac{mass}{volume} = \frac{P \times M}{R \times T}$

   Where: P = pressure
   M = molecular weight
   R = molar gas constant
   T = temperature

5. Concentration (C_a): [percent (ml/100 ml) or ppm (μl/l)]

   $$C_a(\%) = \frac{100 \times V_a}{V_a + V_b + \ldots V_n} = \frac{100 \times P_a}{P_a + P_b + \ldots P_n}$$

   Where: V_a to V_n = volume of components
   P_a to P_n = partial pressures of components

   $$C_a \text{ (ppm)} = \frac{10^6 \times V_a}{V_D + V_a} = \frac{10^6 \times P_a}{P_D + P_a}$$

   Where: V_D = Volume of diluent gas
   P_D = Partial pressure of diluent gas

Gas Mixing Techniques

Static system

1. Gravimetric procedure (mixing by weight): This method is independent of temperature, pressure, and compressibility. It involves weighing components into a gas cylinder.

2. Mixing by volume: Evacuate cylinder to 0.1 mm Hg, flush with diluent gas, evacuate again, inject component gas using a gas-tight syringe, allow diluent gas to pressurize cylinder to the desired pressure.

3. Mixing by pressure: Because the partial pressure of each component equals its mole fraction (MF) times total pressure (P_t) of the mixture, a mix of 10 percent A and 90 percent B at a total cylinder pressure of 2,000 psia can be prepared as follows:

   $P_A = MF_A \times P_t = 0.10 \times 2000 = 200$ psia

   Add 200 psia of A, then 1,800 psia of B

4. Homogenizing the gas mixture:

   ☐ Homogeneity depends upon the densities and the relative amounts of the components

   ☐ Homogenize gas mixtures by rolling cylinders or by thermal convection (temperatures above 50°C [122°F] should be avoided)

   ☐ Once the mixture is homogeneous, it remains so and does not separate except in the case of liquified gases; liquified components may partially condense in the cylinder if subjected to low temperatures

5. Gas mixtures should then be calibrated (analyzed) using chemical and gravimetric techniques (for some primary standards) or other gas analysis methods mentioned later in this chapter.

6. Accuracy, purity, and tolerances: Commercially-available gas mixtures vary in their accuracy (see table,

end of chapter). Even the purest gases and gas mixtures may contain impurities. This is of more concern in research work than in technological practice.

7. Storage and handling of compressed gas cylinders:

    ☐ Gas cylinders should be tested by hydrostatic pressure for their suitability for use with compressed gases

    ☐ For cylinder filling, the pressure limit is 2,000 psi at 21°C (70°F)

    ☐ Cylinder contents and whether the cylinder is full or empty should always be identified clearly

    ☐ Cylinders must be well-secured and are best stored at 21°C (70°F)

    ☐ Proper transportation procedures should be followed

    ☐ Proper valves and regulators relative to the standardized outlets for various families of gases should be used to prevent interchange of regulator equipment between gases that are not compatible

    ☐ Safety procedures required for toxic and/or flammable gases (e.g., CO at 12.5% to 75% and C_2H_4 at 3% to 30%) must be adhered to in handling gas cylinders

Dynamic system

Gases are mixed (continuous flow mixing) as needed by volume at constant pressure and temperature using flow control devices such as capillary tubing and needle valves.

Gas Sampling

Sampling and sample containers

1. Syringes of various volumes—but most commonly used ones are between 1 and 10 ml
2. Plastic film gas-impermeable bags with sealable gas inlet and septum for withdrawing subsamples for analysis
3. Glass containers of various capacities with gas inlet and sampling port
4. Vacuum containers:

    Evacuated 150- to 250-ml cans with septum

    Vacuotainers—evacuated 20-ml test tubes (commonly used for blood sampling)

Important points to consider

- Make sure sample containers are gas-tight to insure no leaks and are clean before use to minimize errors.
- When vacuum containers are used, vacuum should be determined on each container before use, and appropriate correction factors should be applied to the analysis data.
- Samples should be representative of the atmosphere to be analyzed.

Gas Analysis Methods

Methods for immediate and on-the-spot analysis

- Volumetric gas analyzers for O_2 and CO_2 (Orsat, Fyrite, and so on)
- Kitagawa gas sampler and detector tubes for C_2H_4, CO, SO_2, and other gases
- Portable gas analyzers (O_2, CO_2, CO, SO_2, NH_3, and other gases)
- Snoopy Electronic Ethylene Detector

Laboratory gas analysis instruments

The following methods are much more accurate than methods mentioned above for on-the-spot analysis. They can be used to monitor atmospheric composition in controlled atmosphere storage facilities, ripening rooms, and SO_2 fumigation chambers.

Gas	Instruments
O_2	Oxygen analyzers (paramagnetic; polarographic; electrochemical)
	Gas chromatography (thermal conductivity detector)
CO_2	Infrared CO_2 analyzer Gas chromatography (thermal conductivity detector)
CO	Gas chromatography (thermal conductivity detector); Snoopy Gas Detector
C_2H_4	Gas chromatography (flame ionization detector)
SO_2	Infrared SO_2 analyzer

Methods of measuring respiration rates

1. To determine O_2 consumed: For tissue slices or organelles—Warburg method or O_2 electrode; and for intact plant organs—laboratory gas analysis methods for O_2 mentioned above.
2. To determine CO_2 produced: Colorimetric method (Claypool & Keefer, 1942); and methods mentioned above for laboratory gas analysis of CO_2.
3. Results are usually expressed as ml O_2 (or CO_2) per kg-hr and calculated as follows:

$$\frac{\Delta O_2\% \text{ or } CO_2\%}{100} \times \frac{\text{flow rate (ml/hr)}}{\text{sample weight (kg)}}$$

Commercial gas mixtures: Accuracy

Designation	Accuracy limits
Primary standards	Within 0.02% absolute or 1% of the component, whichever is smaller
Certified mixtures	Within 2% to 5% of component
Unanalyzed (commercial grade)	Same as certified but without certificate of analysis

References

1. Barmore, C. R., and T. A. Wheaton. 1978. Diluting and dispensing unit for maintaining trace amount of ethylene in a continuous flow system. *HortScience* 13(2): 169-71.

2. Claypool, L. L., and R. M. Keefer. 1942. A colorimetric method for CO_2 determination in respiration studies. *Proc. Am. Soc. Hortic. Sci.* 40:177-86.

3. Jeffery, P. G., and P. F. Kipping. 1972. *Gas analysis by gas chromatography*. 2nd ed. NY: Pergammon Press. 196 pp.

4. McNair, H. M., and E. J. Bonelli. 1967. *Basic gas chromatography*. Walnut Creek, CA: Varian Aerograph. 306 pp.

5. Nelson, G. O. 1972. *Controlled test atmospheres-principles and techniques*. Ann Arbor, MI: Ann Arbor Sci. Publ. 247 pp.

6. Pratt, H. K., and D. B. Mendoza, Jr. 1979. Colorimetric determination of carbon dioxide for respiration studies. *HortScience* 14:175-76.

7. Pratt, H. K., M. Workman, F. W. Martin, and J. M. Lyons. 1960. Simple method for continuous treatment of plant material with metered traces of ethylene or other gases. *Plant Physiol.* 35:609-11.

8. Saltveit, M. E. 1978. Simple apparatus for diluting and dispensing trace concentrations of ethylene in air. *HortScience* 13:249-51.

9. _____. 1982. Procedures for extracting and analyzing internal gas samples from plant tissues by gas chromatography. *HortScience* 17:878-81.

10. Watada, A. E., and D. R. Massie. 1981. A compact automatic system for measuring CO_2 and C_2H_4 evolution by harvested horticultural crops. *HortScience* 16:39-41.

11. Young, R. E., and J. B. Biale. 1962. Carbon dioxide effects on fruit respiration. 1. Measurement of oxygen uptake in continuous gas flow. *Plant Physiol.* 37(3): 409-15.

13

Ethylene in Postharvest Technology

MICHAEL S. REID

Role of Ethylene

In the postharvest physiology of most horticultural crops, ethylene plays an important role, often deleterious, increasing the rate of senescence and reducing shelf-life, sometimes beneficial, improving the quality of the product by faster and more uniform ripening prior to retail distribution. This chapter is concerned with the properties of this gas, and ways to harness its beneficial effects and avoid its deleterious effects during postharvest handling of perishable commodities.

Properties of Ethylene Gas

Physical properties

Ethylene is the first member of the unsaturated or olefin series of hydrocarbons, and its properties are summarized in table 13.1.

Toxicological properties

Ethylene is a gas with a characteristic suffocating, sweetish odor. It is both an anaesthetic and asphyxiant. High vapor concentrations can cause rapid loss of consciousness and perhaps death by asphyxiation. Removal to fresh air usually results in prompt recovery if the person is still breathing. When the gas is handled in liquefied form, skin and eye burns can result from contact with the liquid. Cases in which liquid ethylene contacts the eye must be seen to by a physician.

FDA status

Use of ethylene gas to promote ripening of fruits and vegetables is sanctioned under FDA Regulation §120,1016. Ethylene is exempted from the requirement of a residue tolerance when used as a plant regulator either before or after harvest.

Explosive properties

Mixtures of ethylene gas and air are potentially explosive when the concentration of ethylene rises above 3.1 percent by volume. This concentration is at least 30,000 times the concentration required for initiating the ripening of most fruits and vegetables. Above 32 percent by volume, ethylene/air mixtures are not explosive.

Ethylene as a Fruit-ripening Hormone

History. The use of ethylene for hastening the ripening of fruits dates back to antiquity. Many examples have been cited, and include the ripening of sorb apples in southern Italy using emanations from ripe quinces and the ripening of mangoes in India in an atmosphere of burning straw. In recent times, the discovery of the hormonal nature of ethylene was associated with the discovery that ethylene would ripen fruits, and the observation that ripe fruits would cause ripening of other fruits.

Physiology. The concentrations of ethylene required for the ripening of various commodities vary (table 13.2), but in most cases are in the range of 0.1 to 1 parts per million (ppm). The time of exposure to initiate full ripening may vary, but normally exposure times of 12 hours or more will have an effect. Full ripening may take several days.

Practical considerations. The effectiveness of ethylene in achieving faster and more uniform ripening depends on maturity stage, temperature and relative humidity of

Table 13.1. Physical properties of ethylene gas

Appearance	Colorless, hydrocarbon gas with a faint, sweetish odor that is easily detected in parts per million concentration.
Molecular weight	28.05
Boiling point:	
at 760 mm Hg	$-103.7°C$
at 300 mm Hg	$-118°C$
at 10 mm Hg	$-153°C$
$\Delta b.p./\Delta p$ at 750 to 770 mm Hg	0.022°C per mm Hg
Freezing point at saturation pressure (triple point)	$-169.2°C$
Surface tension at $-103.7°C$	16.4 dynes/cm
Flammable limits in air:*	
lower	3.1% by vol
upper	32% by vol

*All compositions between the upper and lower limits are flammable and can be explosive.

the ripening room, ethylene concentration and duration of exposure to ethylene. In general, optimum ripening conditions for fruits are:

Temperature: 18° to 25°C (65° to 77°F)

Relative humidity: 90 to 95 percent

Ethylene concentration: 10 to 100 ppm

Duration of treatment: 24 to 72 hours depending on fruit kind and maturity stage

Air circulation: Sufficient to ensure distribution of ethylene within the ripening room

Ventilation: Require adequate air exchanges to prevent accumulation of CO_2 which reduces the effectiveness of C_2H_4

Systems for Ethylene Treatment

Handlers can equip existing rooms at comparatively small cost for use as ripening rooms or they can install specially built chambers with automatic control of temperature, humidity, and ventilation. It is not essential that the rooms be hermetically sealed, but they should be as tight as practicable to prevent leakage of the gas.

Amount of gas needed

The recommended treatment concentration is 100 ppm (1 cubic foot of C_2H_4 in 10,000 cubic feet of room space). Use of higher concentrations will not speed up the ripening process. Adding too much ethylene may result in accumulation of an explosive air-gas mixture.

Temperature

Optimum ripening temperatures are 18° to 25°C (65° to 77°F). At lower temperatures ripening is slowed; at temperatures over 25°C, bacterial growth and rotting may be accelerated, and above 30°C (86°F) ripening may be inhibited.

Table 13.2. Threshold for ethylene action in various fruits

Fruit	Threshold concentration (ppm)
Avocado (var. Choquette)	0.1
Banana (var. Gros Michel)	0.1 - 1
(var. Lacatan)	0.5
(var. Silk fig)	0.2 - 0.25
Cantaloupe (var. P.M.R. No. 45)	0.1 - 1
Honeydew melon	0.3 - 1
Lemon (var. Fort Meyers)	0.1
Mango (var. Kent)	0.04 - 0.4
Orange (var. Valencia)	0.1
Tomato (var. VC-243-20)	0.5

Rooms should be heated with hot water or steam pipe systems or with a gas or electric heater that has been examined and U.L. listed, never with an open flame. Because of the rapid increase in respiratory heat production following ethylene treatment, ripening rooms should be equipped with refrigeration systems adequate to hold the temperature in the desired range. The temperature in the room should be continuously monitored using a distant-reading or thermocouple thermometer.

Safety precautions

Because of the explosion hazard of ethylene mixed with air at concentrations between 3 percent and 30 percent, it is important to stringently follow the rules listed below to prevent buildup of these concentrations, and to prevent ignition if they should form.

1. Do not permit open flames, spark-producing devices, fire, or smoking in or near a room containing ethylene gas, or near the cylinder.

2. Use an approved meter for accurately measuring the gas when discharging ethylene from the cylinder.

3. Ground all piping to eliminate the danger of electrostatic discharge.

4. Store ethylene cylinders in accordance with all instructions and standards of the National Board of Fire Underwriters.

5. All electrical equipment, including lights, fan motors, and switches should comply with the National Electric codes for Class 1, Group D equipment and installation.

6. Instruments can be purchased that detect the concentration of ethylene in air and these can be set to sound an alarm if the concentration approaches explosive levels.

Introducing ethylene into the ripening room

Shot system. In the shot system, ethylene is introduced into the room in accurately measured quantities at regular intervals, using a gauge which registers the discharge of ethylene in cubic feet per minute. The required ethylene application is obtained by adjusting the regulator to give an appropriate flow rate, then timing the delivery of the gas. The piping, which leads into the ripening room, should be grounded.

Thorough ventilation is essential. Because of the inhibiting effects of CO_2 produced by the respiring fruit, it is customary to treat the fruits or fruit vegetables twice each day. The room should be well-ventilated before each application of ethylene, particularly if it is well-sealed. The doors should be opened for about half an hour. In large ripening rooms, a ventilating fan should be provided.

Trickle system. In the trickle or flow-through system, the ethylene is introduced into the room continuously, rather than intermittently, as in the shot system. As the flow of ethylene is very small, it has to be regulated carefully. This is usually done by reducing the pressure using a two-stage regulator and passing the gas into the room through a metering valve and flowmeter (fig. 13.1).

To prevent a buildup in either CO_2 or C_2H_4, fresh air is drawn into the ripening room at a sufficient rate to insure a change of air every six hours. The air is vented through a small exhaust port to the rear of the room. The fan size in CFM (cubic feet per minute) is calculated by:

$$\frac{\text{Volume of room (cubic feet)}}{360}$$

The ethylene flow rate (in CFM) needed to maintain 100 ppm in the room is calculated by:

Ventilation fan delivery (CFM) times 0.0001.

In ml (cc)/min, the flow rate is: ventilation fan delivery (CFM) times 2.8.

Other Technologies for Using Ethylene

The use of ethylene gas as described above may not always suit the technologies available or the handling regime of a particular perishable commodity. Other strategies for ripening fruits have been developed, and some useful methods are summarized below.

Explosion-proof ethylene mixtures

The danger of explosions from oversupply of ethylene to a ripening room can be eliminated by using mixtures of ethylene and inert gases. The proportion of the inert gas should be such that at high concentrations of ethylene there is insufficient oxygen remaining in the ripening space to provide an explosive mixture. As an example, "Ripegas" contains 6 percent C_2H_4 in CO_2 (w/w).

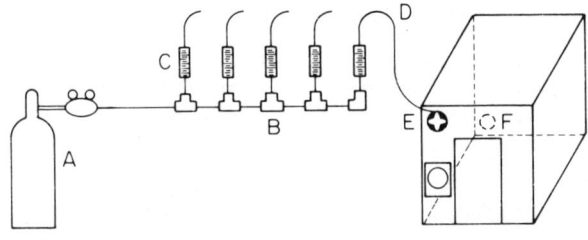

Diagram 1: Flowmeters grouped in one location.

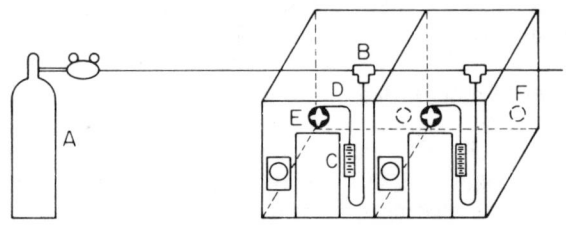

Diagram 3: Flowmeters located at each ripening room.

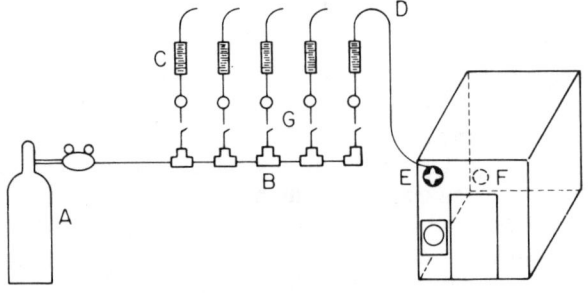

Diagram 2: Flowmeters grouped in one location with optional solenoid and fan switch.

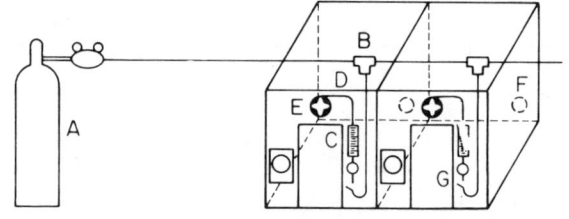

Diagram 4: Flowmeters located at each ripening room with optional solenoid and fan switch.

SYMBOLS:
- A = Ethylene cylinder and regulator
- B = Common line with union tees or appropriate fitting to deliver ethylene to flowmeter.
- C = Flowmeter—must be in a location where it can be read and adjusted if needed.
- D = Tubing to deliver ethylene from flowmeter to ripening room. *Note*—ethylene should be discharged into airstream being introduced into the ripening room.
- E = Blower/fan located somewhere at front of room.
- F = Small exhaust port in back of room.
- G = Optional solenoid and blower/fan switch. *Note*—solenoid should be wired in parallel with blower/fan.

Fig. 13.1. Diagrams of flow-through systems for introducing ethylene in ripening rooms (Sherman and Gull 1983).

Ethylene generators

Ethylene generators, in which a liquid produces ethylene when heated in the presence of a catalyst, are now widely used for supplying ethylene in ripening rooms. The liquid (which is a proprietary product) appears to comprise ethanol and agents that catalyze its dehydration ($C_2H_5OH - H_2O \rightarrow C_2H_4$). The generator combines a simple heating unit with a system for attaching a bottle of the generator liquid. The liquid comes in 1-pint and 1-quart bottles, and is used up by the generator at the rate of 1 pint every 8 hours. The generator delivers about 14 liters (0.5 cubic feet) of ethylene gas per hour, adjustable on newer models. These figures can be used to determine the number of generators to be used in a given size of room, if you know the air leakage or ventilation rate (number of gas exchanges per hour). For example, in a 5,000-cubic-foot ripening room where there is one air exchange per hour, a 1-quart bottle will generate 100 ppm for 16 hours.

Ethephon

Ethephon (2-chloroethane phosphonic acid) is strongly acidic in water solution. When in solutions above a pH of about 5, ethylene is liberated. Ethephon is commercially available (Ethrel, Cepa) and is registered for preharvest use on a variety of crops for controlling developmental processes, or inducing ripening. As a material for enhancing the ripening of harvested fruits, it has the disadvantage that it has to be applied to the fruit in a water solution, either as a spray or as a dip, an extra step in handling, with attendant dangers of microbial infection. In contrast to ethylene treatment, however, no special facilities are required to ripen fruit with ethephon, provided the ambient temperatures are within the range required to ripen the commodity. Ethephon is not yet approved for *postharvest* use on any commodity.

On a small scale, commodities can be treated with ethylene liberated from ethephon. Place the calculated amount of ethephon (approximately 7 fluid ounces of active ingredient to release one cubic foot of ethylene gas) in a stainless steel bowl, then, immediately prior to closing the room, add enough caustic soda pellets (approximately 3 ounces/7 fluid ounces a.i. of ethephon) to completely neutralize the ethephon. CARE! CAUSTIC SODA AND ETHEPHON ARE CORROSIVE. WEAR SAFETY GLASSES AND RUBBER GLOVES.

Calcium carbide

It has long been known that the flowering of pineapples (induced by dusting the plants with calcium carbide) can be synchronized by acetylene gas. This effect is due to the ethylene-like response of plant tissues to acetylene (table 13.3), (or perhaps to impurities of ethylene produced during the hydrolysis of calcium carbide). Calcium carbide, produced by combining calcium oxide and carbon (both readily available materials) in a furnace, releases acetylene when combined with water. Simple generators to provide acetylene for lamps are available and can be used in partially-vented spaces to ripen or degreen fruits under conditions where ethylene is not available.

Table 13.3. Comparative effectiveness of ethylene and related analogues in pea stem-section assay (from Burg and Burg)

Compound	Relative activity: Moles/unit effectiveness
Ethylene	1
Propylene	130
Vinyl chloride	2,370
Carbon monoxide	2,900
Acetylene	12,500
1-Butene	140,000

Table 13.4. Ethylene production rates by fruits at 20°C

Range (μl/kg-hr)	Commodity
0.01 - 0.1	Cherry, citrus, grape, strawberry
0.1 - 1.0	Blueberry, cucumber, okra, pineapple
1.0 - 10.0	Banana, fig, honeydew melon, mango, tomato
10.0 - 100.0	Apple, avocado, cantaloupe, feijoa, nectarine, papaya, peach, pear, plum
>100	Cherimoya, passion fruit, sapote

Use of fruits

Traditionally, the ripening of fruits has often been stimulated by enclosing them with other fruits which are already ripe. This technique, suitably applied, forms the basis for a cheap and simple method of fruit ripening. Table 13.4 shows the range of ethylene production known for some common fruits. Ripe fruits in the high ethylene production categories can be used in very small-scale commercial operations or at home to ripen or degreen other fruits in much the same way as any other ethylene generating system.

Deleterious Effects of Ethylene on Perishables

The potent effects of ethylene on plant growth, development, and senescence, mean that this gas, which is commonly found in the environment, can greatly reduce the life of perishable commodities sensitive to it. Techniques to remove it, or to avoid its effects are, therefore, of considerable importance to the postharvest technologist.

Undesirable effects of ethylene. Important effects of ethylene in hastening the deterioration of perishable commodities include:

1. Accelerated senescence and loss of green color in some immature fruits (cucumbers, squash, and so on) and leafy vegetables

2. Accelerated ripening of fruits during handling and storage

3. Russet spotting on lettuce (fig. 13.2)
4. Formation of bitter principle (isocoumarin) in carrots
5. Sprouting of potatoes
6. Abscission of leaves (cauliflower, cabbage, foliage ornamentals)
7. Toughening of asparagus
8. Abbreviated storage life and reduced quality of flowers (sleepiness, for example, in carnations [fig. 13.3])
9. Physiological disorders in flowering bulbs
10. Abscission of florets from cut and potted flower crops
11. Decreased storage life (or reduced quality) of fruits and vegetables

Sources of ethylene in the environment. The sources of ethylene as an air pollutant during postharvest handling of perishables are manifold but clearly the most important are internal combustion engines, ripening rooms, and ripening fruits. Other sources include: aircraft exhaust, fluorescent ballasts, decomposing produce and sometimes fungi growing on it, cigarette smoke, rubber materials exposed to heat or UV light, and virus-infected plants.

Protection of commodities sensitive to ethylene. A number of techniques have been developed to protect sensitive commodities from the effects of ethylene. Selection of the appropriate method obviously depends on the commodity, and the handling techniques used in its marketing.

Fig. 13.2. Ethylene-induced russet spotting on lettuce.

Fig. 13.3. Sleepiness (failure to open) of carnation flowers can be caused by as little as 5 parts per billion ethylene.

1. Ethylene exclusion and removal.

Where possible, removal of ethylene from the atmosphere surrounding the commodity is the preferable method of preventing deterioration of produce sensitive to this gas. A number of methods are available, and the simplest and cheapest are in many cases the most effective.

- Eliminate sources of ethylene. In a great majority of cases, high levels of ethylene in storage and handling areas can be avoided by removing sources of ethylene. In particular, handling of commodities sensitive to ethylene should be carried out with electric fork lifts. Internal combustion engine vehicles should be isolated from handling and storage areas, and engines should never be left idling in an enclosed space during loading and unloading operations. Rigorous attention to sanitation will remove overripe and rotting produce, a source of ethylene.

- Ventilation. Where the air outside storage and handling areas is not polluted, simple ventilation of these areas can reduce ethylene concentrations. An exchange rate of one air change per hour can readily be provided by installing an intake fan and a passive exhaust. The cost of using such a system (ignoring the small initial capital investment) can be determined using the equation:

Cost/year (dollars) = 0.001 x cooler volume (ft^3) × [outside temperature − cooler temperature (°F)]

This equation assumes a power cost of 7 cents per kilowatt hour.

- Chemical removal. Ethylene can be removed by a number of chemical processes; the most important are described below:

Potassium permanganate. Commercial materials, such as Purafil, utilize the ability of potassium permanganate ($KMnO_4$) to oxidize ethylene to CO_2 and H_2O. The requirements for such materials are a high surface area coated with the permanganate, and ready permeability to gases. Many porous materials have been used to manufacture permanganate absorbers, including vermiculite, pumice, and brick. The type of material utilized may depend on the purpose for which the absorber is required. Small brick pieces impregnated with permanganate are admirable for enclosing in sealed polyethylene bags. For removing ethylene from room air, the absorber should be spread out in a number of shallow trays or air should probably be drawn through the absorber system. Attempts to develop liquid scrubbers utilizing $KMnO_4$ have so far been unsuccessful.

Ozone. Ozone, a very potent oxidizing agent, can also be used to remove ethylene from the atmosphere. The effects of ozone on fruits and vegetables can, however, be very harmful. Some germicidal lamps can destroy ethylene without producing harmful concentrations of ozone, and it is now known that the effect is due to the production of ozone by 184.9 nm emissions from the lamp, oxidation of ethylene by the ozone, and removal of remaining ozone by the predominant 253.7 nm emission. Pilot-scale apparatus has been constructed to take advantage of this effect. Air from the storage room is drawn past lamps producing largely 184.9 nm radiation, then past lamps producing 254 nm radiation. In early prototypes the lamps were shielded from each other, but this does not appear to be important in achieving dramatic reductions in the concentration of ethylene in store atmospheres. This technique is not very effective in low-O_2 controlled atmosphere storage.

Activated or brominated charcoal. Charcoal air purifiers, especially if brominated, can be used to absorb ethylene from air. These systems are largely confined to use in the laboratory, since potassium permanganate absorbers are cheaper and more widely available.

Catalytic oxidizers. If ethylene and oxygen are combined at high temperature in the presence of a catalyst (such as platinized asbestos) the ethylene will be oxidized. Prototype ethylene scrubbers utilizing this effect have been constructed, but the difficulty of heating the incoming air, the need to add oxygen for scrubbing CA storage environments, and the need to cool the air before its return to the storage room may make these scrubbers impractical.

Bacterial systems. It has been calculated that approximately 30,000 metric tons of ethylene are liberated into the atmosphere each day from internal combustion engines, but the concentration of ethylene in air remains very low (nearly undetectable in fresh rural air). This implies that some system operates to remove ethylene from the atmosphere. Recently a bacteria has been isolated from soils which utilizes ethylene as a substrate. It seems possible that a scrubber may be developed in which the bacteria grow at the expense of ethylene in the store atmosphere, but initial attempts have not been successful.

- Hypobaric storage. Removal of endogenous ethylene was the first benefit ascribed to the hypobaric or low-pressure system of storage. It was clearly demonstrated that endogenous levels of ethylene in fruits held in hypobaric storage were greatly reduced, and that the longer storage life obtained in such systems could be reduced by adding ethylene to the atmosphere. There appear to be few commodities where the benefits of reducing the ethylene content of the tissue warrant the use of this cumbersome and expensive apparatus. Many of the results obtained in hypobaric storage are more properly attributed to reduction in the partial pressure of oxygen, which automatically accompanies reduction in the atmospheric pressure.

2. Inhibition of the effects of ethylene.

- Controlled atmospheres. The primary effects of controlling the concentrations of O_2 and of CO_2 in the storage atmosphere are thought to be reduction in the rate of respiration and associated metabolic processes. Reduction of the partial pressure of O_2 and elevation of the partial pressure of CO_2 can inhibit the action of ethylene on tissues sensitive to this hormone. The response of fruits to high CO_2 may be an effect on the response to ethylene as well as a direct effect on the rate of metabolism. For example, it has been shown that bananas packed in polyethylene-lined boxes could be transported at high ambient temperatures in the presence of a potassium permanganate absorber. Even without the absorber, fruit transported in this way was in much better condition than control fruit. It is reasonable to suppose that the main effect of the accumulated CO_2 produced by the fruit was in preventing the action of ethylene on the fruit.

■ **Specific anti-ethylene compounds.** Compounds recently reported to inhibit the action of ethylene include Rhizobitoxin and its analogues and the silver ion. While the possibility of registration of these compounds for use on comestibles is remote, their use with ornamental commodities is already in commercial practice. Of particular interest is a complex between silver and thiosulphate (STS), which has a very low stability constant, and, therefore, moves readily from the vase solution to the head of cut flowers. Flowers pulsed with this material last from two to three times longer than control flowers (fig. 13.4). Potted flowering plants do not lose their flowers during transportation if they are first sprayed with STS.

References

1. Abeles, F. B. 1973. *Ethylene in plant biology*. New York: Academic Press. 302 pp.

2. Anon. 1970. *Ethylene for coloring matured fruits, melons, and vegetables*. New York: Product Information, Union Carbide Corporation 11 pp.

3. Burg, S. P., and E. A. Burg. 1966. Fruit storage at sub-atmospheric pressure. *Science* 153:314-15.

4. deWilde, R. C. 1971. Practical applications of (2-chloroethyl) phosphonic acid in agricultural production. *HortScience* 6:359-64.

5. Kader, A. A. 1985. Ethylene-induced senescence and physiological disorders in harvested horticultural crops. *HortScience* 20:54-57.

6. Liu, F. 1970. Storage of bananas in polyethylene bags with an ethylene absorbent. *HortScience* 5:25-27.

7. McGlasson, W. B. 1985. Ethylene and fruit ripening. *HortScience* 20:51-54.

8. Pratt, H. K., and J. D. Goeschl. 1969. Physiological roles of ethylene in plants. *Annu. Rev. Plant Physiol.* 20:541-85.

9. Reid, M. S. 1985. Ethylene and abscission. *HortScience* 20:45-50.

10. Saltveit, M. E., Jr. 1980. An inexpensive chemical scrubber for oxidizing volatile organic contaminants in gases and storage room atmospheres. *HortScience* 15:759-60.

11. Scott, K. J., and R. B. H. Wills. 1973. Atmospheric pollutants destroyed in an ultra violet scrubber. *Lab. Pract.* 22:103-106.

12. Sherman, M. 1985. Control of ethylene in the postharvest environment. *Hortscience* 20:57-60.

13. Sherman, M., and D. D. Gull. 1983. *A flow-through system for introducing ethylene in tomato ripening rooms*. Univ. Florida, Veg. Crops. Fact Sheet, VC-30. 4 pp.

14. Staby, G. L., and J. F. Thompson. 1978. An alternative method to reduce ethylene levels in coolers. *Flor. Rev.* 163:31, 71.

15. Yang, S. F. 1985. Biosynthesis and action of ethylene. *HortScience* 20:41-45.

Fig. 13.4. Treatment of snapdragon flowers with silver thiosulfate protects them against ethylene effects in inducing abscission of the florets.

14
Principles of Disease Suppression by Handling Practices

NOEL F. SOMMER

Introduction

Fresh horticultural crops are increasingly marketed at great distances from point of production. Long-distance marketing sometimes results in large economic benefits which, however, can often be achieved only by stretching the postharvest life of the commodity to its limit. Diseases and disorders ordinarily manageable during handling and transcontinental transit and marketing may be excessive when transoceanic marine transport of longer duration is involved. Similarly, storage of fruits and vegetables for weeks or months prior to marketing may create added disease problems.

The type of postharvest handling horticultural commodities receive will determine in large measure the loss due to rot. The physiological and physical conditions of these commodities are of great importance. For example, fruits in high vitality exhibit considerable resistance to fungal attack while stressed or senescent fruits are often disease prone. The central features of good postharvest handling procedures are to start with commodities in good condition and to emphasize those methods that will maintain their vitality. Debilitating environments are to be avoided and life-shortening injuries prevented or their effects minimized. At the same time, procedures must limit directly or indirectly the activity of pathogens.

Physiology in Relation to Disease

Sugars translocated from photosynthesizing leaves provide the chemical energy and carbon building blocks for growth and development of plants and plant parts. Within the living cells of fruits, metabolic processes convert sugars (along with minerals and water from the soil) into the myriad compounds (carbohydrates, proteins, and fats) that compose the living cell and its storage reserves. These metabolic reactions require chemical energy obtained from compounds in the cell by respiration, a process which itself requires energy to function.

Respiration is thus a process by which the captured energy of light stored in organic compounds by photosynthesis is released by oxidation (low temperature "burning"). The respiratory process is not entirely efficient because not all the energy produced is usable chemical energy. Some is wasted and given off as heat, the so-called 'vital heat' or 'heat of respiration.'

While a fruit or edible plant part is still attached to the plant, the substances oxidized to CO_2 and H_2O are easily replaced by photosynthates from green leaves or from stored reserves in stems. Once the fruit is separated from the tree it is on its own. The respiratory process must function continuously to produce energy for cellular functions or the fruit tissues die. The only available fuel is that which is stored within the fruit itself.

Since O_2 is consumed and CO_2 is lost during respiratory oxidation, measurements of either can be used as an index of respiration. Analyses of these gases indicate the "rate of living." Perhaps it more properly should be called "the rate of dying," since the fruit or other harvested plant part will eventually consume itself through respiration if it is not earlier destroyed by pathogens. The carbon dioxide wasted to the air in the oxidation of sugars represents an irretrievable loss of carbon from the fruit.

In a large part, postharvest environments are designed to reduce the rate of respiration to the minimum required to maintain vital processes. The stored reserves are thereby conserved and the postharvest life of the fruit is extended to a maximum if it is not attacked by disease organisms.

Fruits harvested and placed in respirometers at about 20°C (68°F) exhibit one or the other of two very different respiratory patterns as shown by CO_2 evolution (fig. 14.1). Sweet cherries show a gradual decrease in the respiration rate, as shown by the dotted line, as fruits ripen and

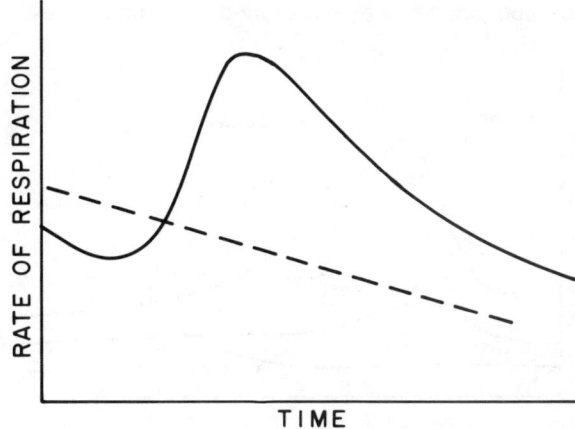

Fig. 14.1. Fruit respiration patterns. Climacteric (*solid line*) respiration of peaches and nonclimacteric (*dotted line*) respiration of sweet cherries.

senesce. When harvested before fully ripe, the cherries will darken in color and become soft during the ripening process. Acids may decrease to result in a sweeter taste but large increases in sugar do not occur because of the absence of large reserves of starch at harvest. In addition to sweet cherries, such fruits as grapes, citrus, strawberries, and pineapples exhibit this nonclimacteric respiratory pattern.

A climacteric respiratory pattern, shown in the solid line in figure 14.1, is exhibited by most deciduous tree fruit species (apples, pears, apricots, peaches, nectarines, plums), many tropical and subtropical fruits (banana, guava, avocado, mango), and some fruit vegetables such as the tomato. During the climacteric rise, fruits soften, and yellow colors intensify through loss of chlorophyll and increase of carotenoid pigments. Anthocyanins (red, blue, and purple colors) may be produced at this time. Ethylene gas evolution increases as do other volatiles, including those associated with fruit aromas. The peak of the respiratory curve approximates the point at which fruits are considered eating ripe. Beyond that point respiration gradually decreases as the fruit senesces during its remaining life. The climacteric rise in respiration roughly coincides with a striking reduction in the fruit's resistance to certain pathogens.

Maximum postharvest life in climacteric fruits can be attained only by harvesting before the start of their respiratory climacteric rise. Studies have shown that the climacteric rise can be initiated prematurely by exposure of the fruit to ethylene gas. Thus, if some of the fruits have started to ripen, the ethylene gas they evolve may trigger the respiratory rise of the remainder. Similarly, rotting or badly bruised fruits may evolve sufficient ethylene to trigger ripening.

It is highly important to maintain the fruit in a vigorous condition through lowering the respiration rate to the minimum that will still permit normal cellular function. With climacteric fruits it is essential to not only reduce the respiratory rate but, if possible, to minimize and delay the climacteric rise and associated ripening processes. Low temperature is the most effective means by which this is accomplished (fig. 14.2), with the possibility of an important assist by use of modified atmospheres.

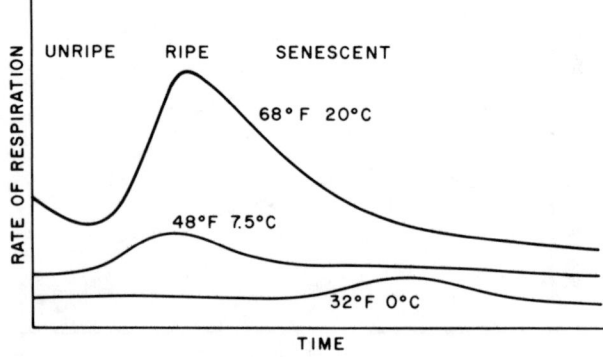

Fig. 14.2. Effect of temperature on suppression and delay of the climacteric rise in rate of respiration. Redrawn from data of Fidler and North 1967.

Resistance to Fungal Attack

Fruits and other detached plant organs resist fungal attack in several ways. The fruit skin (cuticle and epidermis) provides protection against infection in the same manner as does the skin of animals. Most fungi are unable to penetrate the sound fruit skin and must depend upon wounds. Even fungi capable of penetrating the sound skin often do so with considerable difficulty and are highly dependent upon favorable environmental conditions.

Through most of the life of the developing fruit in the orchard a strong healing capability is maintained. Fruits injured by rubbing against a limb or by some other means quickly heal the wound by the formation of resistant cork cells in a periderm. The wound thus effectively regains protection against entrance of microorganisms.

Maturity and biochemical defense

Immature fruits may be penetrated by fungi which are unable to colonize the tissue. Similarly, no disease lesion results if the same immature fruits are punctured with needles contaminated with fungal spores. Commonly, it is observed that a high degree of resistance is maintained until the fruit approaches maturity. Resistance is reduced very noticeably as the fruit begins to ripen. Not only does the fruit become susceptible to its most common disease organisms when it ripens but it succumbs to attack by fungi against which it was formerly resistant. Fruit resistance is further lost during senescence as the fruit nears the end of its postharvest life. Thus, as fruits approach maturity they are increasingly susceptible to attack by *Monilinia fructicola,* the brown rot fungus, or *Botrytis cinerea,* which causes gray mold in refrigerated storage. Other fungi, such as *Rhizopus* spp. or *Penicillium expansum,* causing blue mold rot in storage, are most likely to attack after the fruit is completely ripe or has become senescent.

Nature of biochemical defense

It appears reasonable to believe that all living organisms normally have highly effective mechanisms for disease resistance. Susceptibility, not resistance, appears to be the exceptional condition. If this were not the case, plant and animal species would have disappeared under the onslaught of countless microorganisms. It is unlikely that the resistance resulting from the biochemical action of tissues is the result of a single compound or a single mechanism. Instead multiple systems are evidently involved in complex and sometimes confusing interactions.

Immature fruits are believed to be resistant due in part to fungi-toxic compounds already present at infection. Most often implicated are compounds belonging to a chemical group called polyphenols, sometimes described as tannins. Members of this group are responsible for the browning reaction when fruits are cut or bruised. Analyses of total polyphenols in fruits show a decrease as the fruits approach maturity and ripeness, a decrease which coincides with observed decreased resistance to disease. However, it now appears that the total polyphenol content is not as important as the presence of specific highly fungi-toxic compounds. Furthermore, it has been found that some fungi-toxic compounds not previously present are formed as a consequence of fungal attack. The plant tissue's defense is thus not merely passive but includes the ability to mount a chemical offense against the invading fungus.

Research now underway is beginning to unravel the mysteries of the biochemical defenses of tissues against disease. Almost none of the research has been done on fruits, probably because potato tubers, sweet potato roots, and growing plants are more amenable to study. Nevertheless, the principles being elucidated appear to have widespread applicability throughout the plant kingdom. Resistance mechanisms in fruits are likely to be similar in general, if not in specific details, to mechanisms in other plant tissues.

The evidence is becoming very strong that a major portion of the resistance of tissues results when a penetrating fungus elicits a reaction from nearby cells to produce a compound toxic to the fungus. These fungi-toxic compounds have been termed phytoalexins and are often polyphenolic in nature. As an example, apple fruits have been demonstrated to produce benzoic acid in response to infection by *Cylindrocarpon mali (Nectria galligena)*, an important storage disease of apples in Europe. As different host tissues are studied while under attack by various pathogens, an increasingly long list of compounds is found to be phytoalexinic.

Generally the effect of phytoalexins has been limited to areas near the site of fungal penetration. However, recent instances of systemic protection conferred upon tissues by phytoalexins at considerable distances from the infection have been reported.

Little is known about the effects of postharvest environments on production of phytoalexins. What appears very likely, however, is that the ability to produce phytoalexins is positively correlated to the tissue's vigor.

Although phytoalexins were originally believed to be a specific chemical elicited by one or a group of pathogens, it is now known that at least some can be formed in response to plant wounds. These "wound metabolites" appear to provide a type of wound healing that tends to prevent subsequent fungal colonization of the wound.

Wound healing

Cuts and punctures are common avenues for infection by fungal spores. Conidia of *Monilinia fructicola*, for example, find needed moisture and nutrients for spore germination and growth if deposited in a fresh wound of a stone fruit. Pathogenic growth and rot development follow if temperature conditions are favorable. Fresh wounds that escape colonization by fungi often become less subject to subsequent colonization. In particular, wounds in fruits removed from refrigerated storage may no longer be highly prone to fungal invasion. Some resistance might be explained by a drying of the wound area. However, studies of responses to wounding in a wide assortment of tissue (roots, stems, tubers, leaves, fruits) suggests that a biochemical wound-healing process occurs and that it occurs widely throughout the plant kingdom.

When commodities are cut or otherwise damaged mechanically, they start respiring more rapidly, as measured by oxygen consumption or carbon dioxide evolution, and ethylene production is stimulated. In plant tissues in which the effects of cuts have been studied in detail, the following events have been noted. Cells ruptured by the cut are killed and cellular contents are mixed and exposed in the wound area. Enzymes, such as polyphenol oxidases, which are compartmentalized when the cell is alive, are mixed with the polyphenols in the cell sap. Browning in the wound results from enzymatic oxidation of phenolic compounds. Living cells near the injury are stimulated to become very active metabolically even though they do not themselves show signs of major injury. Repair is set in motion by these stressed but unbroken cells. Polyphenol synthesis may lead to the accumulation of greater quantities of those already present. New compounds, often similar to those which appear following infection, may appear in the wound area. Such substances may also be polyphenols. Some of the compounds produced as a result of wounding are some that are highly toxic to fungi. Germinating spores, which may be deposited in such "protected" wounds, are evidently killed or suppressed.

Temperature Effects on Fungal Rots

Temperature management is so critical to postharvest disease control that all other control methods have sometimes been described as "supplements" to refrigeration. Without intending to minimize the importance of other control measures, one can say without a doubt that temperature management is central to all modern postharvest handling systems, because low temperatures not only slow fungus development but the lowest temperature tolerated by the commodity maximizes its postharvest life potential.

Temperature requirements of postharvest pathogens

The generalized effects of temperatures on growth of postharvest pathogens are shown in figure 14.3. Postharvest pathogens generally grow best at 20° to 25°C (68° to 77°F), depending upon the fungus species, a few responding optimally at slightly higher temperatures. The maximum temperatures for growth are typically about 32° to 38°C (90° to 100°F), but some species can grow at higher temperatures.

Fungi can be conveniently divided into those that have a minimum temperature for growth of about 0°C (32°F) or above and those that grow at lower temperatures. In storage, nonchilling-sensitive horticultural crops can generally best be held at the lowest temperature safe from freezing. At a temperature of −1° to 0°C (30° to 32°F) only a limited group of fungi can be expected to pose difficulties. By far the most important of these is *Botrytis cinerea*, particularly if the storage period extends for 3 to 4 weeks. *Penicillium expansum*, the cause of the blue mold disease of deciduous fruits may also be of concern. Other fungi causing significant rot at 0°C (32°F) include *Alternaria alternata* and *Cladosporium herbarum*. *Monilinia fructicola* grows so slowly at 0°C (32°F) that visible

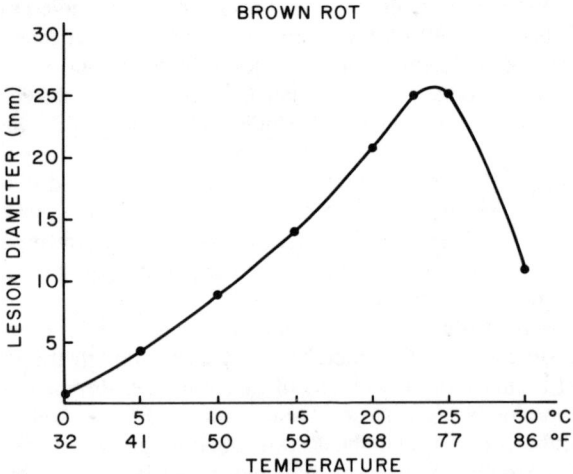

Fig. 14.3. Effects of temperature on growth of *Monilinia fructicola* in peach fruits. Redrawn from Brooks and Cooley 1928.

brown rot of stone fruits can be seen only after excessive storage periods.

Fungi having a minimum temperature for growth of −5° to −2°C (23° to 28°F) cannot be completely stopped by refrigeration without freezing the fruit. The mistaken view is sometimes held that low temperatures are unimportant in the suppression of these fungi. Nothing could be further from the truth. Although the fungi are active, the growth rate is only a minute fraction of that experienced at higher temperatures. Figure 14.4 shows the extent of rot development in peaches after inoculation with spores of *Monilinia fructicola* and being held at temperatures indicated.

Significance of the sigmoid growth curve

When a fungus spore lands on a medium suitable for growth (such as nutrient agar in a Petri dish) the spores of many species will swell and after a few hours germ tubes protrude and growth starts. The time taken to germinate and develop a tiny colony is called the *lag phase*

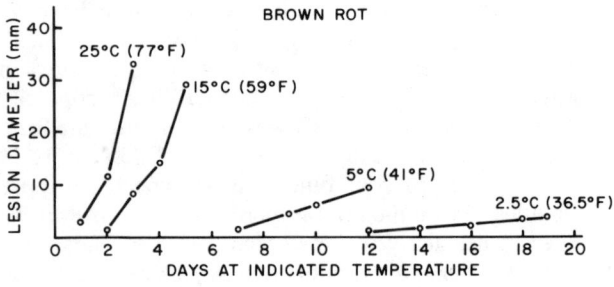

Fig. 14.4 Growth of *Monilinia fructicola* and brown rot developed in peach fruits at selected constant temperatures. Redrawn from Brooks and Cooley 1928.

(fig. 14.5). Growth soon achieves a rapid steady state called the *log phase*, which continues until growth is slowed for some reason, at which point the stationary phase of the curve is entered.

On fruit the lag phase is usually longer than on culture medium because the spore must not only germinate, but also initiate growth in the living tissues of the fruit despite considerable resistance of the tissues. Again, because of host fruit resistance, the rate of growth in the log phase may be slower than in culture, to result in a curve much less steep. The rate of growth slows when much of the fruit has become invaded.

As the temperature is lowered from near optimum to less favorable, growth curves change considerably but retain the general sigmoid (S) shape. A striking feature of the curve is that the time involved in the lag phase becomes greatly extended as a consequence of very slow germination and establishment of the infection. Depending upon the fungus species, the lag phase may lengthen from a few hours to several days at optimum temperatures to weeks or months at temperatures near minimum for fungus growth. Even when a steady state has been reached the very slow growth rate results in a much reduced slope of the log phase. This effect is illustrated by the development of brown rot in peaches (fig. 14.4). The length of the lag phase can be seen by extending the lower point of each curve to zero days.

The importance of the lag phase in fruit handling is further illustrated in figure 14.6. Data show the amount of brown rot that developed in peaches following delays in fruit cooling. After inoculation with spores of the fungus, one group of fruit was placed immediately at 0°C (32°F) while others were placed at that temperature after delays of 24 and 36 hours at 20°C (68°F). After 3 days at low temperature, obvious disease lesions had not developed.

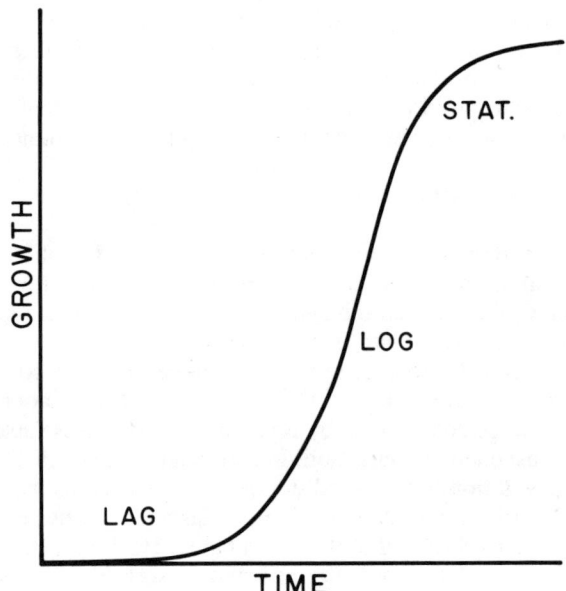

Fig. 14.5. Sigmoid curve of rot development.

Data taken after the fourth and sixth days at 15°C (59°F) show that the effects of delayed cooling extend into the normal marketing period.

A high percentage of spore-contaminated wounds may not develop into lesions if cooling is sufficiently prompt. Spore germination is extremely slow and may fail near the minimum temperature for growth of the fungus. Processes involved in establishing the infection are also likely to be marginally functional near growth-stopping temperatures. Low temperatures while the fungus is still early in the lag phase may, consequently, result in fewer fungal lesions and delay their development.

Fungal sensitivity to cold

Some postharvest pathogens having minimum temperatures for growth of about 5°C (41°F) or higher have developmental stages very sensitive to low temperatures. *Rhizopus stolonifer* and *Aspergillus niger* are examples. Although ungerminated spores are not adversely affected by low temperatures of fruit storages, most spores that have started to germinate may be killed by several days at 0°C (32°F) (fig. 14.7). Generally, very small lesions in the fruit are sensitive to cold also, but in larger lesions the fungus will no longer be killed by the cold temperature. Because only germinating spores are affected, the killing is not complete. Frequently two or three percent of the spores are still viable after several days in the cold. Nevertheless, a dramatic reduction in the incidence of the disease can result from the lethality of the cold if the level of inoculum is not too high. This cold sensitivity, along with wound healing, is believed responsible for the general absence of Rhizopus rot after peach fruits are removed from refrigeration following a good management program. Such fruits tend to remain free of Rhizopus rot even after they are removed from refrigeration, although similar fruits not refrigerated have been heavily attacked.

The ideal objectives in the use of refrigeration for disease control are to lower the temperature (1) below the minimum temperature for growth of the fungus, (2) to a point at which the lag phase has not been completed before the fruit is consumed, or (3) in the case of cold sensitive fungi, kill the spores while they are germinating. These ideals are often unattainable due to better adaptability of the pathogen than the fruit host to low temperatures. The pathogen is often merely suppressed by the low temperatures that are best for maintaining the fruit in good physiological condition.

To obtain the full advantage of refrigeration, it is essential to handle fruit without delays. Field heat should be removed as soon as possible and the fruit temperature lowered to near 0°C (32°F) or the lowest temperature tolerated by the commodity. Even if subsequent transport is at about 5°C (41°F), there are advantages to cooling to 0°C (32°F) because suppression of fungi is likely to be more nearly complete. The lesions of cold sensitive fungi may be permanently halted. Further, to cool to the lowest safe temperature is advantageous from the standpoint of the physiology of the fruit.

Modified or Controlled Atmospheres

Atmospheres around fruits are sometimes provided in which the oxygen (O_2) is low, carbon dioxide (CO_2) high, or both. If a close control of these gases is maintained the synthetic atmosphere is commonly called a controlled

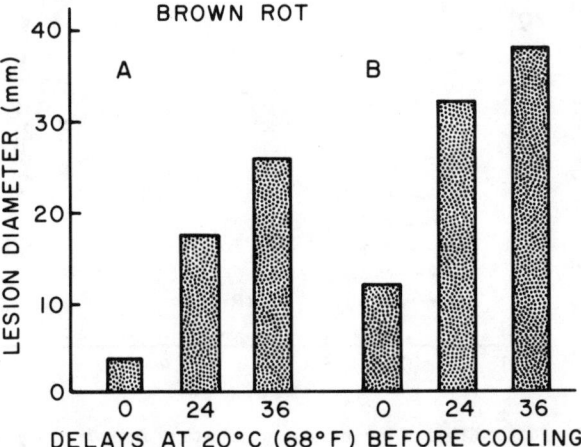

Fig. 14.6. Cooling delays and subsequent rot development. Peaches were stored at 0°C (32°F) immediately or after delays of 24 or 36 hours. Data indicate rot development A) 3 days, and B) 6 days after removal from cold.

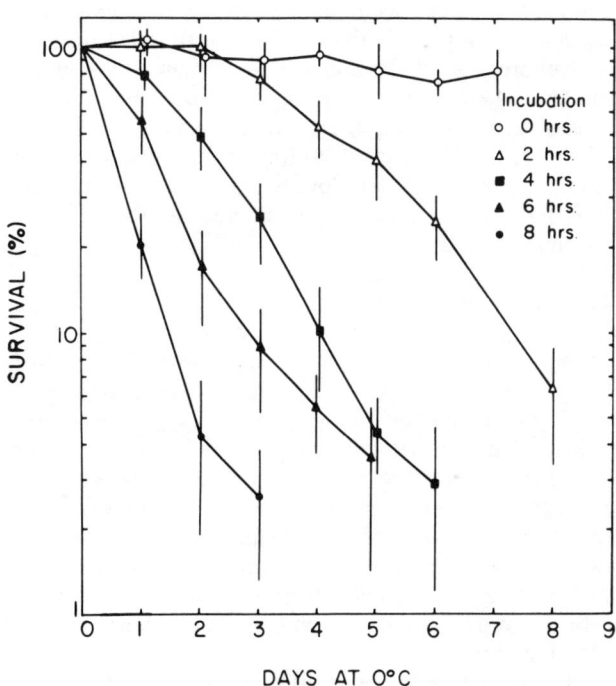

Fig. 14.7. Killing of *Rhizopus stolonifer* spores by exposure to 0°C (32°F) after various periods of incubation at 25°C (77°F). Matsumoto and Sommer.

atmosphere (CA). Modified atmosphere (MA) is a term which may be used to designate any synthetic atmosphere but often is used if there is little or no possibility to make adjustments in gas composition during storage or transportation. The purpose of these atmospheres is usually to extend the fruits' postharvest life as a consequence of a suppression of the rate of respiration. Another objective is to suppress diseases.

The effects of modified atmospheres on postharvest diseases can be either direct or indirect. The maintenance of the fruit in good physiological condition may result in a fruit with considerable disease resistance. A direct effect is also possible. Because the fungus pathogen respires as do the fruits, lowering the oxygen or raising the carbon dioxide can suppress pathogenic growth of the fungus in the fruit host. In this discussion, primary attention is given to the direct effects of atmosphere modification on the fruit pathogen.

Low oxygen

Oxygen is required for normal respiration of both the fruit and its fungus pathogen. The beneficial effects of low O_2 on fruit become evident as oxygen in the atmosphere is decreased to 5 percent or below; benefits are increased at lower O_2 levels. In controlled atmosphere storages the level of O_2 is commonly maintained at about 2 to 3 percent. It is generally believed these levels are the lowest that can prudently be maintained with the control methods usually available in storage.

Anaerobic or fermentative respiration is the consequence of an excessively low O_2 level. The fruit first develops off-flavors as substances, particularly alcohols and acetaldehyde, accumulate in the tissues. Eventually tissues are irreparably damaged and fruit death results.

Suppression of fungi by a 2 percent O_2 atmosphere is modest, often no more than about 15 percent below the rate of growth in air (21% O_2) as shown for *Botrytis cinerea* (figs. 14.8 and 14.9). Important growth reductions result if the O_2 level is lowered to 1 percent but that, unfortunately, is generally considered too low for commodity safety.

Hypobaric atmospheres

In recent years, storage and transport under low pressure or hypobaric (hypo = less than; baric = barometric pressure) has stirred considerable interest. Test vans have been constructed that maintain low pressure by use of vacuum pumps and regulated flow of air through the van. The refrigeration is conventional. When a 0.1 atmosphere is maintained, the available O_2 is reduced from ca. 21 percent of air to about 2 percent. A 0.05 atmosphere is equivalent to 1 percent O_2. An added benefit to the low O_2 is a very effective removal and sweeping out of ethylene produced by commodities.

From the standpoint of disease, very few critical data are available but a comparison of results with *Botrytis cinerea* at 0.1 and 0.05 atmospheres suggests that the suppressive effect is similar to controlled atmospheres at 2 or 1 percent O_2.

Carbon dioxide elevation

Air commonly contains about 0.03 percent CO_2. Elevation of CO_2 above about 5 percent noticeably suppresses fruit respiration. If the concentration of CO_2 is excessive, however, off-flavors develop and fruit injury results. The relationship of CO_2 concentration to fruit injury is time and temperature related. Fruits tolerate very high levels of CO_2 (>20%) for several days at transit temperatures, i.e., 3° to 5°C (38° to 41°F), but few fruits tolerate those elevated concentrations if storage or transit in the modified atmosphere is extended for several weeks. However, species and varietal differences in CO_2 tolerance may be important.

The addition of 10 percent to 15 percent CO_2 at a transit temperature of 5°C (41°F) has commonly affected both host and pathogens in a manner roughly comparable to a temperature of 0°C (32°F) in air. Carbon dioxide added to air has been widely utilized in transport of Bing cherries, (primarily to suppress *Botrytis cinerea* [gray mold] and *Monilinia fructicola* [brown rot]) and strawberries (to suppress *B. cinerea*).

Although fungi are suppressed by elevated (i.e., 10% to 20%) CO_2 levels, many fungi grow poorly in its complete absence. A number of enzymes have been implicated in CO_2 fixation within fungal cells.

Combined low oxygen, high carbon dioxide

The effects of low-O_2, high-CO_2 atmospheres are believed to be additive. Commonly used atmospheres of about 2 percent to 4 percent O_2 and 5 percent to 7 percent CO_2 suppress respiration and delay ripening of fruit, which could not safely be achieved with modification of the atmosphere by single gases. Modification of O_2 alone

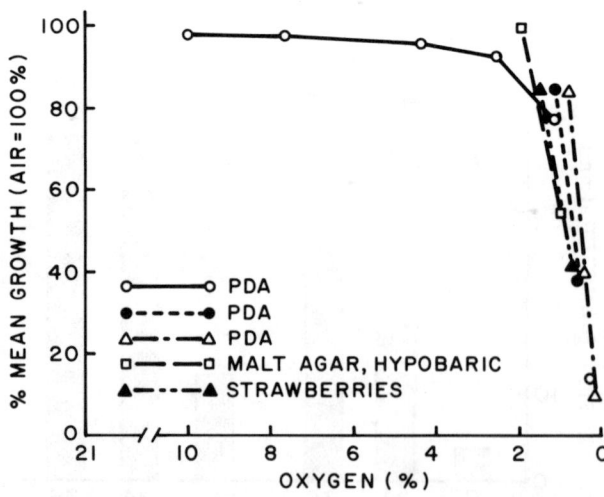

Fig. 14.8. Effect of low oxygen on growth of *Botrytis cinerea* in culture (potato-dextrose agar [PDA], malt agar) or strawberry fruits. All tests were conducted at normal barometric pressure in an atmosphere of low oxygen composition except hypobaric was in air at 0.1 and 0.05 atmospheres. Data from Borecka and Olak 1978; Couey et al. 1966; Follstad 1966; Sommer et al. 1981.

would likely require 1 percent, or less, O_2 to achieve similar effects. CO_2 in air might require 15 percent to 20 percent, or more, to equal the combined effect.

Controlled atmosphere with carbon monoxide

The very modest suppression of fungi by low O_2 and elevated CO_2 concentrations commonly used suggests the desirability of the inclusion of a fungistatic gas. Currently, there is interest in carbon monoxide (CO) for that role. Although tests have been encouraging with some fruits, the commercial use of CO with fruits is still very limited.

Carbon monoxide functions physiologically as an enzyme inhibitor and a competitor of O_2. Carbon monoxide (10%) added to air results in only modest reduction in growth of postharvest pathogens. When added to an atmosphere of low oxygen ($\leq 3\%$), however, much greater suppression has been noted (fig. 14.9). If added to a controlled atmosphere of 2.2 percent O_2 + 5 percent CO_2, even greater suppression is observed because of the additive effects of CO and CO_2.

In general, the suppressive effects of CO increase with lowered O_2 (figs. 14.9, 14.10). Further, the suppression is greater at 5.5°C (42°F) than at higher temperatures 12.5°C (55°F).

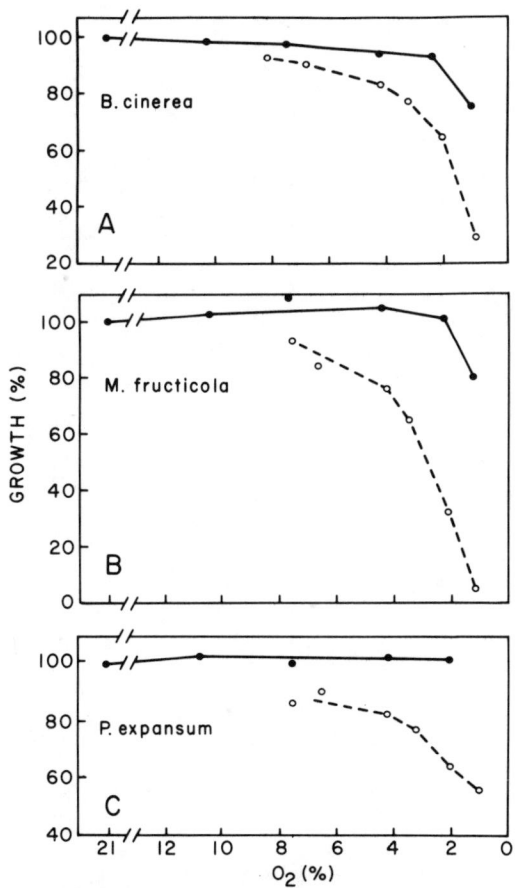

Fig. 14.9. Suppression of stone fruit postharvest pathogens by oxygen alone (*solid lines*) or combined with 10 percent carbon monoxide (*dotted lines*). Sommer et al. 1981.

Humidity

Water vapor is a true gas constituting a very important portion of the atmospheric environment of harvested perishable commodities. Its composition in the atmosphere as a percentage of saturation (relative humidity or RH) varies widely with temperature changes. Although the RH of storage is ordinarily never at saturation, free water on the surface of commodities can occur. Liquid water forms if, at any time in the normal cycling of the temperature within a refrigerated storage, the temperature of the commodity surface falls below the dew point temperature of the atmosphere.

Problems of accurately measuring RH in the microclimate of the surface of fruits and large changes in RH due to minor temperature fluctuations have made studies of humidity effects difficult under postharvest conditions. With pathogens such as *Monilinia fructicola*, germination and direct penetration of commodities are aided by saturated atmospheres or liquid H_2O on the fruit surface. With jacketed storage or packages with moisture barriers of plastic film, the high humidity might be a factor in promoting disease if temperatures were favorable. The disease-enhancing tendency of these high-humidity environments would likely increase if the commodity were wet when packaged.

Cold commodities removed from refrigerated to ambient temperatures condense moisture on their surface from surrounding warm humid air. This sweating continues until the temperature of the fruit has warmed to a temperature exceeding the dew point temperature of the atmosphere. The duration of the sweat depends on such factors as the fruit-air temperature difference, the exposure of the commodity to the air, air movement, and size or bulk of the commodity. Although moisture from sweating at the fruit surface has often been a concern among handlers,

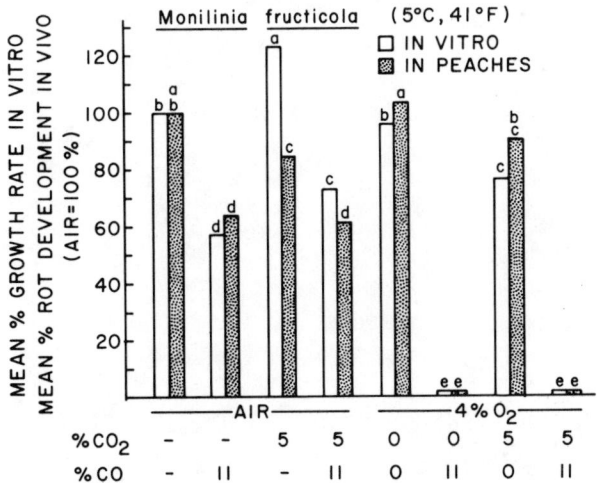

Fig. 14.10. Effects of atmospheres of varying oxygen, carbon dioxide, and carbon monoxide content on suppression of *Monilinia fructicola* in culture and in peach fruits. Redrawn from Kader et al. 1982.

it appears likely that the warm-up period of many commodities may be too short to be an important factor.

The horticultural necessity for maintaining high humidity in commodity environments is primarily to minimize loss of moisture, which results in shrivel and loss of turgidity in tissues. With peaches, for example, a loss of 3 percent to 4 percent of the original weight usually results in noticeable shrivel. The effect, if any, of weight loss on disease resistance requires added studies. In some vegetables, however, weight loss has been associated with increased susceptibility to disease. Carrots in storage were initially resistant to *Botrytis cinerea* but became susceptible after a moisture loss of 8 percent.

References

1. Borecka, H., and J. Olak. 1978. The effect of hypobaric storage conditions on the growth and sporulation of some pathogenic fungi. *Fruit Sci. Rep*. (Poland) 5:39-41.

2. Brooks, C., and J. S. Cooley. 1928. Time-temperature relations in different types of peach rot infection. *J. Agric. Res*. 37:507-43.

3. Couey, H. M., M. N. Follstad, and M. Uota. 1966. Low oxygen atmospheres for control of postharvest decay of fresh strawberries. *Phytopathology* 56:1339-41.

4. Fidler, J. C., and C. J. North. 1967. The effect of conditions of storage on the respiration of apples. I. The effects of temperature and concentrations of carbon dioxide and oxygen on the production of carbon dioxide and uptake of oxygen. *J. Hortic. Sci*. 42:189-206.

5. Follstad, M. N. 1966. Mycelial growth rate and sporulation of *Alternaria tenuis, Botrytis cinerea, Cladosporium herbarum,* and *Rhizopus stolonifer* in low oxygen atmospheres. *Phytopathology* 56:1098-99.

6. Harding, V. K., and J. S. Heale. 1980. Isolation and identification of the antifungal compounds accumulating in the induced resistance response of carrot root slices to *Botrytis cinerea. Physiol. Plant Pathol.* 17:277-89.

7. Kader, A. A., M. A. El-Goorani, and N. F. Sommer. 1982. Effects of carbon monoxide added to controlled atmospheres on postharvest decay, physiological responses, and quality of peaches. *J. Am. Soc. Hortic. Sci* . 107(5): 856-59.

8. Matsumoto, T. T., and N. F. Sommer. 1967. Sensitivity of *Rhizopus stolonifer* to chilling. *Phytopathology* 57:881-84.

9. Sommer, N. F., R. J. Fortlage, J. R. Buchanan, and A. A. Kader. 1981. Effect of oxygen on carbon monoxide suppression of postharvest diseases of fruit. *Plant Dis*. 66(5): 357-64.

15

Strategies for Control of Postharvest Diseases of Selected Commodities

NOEL F. SOMMER

Decay losses or the threat of decay losses influence most aspects of modern fresh horticultural crops' handling systems. Postharvest diseases must be considered in selection of handling practices or methods. Therefore, an understanding of disease organisms, the host commodity, and the relation of handling methods to both is of critical importance. Interrelated with the pathogen and the host commodities are the varied environments and handling stresses to which fruits are subjected.

Handling practices may affect the commodity's disease susceptibility as a consequence of stage of maturity, ripeness, or senescence. Cuts, bruises, and punctures may facilitate the entrance of the pathogen into the commodity. Stresses, such as high or low temperatures, may result in altered physiology leading to increased susceptibility of the commodity to certain pathogens.

Environmental factors of primary concern in postharvest diseases of horticultural crops are temperature, relative humidity (vapor pressure deficit), and atmosphere composition. The presence of pathogens completes the disease triad (organism-host-environment). The pathogen's level of pathogenicity and the incidence (population) of fungus spores are important determinants of disease incidence and severity.

The Pathogen

Fungi are overwhelmingly important in postharvest diseases of both fruits and vegetables. Bacteria frequently cause disease in certain vegetables but are generally of minor importance or rare in tree fruits and berries. Virus diseases may sometimes develop or intensify symptom expression in certain root or tuber crops but are unimportant in fruits after harvest.

Fungal pathogens are most commonly members of the Class Ascomycetes and the associated Fungi Imperfecti. Phycomycetes are represented by members of the genus *Rhizopus* and near relatives, and by the genera *Phytophthora* and *Pythium*. Basidiomycetes, with very few exceptions, are not postharvest pathogens.

Among the Ascomycetes, pathogens are usually encountered in the asexual or vegetative conidial (imperfect) state in postharvest diseases. The sexual state is rarely or never seen in culture or in diseased commodities. In some species the sexual state is extremely rare in nature. As a practical response to the need to utilize the vegetative conidial state for purposes of identification, the imperfect binomial is commonly applied to Ascomycetes as well as Fungi Imperfecti. Exceptions are certain well-known fungi, such as *Monilinia fructicola,* and fungi such as *Whetzelinia sclerotiorum* that lack a vegetative spore state.

The Infection Process

Fungal Spore

The propagule functioning to disperse fungi is commonly a spore, but other propagules exist. In fact, most living parts of the fungus are capable of growth and disease development if conditions are favorable.

Spores of postharvest fungi exist in a diversity of sizes and shapes. They may be either vegetative or sexual. Sexual spores may be very important in the life cycle of the fungus and, sometimes, may serve to permit survival of the fungus through the adversities of drought or winter cold. The major means of spread of diseases as fruits approach maturity and after harvest, however, is the vegetative spore.

Spore germination

Vegetative spores in the inactive condition are usually not dormant. If a spore is deposited in a moist, fresh wound, for example, it will start germinating immediately if temperature and atmospheric conditions are favorable. Certain substances, in addition to water, are required. Oxygen is almost certainly a requirement for germination of spores of postharvest fungi if not for all fungi. However, very low O_2 tensions suffice. Commonly, spores will not germinate well in the complete absence of CO_2 which may be fixed during germination. Further, the presence of metabolizeable organic compounds in the liquid enhance germination, and may sometimes be required.

Moisture absorption is an initial stage of germination and is usually accompanied by spore swelling. Some spores increase their volume many fold while others swell relatively little. As swelling becomes noticeable a sharp increase in O_2 consumption or CO_2 evolution marks an enhanced metabolic rate.

Before germination is triggered, spores exhibit only minimal metabolic turnover of structural and informational molecules. The initiation of germination is associated with a rapid increase in the synthesis of RNA, DNA, and protein. Increases are also seen during germination in the amount of endoplasmic reticulum and the number of mitochondria.

Spores are characteristically covered by a relatively thick spore coat. After swelling, a germ tube protrudes out through the spore coat. The wall of the germ tube is continuous with and a part of the innermost layer of the spore coat. Germ tube protrusion, and possibly much of the prior spore swelling, is believed dependent upon protein synthesis. As the germ tube lengthens through polar growth, side branches are soon initiated (fig. 15.1).

Spore germination is a hazardous period in the life of the fungus. During the swelling period spores are increasingly susceptible to lethal effects of such stresses as gamma and ultraviolet irradiation, low temperatures, high temperatures, absence of O_2, and exposure to toxic chemicals. Once underway, the germination process evidently cannot be stopped for very long without loss of the ability to resume normal growth.

Spore dissemination

Many spores produced in exposed structures are powdery and ideally suited to wind transport. Other spores, produced in more or less enclosed structures, may be exuded to the surface in a gelatinous or mucilaginous substance. Such spores are dispersed by rain and often for considerable distances by windborne mist.

A few postharvest pathogens, primarily in the genus *Phytophthora* produce structures called *sporangia,* which may germinate by protrusion of germ tubes as if they were spores. However, under favorable conditions, motile spores may be formed within the sporangium instead. Upon emergence, the motile spores are capable of swimming in soil water before coming to rest. When conditions are favorable, germination is by formation of a germ tube. These sporangia or motile spores are usually deposited on fruits in the orchard with soil splashed on lower fruits as a consequence of driving rains. Spread and deposit through sprinkler irrigation systems have been observed.

Spore dissemination by insects and other small animals, birds, and humans may often play an important role.

Initial fruit penetration

Postharvest pathogens are divided into two groups with regard to their ability to gain entrance through the fruit skin into the fruit. One group is able to bypass the protective skin only through wounds (wound infections). The other group has the ability to form special morphological structures called appressoria that permit the fungus to penetrate the fruit cuticle and epidermis (direct infection).

With wounds, the germinating spore may grow and colonize in the exposed fruit tissue, provided the fungus is pathogenic (see wound healing, below). Wounds are the most common avenue of entrance of postharvest decay fungi. This is true even in the case of most fungi capable of penetrating the fruit cuticle and epidermis. Bruises, stem punctures, cuts, limb-rubbed areas, abrasions, and insect punctures are common wounds leading to entrance by decay fungi.

Spores of all postharvest pathogens require high humidity or free water for a number of hours for successful spore germination. Frequently, the atmosphere on the surface of the fruit is too dry for germination but if the spore is in a wound, the fruit juice present permits germination.

Serious stem-end rots result from infection of the wound caused when the stem is severed at harvest. *Diplodia natalensis* causes serious stem-end rots of citrus fruits, mangos, and papayas. *Thielaviopsis paradoxa* and *Botryodiplodia theobromae* similarly invade stem tissues of banana fingers, and the former invades pineapple fruits. Fleshy-stemmed varieties of apple and pear may be invaded by *Botrytis cinerea* or *Penicillium expansum* in long-term refrigerated storage at −1°C (30°F). In some fruit species, stems may be the first part of the fruit to become senescent. Senescence is likely a factor in Alternaria stem-end rots of citrus and papaya fruits. Some varieties of apples suffer chilling injury if stored for many months below about 3.5°C (38°F). Fruits so chilled may suffer serious losses from Alternaria stem-end rot.

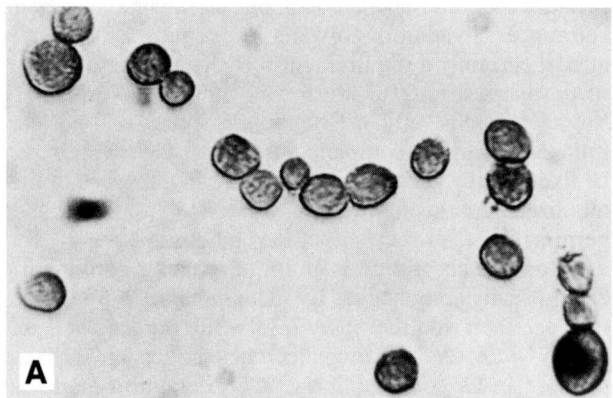
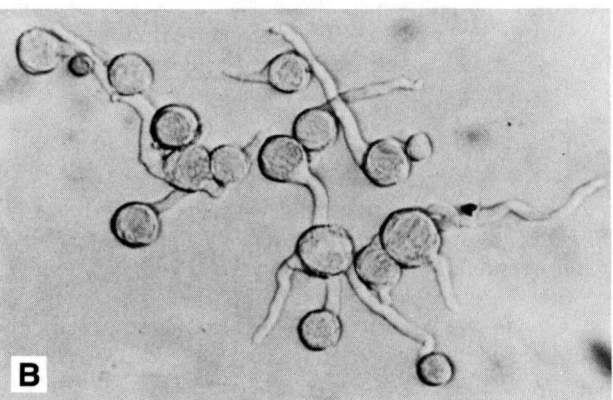

Fig. 15.1. Sporangiospores of *Rhizopus stolonifer.* A) Ungerminated spores. B) Germinated spores. Magnification 400×.

Fungi may colonize senescent floral parts at blossoming and only much later grow into and rot the fruit proper. California Bartlett pears are an example in which floral parts are infected by *Botrytis cinerea* near the end of the blossoming period in March. Dead styles and stamens colonized by the fungus are retained within the floral cavity near the core area. No rot will have occurred at harvest during August nor during much of the storage period. Only as the fruits become senescent near the end of their storage life does the fungus successfully invade the fruit flesh.

Direct penetration of fruit skin by a fungus occurs according to the following scenario. Spores landing upon a fruit, provided temperature and humidity conditions are satisfactory, will germinate within a few hours by sending out a germ tube. After the germ tube has become well-formed, a thick-walled structure (appressorium) is formed (fig. 15.2). The appressorium and germ tube adhere tightly to the fruit surface by mucilaginous material produced by the germ tube. The heavy thickening of appressoria walls is complete except for a pore on the underside against the fruit surface. The pore is covered only by the relatively thin germ tube wall. It is now believed that certain enzymes are excreted through the pore to the fruit surface. Among these enzymes is a cutinase capable of hydrolyzing the cutin which overlays the epidermis. Through the pore of the appressorium, a very fine germ tube-like protrusion called an infection peg penetrates the cuticle, which has been weakened by enzymatic action (fig. 15.3). Penetration of the infection peg is aided by the considerable pressure the appressorium is capable of exerting. After penetration, the infection peg regains the normal size for mycelia of the fungus. It proceeds to branch and rebranch to thoroughly invade the fruit flesh.

Latent infections result as a consequence of an interruption in infection following direct penetration. If the infection peg penetrates but is unable to overcome host resistance, the infection may remain quiescent until reduced resistance of the fruit permits it to proceed. An example is found in *Colletotrichum gloeosporioides,* a fungus causing anthracnose of many fruits (avocados, citrus, mangos, papayas, and others). Fruits are commonly penetrated while they are developing in the orchard. Before the fruit ripens it is highly resistant to this fungus, but it becomes susceptible upon ripening. Typically, anthracnose is a rot of ripe fruit.

Tissue invasion and rotting

Once penetration has succeeded, if fruit resistance is successfully overcome, the mycelium grows and branches to thoroughly invade the fruit flesh. The advancing mycelium excretes toxins into the fruit tissue that kill the cells. Extracellular enzymes are produced that break down constituents. The enzymes perform the functions of digestion by degrading complex substances into low-molecular-weight compounds that can enter the fungus cell. After entering the fungus, the compounds provide the building blocks for the synthesis of substances required for growth, plus the energy for growth and other life processes.

Sporulation

The production of spores is the common final step in the fruit rotting process and completes the vegetative spore cycle.

Disease Control

A discussion of control measures, commodity by commodity, follows. The appropriate measures may vary from one growing region to another depending upon the nature and seriousness of the threatening disease. Distance and time of transportation to market affect disease pressures. Finally, the capital investment required for handling facilities influences greatly the choice of control strategy. Thus, there is often no one best approach to disease control. Common postharvest diseases with names of pathogens and appropriate minimum temperatures for growth are listed in the table.

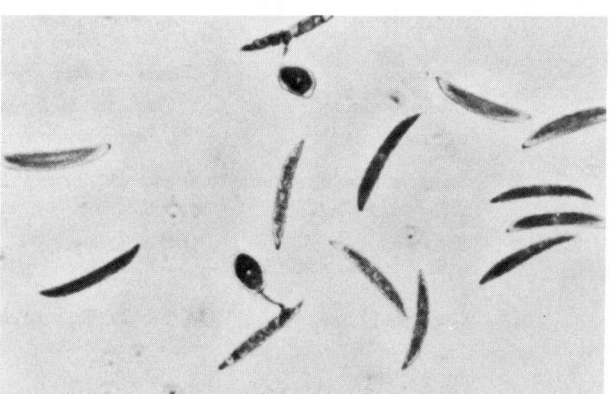

Fig. 15.2. Appressoria produced on short germ tubes of elongated conidia of *Colletotrichum* sp. Magnification approx. 700×.

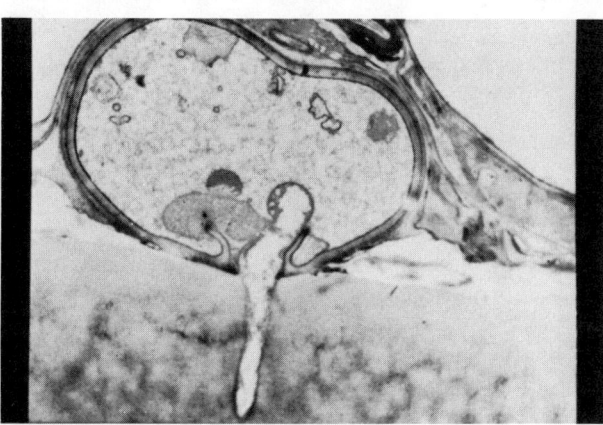

Fig. 15.3. Electron micrograph of *Colletotrichum gloeosporioides* penetrating the host. Courtesy of Dr. G. Eldon Brown, University of Florida, Lake Alfred Research and Extension Center. Magnification approx. 7,000×.

(Continued on page 89)

Table. Principal postharvest diseases of fruits of commerce

Disease	Pathogen Conidial state	Pathogen Sexual state	Estimated minimum temperature for disease
A. Temperate zone fruits			
1. Pome fruits (apples, *Malus pumila* Mill., and pears, *Pyrus communis* L.)			
Gray mold	*Botrytis cinerea* Pers. ex Fr.	*Botryotinia fuckeliana* (d.By.) Whetzl.	− 2°C
Blue mold	*Penicillium expansum* (Lk.) Thom		− 2°C
Anthracnose rots			
Bull's-eye rot	*Cryptosporiopsis curvispora* (Peck) Grem. = *C. malicorticis* (Cordl.) Nannfl. = *Gloeosporium perennans* Zeller and Childs	*Pezicula malicorticis* (Jacks.) Nannfld. = *Neofabria malicorticis* Jacks.	− 4°C
Lenticel rot	*Phlyctaena vagabunda* Desm. = *Gloeosporium album* Osterw.	*Pezicula alba* Gutherie	> 0°C
Bitter rot	*Colletotrichum gloeosporioides* (Penz.) Arx	*Glomerella cingulata* (Stonem.) Spauld. and Schr.	+ 3° to 9°C
Black rot	*Sphaeropsis malorum* Pk.	*Physalospora obtusa* (Schw.) Cook = *Botryosphaeria obtusa* (Schw) Shoem.	+ 2°C
White rot	*Dothiorella gregaria* Sacc.	*Botryosphaeria dothidea* (Moug. ex Fr.) Ces. and de Not.	+ 2°C
Alternaria rot	*Alternaria alternata* (Fr.) Keissler = *A. tenuis* auct.		− 3°C
Pleospora rot	*Stemphylium botryosum* Wallr.	*Pleospora herbarum* (Fr.) Rab.	− 3°C
Cladosporium rot	*Cladosporium herbarum* Lk. ex Fr.		− 5°C
Eye rot	*Cylindrocarpon mali* (Alles.) Wollenw.	*Nectria galligena* Bres.	
2. Stone fruits (apricots, *Prunus armeniaca* L., cherries, *P. avium* L., nectarines and peaches, *P. persica* (L.) Batsch., plums, *P. salicina* Lindl. and *P. domestica* L.)			
Brown rot	*Monilia* spp.	*Monilinia fructicola* (Wint.) Honey = *Sclerotinia fructicola* (Wint.) Rehm.	0°C
Gray mold	*Botrytis cinerea* Pers. ex Fr.	*Botryotinia fuckeliana* (Jack.) Nannfld.	− 2°C
Rhizopus		*Rhizopus* spp.	+ 2°C
Blue mold	*Penicillium expansum* (Lk.) Thom		− 2°C
Cladosporium rot	*Cladosporium herbarum* Lk. ex Fr.		− 5°C
Alternaria rot	*Alternaria alternata* (Fr.) Keissler = *A. tenuis* auct.		− 3°C
3. Strawberries, *Fragaria* spp.			
Gray mold rot	*Botrytis cinerea* Pers. ex Fr.	*Botryotinia fuckeliana* (d.By.) Whetz.	− 2°C
Rhizopus rot		*Rhizopus stolonifer* (Ehr. ex Fr.) Vuill.	+ 2°C
Leather rot		*Phytophthora cactorum* (Leb. and Cohn) Schtr.	+ 2°C
Anthracnose	*Colletotrichum gloeosporioides* (Penz.) Arx = *Gloeosporium fragariae* (Lib.) Mont.	*Glomerella cingulata* (Stonem.) Spauld. and Schr.	+ 3° to 9°C
4. Grapes, *Vitus vinifera* L.			
Gray mold rot	*Botrytis cinerea* Pers. ex Fr.	*Botryotinia fuckeliana* (d.By.) Whetz.	− 2°C
Alternaria rot	*Alternaria alternata* (Fr.) Keissler = *A. tenuis* auct.		− 3°C
Cladosporium rot	*Cladosporium herbarum* Lk. ex Fr.		− 5°C
Blue mold rot	*Penicillium expansum* (Lk.) Thom		− 2°C

Continued on next page

Table—Continued

Disease	Pathogen		Estimated minimum temperature for disease
	Conidial state	Sexual state	
B. Subtropical fruits			
1. Citrus fruits (oranges, lemons, grapefruit, limes, *Citrus* spp.)			
Blue mold rot	*Penicillium italicum* Wehm.		0°C
Green mold rot	*Penicillium digitatum* Sacc.		+ 3°C
Brown rot		*Phytophthora citrophthora* (Sm. and Sm.) Leon.	
Diplodia rot	*Diplodia natalensis* P. Evans	*Physalospora rhodina* (Berk. and Curt.) Ck.	+ 2°C
Phomopsis rot	*P. citri* Faw. Penz. and Sacc.	*Diaporthe medusaea* Nit. = *D. citri* Wolf	
Alternaria rot	*Alternaria citri* Ell. and Pierce		− 2°C
Trichoderma rot	*Trichoderma lignorum* (Tode) Harz.		+ 15°C
Sour rot	*Geotrichum candidum* Lk. ex Pers.		+ 2°C
2. Avocado, *Persea americana* Mill.			
Dothiorella stem-end rot	*Dothiorella gregaria* Sacc.	*Botryosphaeria dothidea* (Moug. ex Fr.) Ces. and de Not.	+ 2°C
Diplodia stem-end rot	*Diplodia natalensis* P. Evans	*Physalospora rhodina* (Berk. and Curt.) Ck.	+ 2°C
Alternaria stem-end rot	*Alternaria* spp.		
Anthracnose	*Colletotrichum gloeosporioides* (Penz.) Arx	*Glomerella cingulata* (Stonem.) Spauld. and Schr.	+ 3° to 9°C
Phomopsis stem-end rot	*Phomopsis citri* Faw.	*Diaporthe citri* (Faw.) Wolf	− 2°C
3. Fig, *Ficus carica* L.			
Endosepsis	*Fusarium moniliforme* Shel.	*Gibberella* spp.	+ 6°C
Gray mold	*Botrytis cinerea* Pers. ex Fr.	*Botryotinia fuckeliana* (d.By.) Whetz.	− 2°C
Souring	Various yeasts		
Smut	*Aspergillus niger* V. Tiegh.		+ 11°C
4. Kiwifruit (or Chinese gooseberry), *Actinidea chinensis* Plach.			
Gray mold	*Botrytis cinerea* Pers. ex Fr.	*Botryotinia fuckeliana* (d.By.) Whetz.	− 2°C
Surface mold	*Alternaria alternata* (Fr.) Keissler = *A. tenuis* auct.		− 3°C
Phomopsis rot	*Phomopsis* spp.	*Diaporthe actinidiea* Som. & Ber.	− 1°C
C. Tropical fruits			
1. Banana and plantain, *Musa* spp.			
Anthracnose	*Colletotrichum musae* (Berk. and Curt.) Arx = *Gloeosporium musarum* Cke. and Mass.	*Glomerella cingulata* (Stonem.) Spauld. and Schr.	+ 7° to 9°C
Thielaviopsis finger-stem rot	*Thielaviopsis paradoxa* (de Seynes) Hohnel = *Ceratostomella paradoxa* (de Seynes) Dade	*Ceratocystis paradoxa* Dade	+ 5°C
Botryodiplodia finger-stem rot	*Botryodiplodia theobromae* Pat. = ? *Diplodia natalensis* ?		+ 8°C
Fusarium finger-stem and spot rot	*Fusarium roseum* Lk. amend. Snyd. and Hans.		+ 3°C

Continued on next page

Table—Continued

Disease	Pathogen Conidial state	Pathogen Sexual state	Estimated minimum temperature for disease
C. Tropical fruits (cont.)			
Cigar-end rot	*Verticillium theobromae* (Turc.) Hughes = *Stachylidium theobromae* Turc. = *Verticillium staphylidium* Holmes *Trachysphaera fructigena* Tabor & Bunt.		+ 7°C
Squirter rot	*Nigrospora sphaerica* (Sacc.) Mason		+ 5°C
Crown rot	*Fusarium roseum* often with *T. paradoxa*, *B. theobromae*, *C. musae*, *N. sphaerica*, and *V. theobromae*		
2. Guava, *Psidium guajava* L.			
Anthracnose	*Colletotrichum gloeosporioides* (Penz.) Arx = *Gloeosporium psidii* Delacr.	*Glomerella cingulata* (Stonem.) Spauld. and Schr.	+ 3° to 9°C
Phoma rot	*Phoma psidii* P. Hen.		+ 3°C
Pestalotia rot	*Pestalotia psidii* Pat.		
Phytophthora rot		*Phytophthora nicotiana* var. *parasitica* (Dast.) Waterh.	+ 10°C
Botryodiplodia stem-end rot	*Botryodiplodia theobromae* Pat.		+ 8°C
3. Pineapple, *Ananas comosus* L.			
Thielaviopsis soft rot = water blister = black rot	*Thielaviopsis paradoxa* (de Seynes) Hohnel	*Ceratocystis paradoxa* Dade	+ 5°C
Bacterial brown rot	*Erwinia ananas* (E.F.Sm.) Serr.		
Pink disease	Unidentified bacteria		
Fruitlet core rot	*Penicillium funiculosum* Thom *Fusarium moniliforme* Shel.		+ 6°C
Fruit rot and fruitlet core rot	*Fusarium moniliforme* var. *subglutinans* Wr. and Reink.		
4. Papaya, *Carica papaya* L.			
Anthracnose	*Colletotrichum gloeosporioides* (Penz.) Arx	*Glomerella cingulata* (Stonem.) Spauld. and Schr.	+ 3° to 9°C
Stem-end rots			
Ascochyta	*Ascochyta caricae-papayae* Pat.	*Mycosphaerella* sp.	+ 2°C
Phomopsis	*Phomopsis caricae-papayae* Petri & Cif.		
Phytophthora	*Phytophthora nicotiana* var. *parasitica* (Dast.) Waterh.		+ 10°C
Botryodiplodia	*Botryodiplodia theobromae* Pat.		+ 8°C
Rhizopus rot		*Rhizopus* spp.	0° to 4°C
Alternaria rot	*Alternaria alternata* (Fr.) Keissler = *A. tenuis* auct.		− 3°C
D. Tomatoes (*Lycopersicon esculentum* Mill.)			
Alternaria rot	*Alternaria alternata* (Fr.) Keissler = *A. tenuis* auct.		− 3°C
Pleospora rot	*Stemphylium botryosum* Wallr.	*Pleospora lycopersici* El. & Em. Marchall	− 3°C
Buckeye rot		*Phytophthora parasitica* Dast. *P. capsici* Leonin, *P. drechsleri* Tucker	+ 10°C
Botrytis rot	*Botrytis cinearea* Pers. ex. Fr.	*Botryotinia fuckeliana* (d. by) Whetzl.	− 2°C
Rhizopus rot		*Rhizopus* spp.	+ 2°C
Sour rot	*Geotrichum candidum* Lk. ex. Pers.		+ 2°C

(Continued from page 85)

Apples *(Malus pumila* Mill.) and Pears *(Pyrus communis* L.)

For best storage, fruits must be placed in storage while preclimacteric. Depending upon variety, fruits have maximum storage lives of about 2 to 12 months or longer. Storage temperature depends upon varieties. If not chilling sensitive, they should be stored as near the point of fruit freezing as can safely be done with the storage facilities available, usually $-1°$ to $0°C$ ($30°$ to $32°F$). Some apples are chilling sensitive and must be stored at $3°$ to $5°C$ ($37°$ to $41°F$). To increase their storage life, chilling-sensitive apples have often been stored in controlled atmospheres (CA) containing about 2 percent to 3 percent oxygen + 5 percent carbon dioxide (the remainder of the atmosphere is nitrogen). Increasingly, nonchilling-sensitive apple varieties are also being placed in CA storage to improve quality of apples held for very long storage.

Direct suppression of Penicillium blue mold *(Penicillium expansum)* (fig. 15.4) and Botrytis gray mold *(Botrytis cinerea)* (fig. 15.5) is achieved by low temperature and with the carbon dioxide of CA. Indirect suppression is obtained by maintaining fruits in their best physiological condition through use of good temperature management and controlled atmospheres. A high level of disease resistance will thereby be maintained.

In some growing areas bull's-eye rot *(Pezicula malicorticis)* (fig. 15.6) and/or lenticel rot *(Pezicula alba)* are more serious than blue and gray mold rots. Bitter rot *(Colletotrichum gloeosporioides)* is a problem in the orchard and in nonrefrigerated storage and marketing but not in refrigerated storage. Because bull's-eye rot and lenticel rot are present as diseases in the orchard where they produce limb cankers (fig. 15.7), control measures are often required in the orchard.

Black rot *(Sphaeropsis malorum)* and white rot *(Dothiorella gregaria)* (fig. 15.8) are found most commonly in the eastern United States and other humid growing areas. Both diseases have a more or less serious limb-canker phase in the orchard. Sporulation of the organisms on the cankers provides spores that infect the fruits.

Alternaria rot *(Alternaria alternata)*, often as a stem-end rot, has been observed in very high frequency in chilling-sensitive varieties that have been subjected to marginal chilling (fig. 15.9). Susceptibility evidently occurs before physiological chilling symptoms are readily detectable in the fruit. Alternaria rot frequently also develops in fruits that have been sunburned. This phase of the disease is especially troublesome in apples subject to delayed sunscald, in which no evidence of injury can be seen at harvest but after a few weeks in storage a yellowing or browning appears on the side of the fruit exposed to the sun. *Stemphyllium botryosum* and *Cladosporium herbarum* rots are often found associated with Alternaria rot in sunburned fruit.

Apples may be dipped in a broad-spectrum fungicide such as sodium orthophenyl phenate and/or benomyl before storage if experience has shown that danger of loss is high. In California, often no prestorage fungicide is applied. After storage in bins, any rotted fruits are eliminated during sorting prior to packing in shipping containers. Wax, sometimes containing a fungicide such as benomyl, may be applied immediately prior to packaging.

Pears are generally packed for shipping before storage, rather than after, as with apples. The reason for this handling difference is that pears become very sensitive to skin injuries from handling after only a few weeks in storage. Since there is no poststorage sorting of pear fruits, some rotted fruits may occur in packed containers after many months of storage. Occasional rotted fruits are tolerated by handlers and are eliminated in markets.

Stone Fruits *(Prunus* spp.)

Fruits may be infected with brown rot *(Monilinia fructicola)* in the orchard but the infection may remain quiescent until about time of harvest. Further infections may result from fruit injuries occurring during harvest and subsequent handling. Rhizopus rot *(Rhizopus* spp.) ordinarily results entirely from the infection of wounds. Both fungi, however, may grow from diseased to healthy fruit in packed containers by direct penetration of mycelia (fig. 15.10).

Brown rot requires appropriate control measures in the orchard and these differ with the seriousness of the disease in different areas. For peaches, nectarines, and plums, a postharvest application of benomyl and botran in wax is commonly made after harvest in California.

Atmospheres of high carbon dioxide have been used in transit vehicles for disease suppression in sweet cherries for many years. Traditionally, dry ice (solid carbon dioxide) was placed in the refrigerated rail car. When the dry ice sublimed, an initial atmospheric concentration of carbon dioxide of about 20 percent was sought. That concentration of carbon dioxide was potentially fruit-damaging, but a safe level was reached within a day or so due to dilution by air leakage in and out of the car. Wrapping the dry ice with several layers of kraft paper slowed the rate of sublimation and, thereby, maintained an elevated level of carbon dioxide for a longer period.

Modern refrigerated marine containers and highway vans are available that are designed to utilize modified atmospheres if needed. Although use of benomyl has reduced the need for modified atmospheres to slow Botrytis gray mold *(B. cinerea)* in sweet cherries, it remains a viable option for shipment to markets that do not permit such fungicidal applications, or it may be needed if fungicidal effectiveness is lost as a consequence of the development of resistant strains of the pathogen. Carbon dioxide has never been widely used for peaches, nectarines, or plums. Some commodities benefit from modified atmospheres by delayed ripening as well as for disease suppression, but others are injured.

Blue mold *(Penicillium expansum)* and Cladosporium *(C. herbarum)* and Alternaria *(A. alternata)* rots are seldom found in high levels unless fruits have been held in low temperature storage $0°C$ ($32°F$) for an excessive period (4 to 5 weeks).

Strawberries (*Fragaria* spp.)

Botrytis gray mold rot (*B. cinerea*) is by far the most important postharvest disease of strawberries handled under conditions of good low-temperature management. Strawberries may become infected in the field when senescing blossom parts are colonized followed by growth of the fungus into the flesh as the fruit starts to ripen. Alternatively, the fruit touching the ground is invaded by mycelium of the fungus growing in organic matter in the soil. Fruit in packed containers may be infected when fungal mycelium grows from an infected berry to adjacent healthy fruits (fig. 15.11).

Field sprays are used to suppress the disease in the field. Extensive use of benomyl, however, has resulted in the development of strains of *B. cinerea* very resistant to that fungicide.

The use of corrugated paper containers for field packing fruits precludes application of fungicidal dips or sprays after harvest. Postharvest disease control consists of (1) care to avoid packing infected fruit to the extent possible, (2) care to avoid injuries to the fruit during harvesting and handling, (3) cooling promptly to near 0°C (32°F), and (4) use of high-speed refrigerated trucks to deliver fruits to North American markets or airplanes for transoceanic shipment.

Atmospheres modified by high levels of carbon dioxide (10% to 15%) are often used in either air or surface transport. The atmosphere may be established in the van by introduction of gaseous carbon dioxide or by sublimation of dry ice. Alternatively, plastic pallet covers may be sealed to permit a modified atmosphere to be introduced or dry ice inserted into a stack of packages on a single pallet.

Table Grapes (*Vitis vinifera* L.)

Botrytis cinerea is usually the most serious rot problem of table grapes after harvest. The fungus may infect berries in the vineyard, particularly if there are extended periods of rainy weather before harvest. However, wounds are a common entry point for the fungus. A frequently invaded wound is at the point where berries are joined to the stem. Berries may be partially loosened as a consequence of harvesting and handling.

Commercial control of the disease is obtained by fumigating the grapes while in storage at 0°C (32°F) using 0.5 percent sulfur dioxide gas for about 20 minutes and then clearing the air of the gas. This initial fumigation should be within 12 hours of harvest. If the grapes are stored, treatment needs to be repeated about every 7 days at a concentration of 0.05 percent to 0.2 percent for 30 to 45 minutes. If the disease potential is expected to be high, as a result of rain before harvest or other causes, sulfur dioxide concentrations may be at the upper levels. If storage temperatures are higher than 0°C, increased sulfur dioxide concentrations and more frequent fumigations may be required.

The sulfur dioxide has little effect against fungal lesions established in the vineyard before harvest. The storage fumigation treatments do serve the extremely important role of preventing fruit-to-fruit contact infection and nesting of the fungus within packed containers.

Some injury is associated with fumigation when the sulfur dioxide is absorbed into wounds in the berry. The injury increases with each successive fumigation. The most obvious symptom of injury is bleaching of berry tissues surrounding a wound or near the point of attachment of the berry to the stem.

A slow release of sulfur dioxide has sometimes been used for low-level fumigation within packages. Compounds such as sodium bisulfate release sulfur dioxide gas under high humidity conditions within fruit packages. The chemicals have been added to sawdust in export chests containing grapes. Alternatively, the chemicals are contained within plastic pouches or are absorbed in paper pads and placed within packages.

Avocados (*Persea americana* Mill.)

Anthracnose (*Colletotrichum gloeosporioides*) is generally found in humid growing areas. Blossoms, leaves, twigs, and young fruits may be attacked. Especially the dried tips (tip burn) of leaves appear to produce abundant inoculum by sporulation of the fungus. Careful studies in Israel have shown that spores of the fungus, deposited at any time during fruit development, germinate on the fruit surface when temperature and humidity conditions are favorable. Upon germination a thick-walled appressorium is produced and rests in the waxy surface of the fruit. When the fruit softens, the appressorium germinates and the infection hyphae penetrates the skin of the fruit after which mycelium develops in the flesh (fig. 15.12).

Diplodia stem-end rot (*Diplodia natalensis*) resulting from latent or quiescent infections of fruits in the orchard is common, especially in humid growing areas. Infections of the fruit stem could occur, under Israeli conditions, whenever temperature and moisture were favorable during fruit development. The rot did not occur while the fruit was still attached to the tree. After harvest the fungus colonizes first the stem and then the flesh near the stem.

Temperature management is important in delaying the onset of rot. The Diplodia rot did not develop at temperatures below 10°C (50°F). Anthracnose did not develop below 6°C (43°F). Consequently, fruits are stored at 6° to 8°C (43° to 46°F) and ripened at about 12°C (54°F).

Orchard control measures may be instituted when justified by the severity of the disease. Bordeaux mixture or other copper sprays have traditionally been used. Three to six applications applied during fruit development have nearly eliminated anthracnose from the fruits and reduced the incidence of Diplodia stem-end rot. A postharvest dip in Thiabendazole (TBZ) largely eliminated Diplodia rot.

The replacement of copper orchard sprays with about the same number of benzimidazole-type sprays (TBZ or benomyl) have largely eliminated Diplodia rot. However, Alternaria stem-end rot became a major concern although it had only occasionally been seen when copper sprays were used. *Alternaria* spp. have usually been shown to be resistant to benzimidazole-type fungicides. Of further

(Continued on page 95)

Fig. 15.4. Blue mold (*Penicillium expansum*) in ripe Bartlett pears.

Fig. 15.5. Gray mold (*Botrytis cinerea*) of d'Anjou pears stored at −1°C. Sommer 1982.

Fig. 15.6. Bull's eye rot (*Pezicula malicorticis*) of Bartlett pear. Sommer 1982.

Fig. 15.7. Limb cankers of apple, caused by *Pezicula malicorticis*, are sources of spores causing fruit bull's eye rot. Courtesy of Dr. Audus W. Helton, University of Idaho, Moscow.

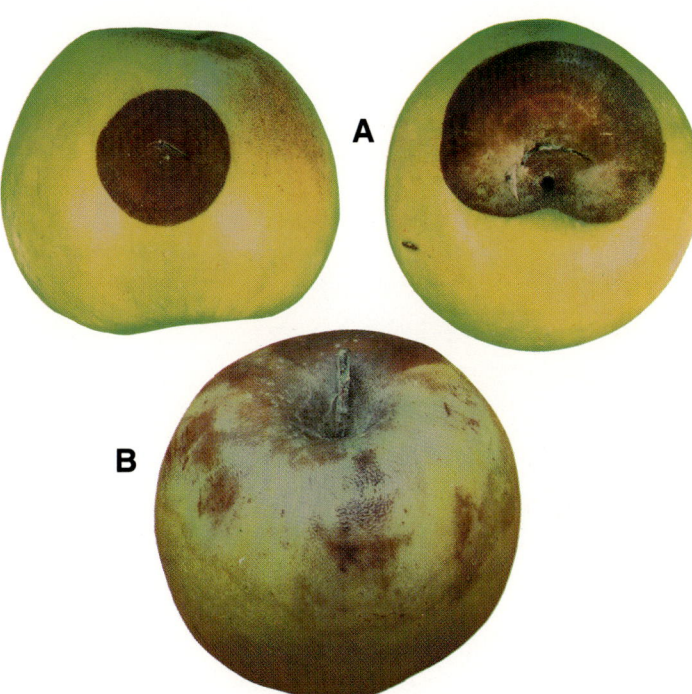

Fig. 15.8. Black and white rots of Yellow Newtown apples. A) Black rot caused by *Physalospora obtusa*. B) White rot caused by *Botryosphaeria dothidea*.

Fig. 15.9. Alternaria rot (*A. alternata*) in chilling injured Yellow Newtown apple. Sommer 1982.

Fig. 15.10. Rhizopus rot (*R. stolonifer*) (*black*) and brown rot (*Monilinia fructicola*) (*tan*) nests caused by mycelial infection of packed nectarines. Sommer 1982.

Fig. 15.13. Nest resulting from mycelial infection of kiwifruits by *Botrytis cinerea* during storage at 0°C.

Fig. 15.11. Gray mold (*Botrytis cinerea*) nest in strawberries. Sommer 1982.

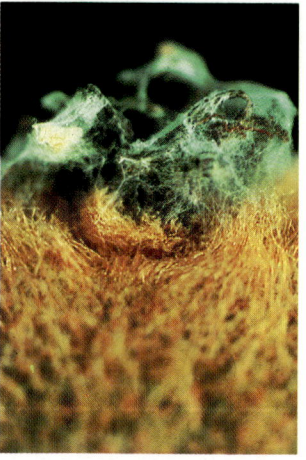 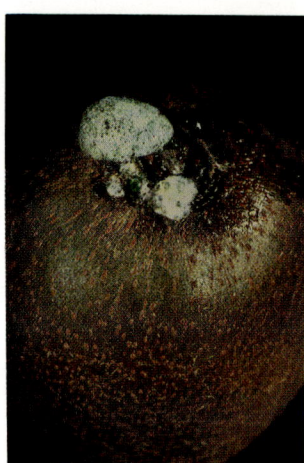

Fig. 15.14. *Alternaria alternata* sporulating on adhering styles of kiwifruits in 0°C storage. Sommer 1983.

Fig. 15.15 Phomopsis rot (*Diaporthe actinidea*) of New Zealand kiwifruit. Notice wetness of stem end. Froth from gas production is sometimes observed.

Fig. 15.12. Anthracnose (*Colletotrichum gloeosporioides*) in avocado.

Fig. 15.16. Crown rot of bananas caused by several fungi.

Fig. 15.17. Water blister (*Thielaviopsis paradoxa*) of pineapple fruit.

Fig. 15.18. Floret rot of pineapple fruits, usually caused by *Penicillium funiculosum* or *Fusarium moniliforme*.

Fig. 15.19. Anthracnose (*Colletotrichum gloeosporioides*) of ripe papaya fruits.

Fig. 15.20. Ascochyta stem-end rot (*Ascochyta caricae*-rot *Ascochyta caricae*-perfect state; *Mycosphaerella* sp.).

Fig. 15.21. Alternaria rot of chilled papaya fruits caused by *Alternaria* spp. and *Stemphyllium* spp. *Cladosporium* spp. and other fungi may also be present.

Fig. 15.22. Penicillium rot of citrus fruits. *Left*, blue mold (*P. italicum*). *Right*, green mold (*P. digitatum*).

Fig. 15.23. Diplodia stem-end rot of orange fruit (*Diplodia natalensis*).

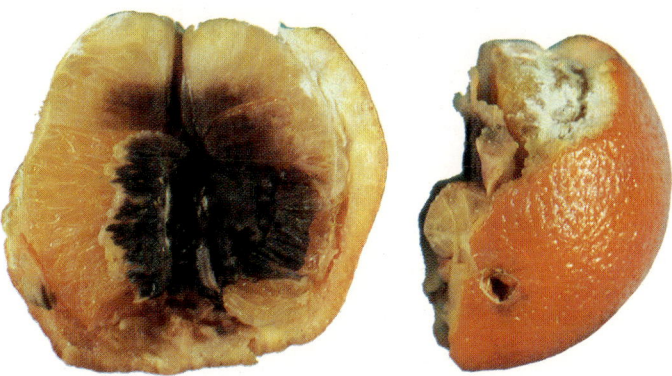

Fig. 15.24. Alternaria stem-end rot (*A. citri*).

Fig. 15.25. Anthracnose (*Colletotrichum gloeosporioides*) of mango.

Fig. 15.26. Diplodia stem-end rot of mango (*Diplodia natalensis* or *Botryodiplodia theobromae*).

Fig. 15.27. Alternaria rot of chilled tomato (*A. alternata*).

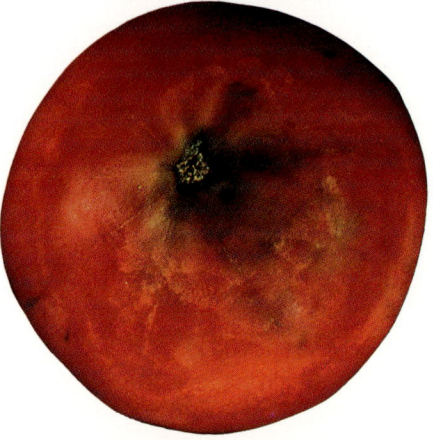

Fig. 15.28. Buckeye rot of tomato (*Phytophthora* spp.).

Fig. 15.29. Rhizopus rot of tomato (*R. stolonifer, Rhizopus* spp.).

Fig. 15.30. Sour rot of tomato (*Geotrichum candidum*).

(Continued from page 90)

concern has been the reduced effectiveness of TBZ against anthracnose, evidently as a consequence of the development of new, thiabendazole-resistant strains of the causal fungus, *Colletotrichum gloeosporioides.*

In California-grown avocados the Diplodia stem-end rot is rarely seen, but occasionally Dothiorella stem-end rot is of importance.

Kiwifruit (*Actinidia chinensis* Planch.)

Kiwifruits, or Chinese gooseberries, are rotted in storage primarily by gray mold (*Botrytis cinerea*) (fig. 15.13). Alternaria rot (*Alternaria alternata*) occurs primarily in fruits exposed to sunscald injury. A so-called Alternaria surface mold occurs in high-humidity storage when adhering styles, which have been colonized by the fungus after flowering, produce aerial mycelium and sporulates during storage (fig. 15.14). Sometimes calyces also adhere and Alternaria develops in them. Ordinarily no rot results from Alternaria surface mold activity but it is unsightly and is a factor in marketing.

A Phomopsis stem-end rot (*Diaporthe actinidiea*) has been observed in New Zealand-grown fruits marketed in the United States since the 1960s (fig. 15.15). The disease was first observed in California-grown fruit in 1981. A *Dothiorella* sp., probably *Dothiorella gregaria*, the imperfect state of *Botryosphaeria dothidea*, and *Botrytis cinerea* are also commonly found in New Zealand kiwifruits. *Botrytis cinerea* is the major fruit rot pathogen in California kiwifruits.

A postharvest treatment of sodium orthophenylphenate and botran in wax may be used to suppress Botrytis rot and largely prevent nesting of *B. cinerea* in packed fruit. Unfortunately, some important export markets do not accept that treatment.

Kiwifruits under California conditions are evidently resistant to *Botrytis cinerea* until fruits are ripe as shown by a reading of 1 1/2 or less pounds, with fruit skin removed, using a penetrometer with a 8 mm (5/16-inch) diameter tip. Phomopsis and Dothiorella rots also appear to be prevalent after fruits have ripened.

Kiwifruits ripen in a few weeks in 32°F (0°C) storage if exposed to ethylene before or during storage. Substitution of electric forklift trucks for internal combustion types has eliminated an important source of inadvertant ethylene exposure. Rigorous avoidance of ethylene has materially reduced the incidence of fruit rot.

Banana (*Musa* spp.) (Also see Chap. 27.)

For long-distance banana transport it is essential that fruits be pre-climacteric (i.e., green). While still-green fruits are essentially immune to loss by diseases of ripe fruit such as anthracnose (*Colletotrichum gloeosporioides*), Thielaviopsis rot (*Thielaviopsis paradoxa*) and Botryodiplodia rot (*Botryodiplodia theobromae*). Those factors which would stimulate the production of ethylene and thereby cause the fruits to ripen must be carefully avoided. Increased ethylene results from fruit injuries, the activity of green-fruit rotting organisms, or the presence of fruits of advanced maturity.

Fruits of proper maturity should be handled carefully to avoid injury. The time between harvest and cooling should be reduced to a minimum. Temperatures of 12° to 14°C (54° to 57°F), depending upon the variety, should be maintained in storage and transit.

Crown rot (fig. 15.16) is the result of fungal invasion of the hand tissue to which fruit fingers are attached. The disease is largely a consequence of an important technological change in which hands are cut from stems and packaged in fiberboard cartons. Formerly, bananas were shipped on the stem. Crown rot has been associated with *Fusarium roseum*, *Thielaviopsis paradoxa*, *Verticillium theobromae*, *Nigrospora sphaerica*, and *Colletotrichum musae*.

When cutting the hands from the stems and in washing to remove the resulting latex, spores evidently may be pulled into the xylem vessels if wash water is contaminated. Thiabendazole (TBZ) at 200 to 500 ppm in the wash water has been reported to effectively control crown rot. Alternatively, wash water may be maintained free of viable fungus spores by the inclusion of hypochlorite. Formalin and quarternary ammonium chlorides have been reported to be superior to hypochlorite for that purpose. The cleansed fruit is subsequently dipped in TBZ solution at ca. 200 to 500 ppm.

Treatment for crown rot also suppresses losses from anthracnose and Thielaviopsis, Botryodiplodia, and Fusarium rots.

Squirter rot is known primarily in Australia where single fingers are marketed. The very soft rotting of the interior of the finger results in the liquified contents squirting out if the finger is squeezed. The causal organism, *Nigrospora sphaerica*, is present in American producing areas but is not known to produce the squirter disease. The disease is found primarily in fruit produced in New South Wales and Queensland during the Australian winters. It appears possible that the fruit has been rendered susceptible to the disease by low winter temperatures in these growing areas which are marginal for banana production. Chemical applications are made during the period starting with May and concluding in November. Dips of salicylanilide have largely been replaced by TBZ.

Pineapple (*Ananas comosus* L.)
(Also see Chap. 27.)

Pineapple fruit is nonclimacteric and must be harvested when near ripe in order to market an acceptable quality.

The major fruit disease is Thielaviopsis soft rot (water blister, black rot) caused by *Thielaviopsis paradoxa* (fig. 15.17). The same organism infects planting material to cause the heart-rot disease in the field.

Entrance of the organism into the fruit is via the cut stem at the base of the fruit. The fungus spores germinate and grow up the central stem to thoroughly invade the interior of the fruit. Puncture wounds inflicted during harvesting and handling may result in rots starting at the side of the fruit.

For longest fruit life the fruit should be cooled to 8° to 10 °C (46° to 50°F). *Thielaviopsis paradoxa* grows very slowly at 11°C (52°F) and rotting is reported to essentially cease at 8°C. Consequently, it is essential to cool the fruits without delay and hold them at 10°C, or slightly lower if experience has shown that damage from chilling injury will not be incurred.

Fruits should be treated within about 6 hours after harvest with a fungicide. The application of salicylanilide (Shirlan WS at 1%) on the cut stem surface has long been the control of choice for the disease. More recently, application by dip or spray of benzimidazole fungicides has gained favor. Imazalil (0.2%) and bayleton (0.4%) have been reported effective in preventing invasion at the base but also have controlled lateral rots initiated in wounds if the fruits were allowed to soak in the materials for 1 minute.

Sheds and packing equipment can be effectively cleaned by use of 3 percent formalin or quarternary ammonium chloride at 3,000 ppm.

A fruitlet core rot and fruit rot disease has been reported from Brazil caused by *Fusarium moniliforme* var. *subglutinans*. Initially the symptoms resemble those of the common fruitlet core rot (fig. 15.18) usually caused by *F. moniliforme* or *Penicillium funiculosum* Thorn. Rot is usually localized to the fruitlet area. However, *F. moniliforme* var. *subglutinans* is a much more vigorous pathogen and causes complete rot of the fruit and may rot leaves and sprouts as well.

Papaya *(Carica papaya* L.) (Also see Chap. 27.)

Anthracnose (*Colletotrichum gloeosporioides*) (fig. 15.19) is almost always a serious problem in high-humidity conditions under which papaya fruits are most commonly grown. On papayas the disease behaves quite similarly to the same disease of avocados grown in humid areas. Trials directed toward control of the disease by orchard spray applications using Dithane M-45 and benzimidazole-type fungicides have not been sufficiently effective to permit elimination of postharvest heat treatments of the fruit which are required to kill the fungus in latent infections.

A high incidence of stem-end rots is sometimes seen in Hawaiian papayas marketed in mainland U.S. Usually most common has been Ascochyta stem-end rot (*Ascochyta caricae-papaya*) (fig. 15.20). Less frequently one sees Phomopsis stem-end rot (*Phomopsis caricae-papaya*) and Phytophthora stem-end rot (*Phytophthora nicotiana* var. *parasitica* or, according to some, *Phytophthora palmivora*). In certain growing areas *Botryodiplodia theobromae* or *Diplodia natalensis* may cause a stem-end rot in considerable frequency. All these stem-end rot organisms may also cause lateral fruit rots when they invade wounds. Rarely, they may enter the fruit at the stylar end.

Papayas suffer from chilling injury if held below about 12°C (54°F). Chilling injury is time-temperature related—the lower the temperature, the shorter the exposure period required to induce injury. Stage of ripeness is evidently also a factor, with fruits becoming less sensitive as they ripen.

Papayas exposed to chilling injury have been shown to lose resistance to Alternaria rot (*A. alternata* and *Stemphyllium* spp.) before visible evidence of injury is seen. Such fruits develop a multitude of lesions over the fruit surface (fig. 15.21). The greatest incidence has been observed on the side of the fruit exposed to sunlight while on the tree. Without chilling injury few, if any, Alternaria lesions develop anywhere on the fruit.

Heat treatments have provided reasonable disease control when fruits have been transported by air and marketing has been expedited. Transportation by ship involves longer transportation periods with increased incidence of all rots, especially Ascochyta rot (*A. caricae*). As with most fruits it is highly important to promptly cool papayas to the lowest nonchilling temperature before storage or transport.

The heat treatment for papayas consists of submerging a batch of fruit in water at about 49°C (120°F) for 10 to 20 minutes. There have been many years' experience with the treatment under Hawaiian conditions. Although the 49°C temperature is sufficient to cause fruit injury, the relatively cool fruits quickly reduce the temperature of the hot water to a safe level.

Attempts have been made to place the heat treatment in a packinghouse line as a continuous, rather than batch, operation. The design of compact equipment for this purpose has generally required drastic reduction of the fruit's immersion time in the hot water. In some treatments the water temperature has been as high as 60°C (140°F) and the immersion time as little as 1/2 minute. Obtaining sufficient heat penetration in a short time to kill the latent infections buried in the fruit skin requires very high temperatures when the exposure time is so short. Very short increases in immersion time could result in heat injury resulting in the failure of fruit to degreen or otherwise ripen normally. Underexposure of the fruit to the hot water results in failure to achieve sufficient heat penetration to kill the latent anthracnose infections. Obviously needed is a positive flow-through system by which fruits would move through the hot water during a precise period.

Because of limited heat penetration, questions have been raised about the effectiveness of treatments involving extremely short immersion periods, particularly with regard to control of stem-end rots. The effectiveness of the heat treatment against stem-end rots takes on added importance if transportation is via slower surface means rather than by air.

Citrus Fruits (*Citrus* spp.)

Blue (*Penicillium italicum*) and green (*Penicillium digitatum*) mold rots (fig. 15.22) are generally present wherever citrus fruits are produced. Severe losses may be incurred if uncontrolled. With both diseases the causal organisms are essentially strictly postharvest pathogens and normally invade fruits via wounds. Some fruit-to-fruit contact infection occurs, particularly with *Penicillium italicum*.

Diplodia (*Diplodia natalensis*) (fig. 15.23) and Phomopsis (*Phomopsis citri*) stem-end rots are prevalent in some humid production areas but are less likely to be troublesome in semiarid areas. Tree diseases are caused in the orchard by both organisms, and fruits may be infected before harvest by spores produced on tree cankers. Alternaria (*Alternaria citri*) (fig. 15.24) is found in semiarid locations as well. Although the incidence of the latter disease may be very low, it is serious because the damage is largely within the fruit, sometimes showing little or no exterior indication of the extent of the damage.

Brown rot (*Phytophthora citrophthora*) causes cankers on the trunk and roots of the tree. During rainy weather water splashing from the ground deposits spores on the lower fruits. Infections occur if rainy weather continues for a sufficient period for spore germination and infection to occur. Fruits with incipient brown rot infections may be included among harvested fruits. They rot and in so doing may also contaminate wash water or involve sound fruit by contact infection.

Sour rot (*Geotrichum candidum*) is a soft, very wet decay with a putrid, sour odor attractive to fruit flies which breed in the rotted flesh. Spread from rotting to sound fruit may occur by contact infection and nesting. Injuries and devitalized calyces (buttons) contribute to infection.

Trichoderma rot (*Trichoderma lignorum*) may cause considerable loss during the storage of lemons. As with sour rot, Trichoderma rot gains entrance into the fruits via injuries or by entrance at the stem of devitalized fruits.

Careful handling to avoid injuries is an important feature of decay control. Turgid fruit, particularly fruit harvested following rains, are particularly susceptible to injuries which release oil from glands of the rind causing oleocellosis. Susceptible fruits may be held for several days to reduce turgidity before subjecting fruits to packinghouse operations. Contaminated bins and equipment should be periodically cleaned.

Water dumps should contain chlorine or another fungicide to prevent a buildup of disease inoculum. A broad-spectrum fungicide such as sodium orthophenylphenate is commonly applied (1% at pH 11.5 to 12.0 followed by a water rinse). Hexamine can be added to reduce danger of phytotoxicity. Subsequently, benomyl, TBZ or 2-aminobutane may be applied in a wax application. Sometimes 2,4-dichlorophenoxyacetic acid (500 ppm) is added to the wax. That plant growth hormone has been shown to delay senescence of calyces, thereby maintaining resistance to various fungi which may enter at the stem end. Fungi commonly entering the stem when it is devitalized include *Alternaria citri*, *Geotrichum candidum*, and *Trichoderma lignorum*.

Biphenyl is commonly used to suppress fungal development within packages. Paper sheets (ca. 24 by 40 cm) are impregnated with about 2.35 g biphenyl. One sheet is placed near the bottom and another near the top of fiberboard citrus cartons of about the following dimensions: 27.5 by 27.5 by 42.5 cm. The crystaline compound sublimes in the packed carton to fumigate the fruit throughout the transportation and market handling period.

The biphenyl reduces the number of decayed fruit and prevents sporulation of *Penicillium* spp. on fruit surfaces. Diplodia and Phomopsis stem-end rots are suppressed in addition to the Penicillium blue and green molds. Alternaria and sour rots, however, are relatively unaffected by the biphenyl fumigation.

Citrus fruit handling and marketing is highly dependent upon effective fungicides. The development of fungal strains resistant to fungicides has, therefore, caused considerable consternation.

Resistance to both biphenyl and the chemically related sodium orthophenylphenate has been found in several species. The ease at which resistance develops to benomyl and TBZ has caused a search for replacement fungicides. Imazalil is one of the most promising of these.

Mango *(Mangifera indica* L.) (Also see Chap. 27.)

Anthracnose (*Colletotrichum gloeosporioides*) (fig. 15.25) is generally the most important postharvest disease of the mango. It may be rampant in the orchard where flowers, young fruits, and succulent leaves and twigs are attacked. Depending upon the severity of the fungal attack, flowers may be destroyed before fruit set, or young fruits may be infected. Sporulation of the fungus on leaves and twigs serves to provide inoculum for infections of developing fruits. Spores of the fungus are distributed from leaf and twig lesions largely by dripping rain water, splashing drops or by wind-driven mist. Infections of developing fruit may remain invisible as latent infections. Lesion development generally commences after harvest at about the time fruits ripen. The infections become visible as enlarging black spots on the fruit skin.

Diplodia (or Botryodiplodia) stem-end rot caused by *Diplodia natalensis* or *Botryodiplodia theobromae* (fig. 15.26) is frequently a problem in humid producing areas. A dieback disease may result from tree infections but loss of fruits is generally the most important aspect of the disease. Disease lesions in trees serve to provide spores which infect fruits and especially stems. Fruit infections typically remain quiescent until ripening commences.

Disease control in humid areas where mangoes are most frequently grown generally requires orchard sprays near time of bloom. Protection of flowers and foliage has traditionally been accomplished by Bordeaux mixture or other copper sprays such as copper oxychloride. Captan, maneb, and zineb have also proved beneficial in orchard control. In recent years, benomyl (8 oz/100 gal) has increasingly been utilized during the bloom period and sometimes in succeeding sprays to protect the fruit. Unfortunately, such orchard applications, particularly if repeated throughout the fruit development season, invite the development of resistant fungal strains against which the fungicide may be of limited effectiveness.

Hot water treatments similar to those used for papayas have been suggested as a postharvest control for both anthracnose and Diplodia stem-end rot of fruits. Dipping fruits in benomyl (500 ppm) or thiabendazole (900 ppm) has been reported effective against latent infections of anthracnose in mango fruits. Improved effectiveness in killing latent infections of both anthracnose and Diplodia stem-end rot has been reported if the fungicide is included in the water during heat treatments.

Tomato (*Lycopersicon esculentum* Mill.)

Alternaria rot is caused by *Alternaria alternata,* a fungus that is present wherever tomatoes are grown. It lives on dead or dying plant tissues. Spores of the fungus are commonly an important component of the microflora of the air.

Lesions of Alternaria rot in tomato are flat at the fruit surface or sunken. The lesion is usually covered by the sporulating black mycelium of the fungus (fig. 15.27). The lesion extends into the flesh where it produces a firm, dry, blackened mass of tissue thoroughly ramified by mycelium of the fungus.

A. alternaria is believed to produce latent infections in the developing fruits in the field by direct penetration of the cuticle. Most such infections seldom develop into disease lesions unless the fruit has been chilled. Upon chilling, however, rot lesions may develop at any point on the fruit surface. Mechanical injuries provide the fungus ready access to internal fruit tissues. A circle of lesions on the shoulders of the fruit around the stem area result from the tendency of fruit shoulders to be abraded during handling and transport of packed fruits.

The dramatic loss of resistance after fruits have been held for extended periods at <ca. 12.8°C (55°F) and particularly <5°C (41°F) necessitates avoidance of such temperatures for more than a few days. Mature-green fruits are most sensitive to low temperatures, followed by pink fruits or those of higher maturity. Resistance to Alternaria rot can be lost without the appearance of obvious physiological symptoms of chilling injury.

Control of the disease consists of avoiding stress that will reduce the fruit's natural resistance such as chilling injury or excessively high temperatures. As chilling injury can occur before harvest, it cannot always be avoided. If it is known that fruits have been chilled, losses can sometimes be minimized by prompt ripening at about 18° to 22°C (ca. 65° to 70°F) and shipping to nearby markets.

A disease with similar symptoms is caused by a fungus having *Pleospora lycopersici* El. and Em. Marchall as the sexual state. The conidial state is thought to be *Stemphylium botryosum* Wallr., a fungus with possible close relationships to *Alternaria alternata*. Ordinarily the two diseases can be readily distinguished by fruit examination with the naked eye only if perithecia (sexual spore bearing bodies) of *P. lycopersici* are present.

Buckeye rot is generally attributed to attack by *Phytophthora parasitica* Dast. (fig. 15.28). *P. capsici* Leonin and *P. drechsleri* Tucker have also been found to cause the disease. Very likely additional *Phytophthora* spp. may sometimes be involved. The fungi are soil inhabitants. During warm (i.e., 18° to 22°C − [ca. 65° to 70°F]) wet weather, sporangia are produced which give rise to motile swimming spores which infect those fruits in contact with the soil. Splattering raindrops and wind-driven mist can cause infection in fruits not in contact with the soil. The disease is minimized if plants are staked and fruits are held at some distance above the ground.

Control consists of staking to keep fruits from contact with soil. A 24-hour delay in packing tomatoes will permit sorting out fruits having lesions invisible when picked.

Gray mold rot caused by *Botrytis cinerea* is common where tomatoes are grown during cool wet weather. Late fall California shipments or winter shipments from Florida may suffer significant losses from the disease.

The causal fungus is a common postharvest pathogen of many fruits, vegetables, and flowers in addition to tomato fruits. The optimum growth of the fungus occurs at 25°C (77°F) but it can rot produce at a reduced rate at temperatures as low as −2° or −3°C (ca. 28° or 26°F). Consequently, disease develops readily at the lowest temperature tomato fruits can tolerate. The fungus can penetrate the fruit skin while still in the field. The more common loci of infection are mechanical injuries or growth cracks which often occur in the vicinity of the stem scar.

The most common symptom of the disease is the "dirty-white" color of mycelium over the lesion. When the fungus is sporulating the color darkens to a gray or gray-brown. Tissues appear water-soaked when invaded and are soft and watery.

As fruits ripen they become susceptible to attack by *Rhizopus* spp. (fig. 15.29) and sour rot (*Geotrichum candidum*) (fig. 15.30). These diseases are often prevalent in processing tomatoes that have ripened in the field.

Summary

Postharvest handling methods leading to the maximum physiological life of a commodity are often those which minimize fungal rots. Maintaining a fruit at high vitality enhances its natural disease resistance and ability to heal wounds. The following measures are important in relation to minimizing postharvest diseases and should be followed to the extent possible.

1. Harvest the commodity at its optimum maturity as determined experimentally or by experience.

2. Avoid cuts or punctures of the commodity's skin that provide avenues for entrance of pathogens. Wounds can also stimulate respiration and ethylene evolution, which may trigger ripening.

3. Cool promptly to ensure a high commodity vitality and natural disease resistance throughout the postharvest period.

4. Use the lowest feasible temperature not damaging to the crop and maintain proper relative humidity (90% to 95% for most commodities).

5. Use controlled or modified atmospheres if tolerated by the commodity and if transport and marketing periods approach the maximum postharvest life of the fruit.

References

1. Bompeix, G., and J. F. Bousquet. 1974. Les composes phenoliques dans la resistance des pommes a l'infection par les *Pezicula malicorticis et. P. alba. Fruits* 29:693-96.

2. Boreck, H., and J. Olak. 1978. The effect of hypobaric storage conditions on the growth and sporulation of some pathogenic fungi. *Fruit Sci. Rep.* (Poland) 5:39-41.

3. Brooks, C., and J. S. Cooley. 1928. Time-temperature relations in different types of peach rot infection. *J. Agric. Res.* 37:507-43.

4. Brown, G. E. 1975. Factors affecting postharvest development of *Colletotrichum gloeosporioides* in citrus fruits. *Phytopathology* 65:404-9.

5. Byrde, R. J. W., and C. V. Cutting. 1973. *Fungal pathogenicity and the plants response. Third Long Ashton Symposium, 1971.* New York: Academic Press.

6. Couey, H. M., M. N. Follstad, and M. Uota. 1966. Low-oxygen atmospheres for control of postharvest decay of fresh strawberries. *Phytopathology* 56:1339-41.

7. Dianese, J. C., H. A. Bolkan, C. B. daSilva, and F. A. A. Couto. 1981. Pathogenicity of epiphytic *Fusarium moniliforme* var. *subglutinans* to pineapple. *Phytopathology* 71:1145-49.

8. Eckert, J. W. 1979. Pathological diseases of fresh fruits and vegetables. *J. Food Biochem.* 2:243-49.

9. Eckert, J. W., and N. F. Sommer. 1967. Control of diseases of fruits and vegetables by postharvest treatment. *Annu. Rev. Phytopathol.* 5:391-432.

10. El-Goorani, M. A., and N. F. Sommer. 1981. Fungistatic effects of modified atmospheres in fruit and vegetable storage. *Hortic. Reviews* 3:412-61.

11. Fidler, J. C., B. G. Wilkinson, K. L. Edney, and R. O. Sharples. 1973. *The biology of apple and pear storage.* Res. Review No. 3, Commonwealth Agr. Bureaux, Slough, England.

12. Follstad, M. N. 1966. Mycelial growth rate and sporulation of *Alternaria tenuis, Botrytis cinerea, Cladosporium herbarum,* and *Rhizopus stolonifer* in low oxygen atmospheres. *Phytopathology* 56:1098-99.

13. Gibb, Ellen, and J. H. Walsh. 1980. Effect of nutritional factors and carbon dioxide on growth of *Fusarium moniliforme,* and other fungi in reduced oxygen concentrations. *Trans. Brit. Mycolol. Soc.* 74:111-18.

14. Haard, N. F., and D. K. Salunkhe. 1975. *Symposium: Postharvest biology and handling of fruits and vegetables.* Westport, CT: AVI Publ. Co. 193 pp.

15. Harvey, J. M. 1978. Reduction of losses in fresh market fruits and vegetables. *Annu. Rev. Phytopathol.* 16:321-41.

16. Isenberg, F. M. 1979. Controlled atmosphere storage of vegetables. *Hortic. Review* 1:337-94.

17. Ismail, M. A., and G. E. Brown. 1975. Phenolic content during healing of 'Valencia' orange peel under high humidity. *J. Am. Soc. Hortic. Sci.* 100:249-51.

18. _____. 1979. Postharvest wound healing in citrus fruit: Induction of phenylalanine ammonia-lyase in injured 'Valencia' orange flavedo. *J. Am. Soc. Hortic. Sci.* 104(1):126-29.

19. Ismail, M. A., R. L. Rouseeff, and G. E. Brown. 1978. Wound healing in citrus: Isolation and identification of 7-hydroxycourmarin (Umbelliferone) from grapefruit flavedo and its effect on *Penicillium digitatum* Sacc. *HortScience* 13:358.

20. Kader, A. A., M. A. El-Goorani, and N. F. Sommer. 1982. Effects of carbon monoxide added to controlled atmospheres on postharvest decay, physiological responses, and quality of peaches. *J. Am. Soc. Hortic. Sci.* 107:856-59.

21. Kolattukudy, P. E. 1978. Chemistry and biochemistry of the aliphatic components of suberin. In *Biochemistry of wounded plant tissues*, ed. G. Kahl, 43-84. New York: Walter de Gruyter & Co.

22. Kosuge, Tsune. 1969. The role of phenolics in host response to infection. *Annu. Rev. Phytopathology* 7:195-222.

23. Kuc, J. 1972. Phytoalexins. *Annu. Rev. Phytopathol.* 10:207-32.

24. _____. 1979. Biochemicals and plant protection. Proceedings: Opening session and plenary session symposium. In *IX Intern. Cong. Plant Protection*, 21-26. Washington, DC.

25. Kuc, J., and N. Lisker. 1978. Terpenoids and their role in wounded and infected plant storage tissue. In *Biochemistry of wounded plant tissues*, ed. G. Kahl, 203-242. New York: Walter de Gruyter & Co.

26. Matsumoto, T. T., P. M. Buckley, N. F. Sommer, and T. A. Shalla. 1969. Chilling-induced ultrastructural changes in *Rhizopus stolonifer* sporangiospores. *Phytopathology* 59:863-67.

27. Matsumoto, T. T., and N. F. Sommer. 1967. Sensitivity of *Rhizopus stolonifer* to chilling. *Phytopathology* 57:881-84.

28. McCracken, A. R., and J. R. Swinburne. 1980. Effect of bacteria isolated from surface of banana fruits on germination of *Colletotrichum musea* conidia. *Trans. Bact. Mycol. Soc.* 74:212.

29. Nelson, K. E. 1979. *Harvesting and handling California table grapes for market.* Univ. Calif. Agric. Sci. Publ. No. 4095. 67 pp.

30. Plant, J. E. van der. 1975. *Principles of plant infection.* New York: Academic Press.

31. Rossall, S., J. W. Mansfield, and R. A. Hutson. 1979. Death of *Botrytis cinerea* and *B. fabae* following exposure to wyerone derivatives in vitro and during infection development in broad bean leaves. *Physiol. Plant Path.* 16:135.

32. Sitterly, W. R., and J. R. Shay. 1960. Physiological factors affecting the onset of susceptibility of apple fruit to rotting by fungus pathogens. *Phytopathology* 50:91-93.

33. Slabaugh, W. R., and M. D. Grove. 1982. Postharvest diseases of bananas and their control. *Plant Disease* 66:746.

34. Smith, J. E., and D. R. Berry. 1974. *An introduction to biochemistry of fungal development.* New York: Academic Press.

35. Smock, R. M. 1979. Controlled atmosphere storage of fruits. *Hortic. Review* 1:301-36.

36. Sommer, N. F. 1982. Postharvest handling practices and postharvest diseases of fruit. *Plant Disease* 66:357-64.

37. Sommer, N. F., R. J. Fortlage, J. R. Buchanan, and A. A. Kader. 1981. Effect of oxygen on carbon monoxide suppression of postharvest diseases. *Plant Disease* 65:347-49.

38. Sommer, N. F., R. J. Fortlage, and D. C. Edwards. 1983. Minimizing postharvest diseases of kiwifruit. *Calif. Agric.* 37(1-2): 16-18.

39. Stanghellini, M. E., and M. Aragaki. 1966. Relation of periderm formation and callose deposition to anthracnose resistance in papaya fruit. *Phytopathology* 56:444-50.

40. Swinburne, T. R. 1974. The effect of store conditions on the rotting of apples, cv. Bramley's seedling, by *Nectria galligena*. *Ann. Appl. Biol.* 78:39-48.

41. Swinburne, T. R., and A. E. Brown. 1975. The effect of carbon dioxide on the accumulation of benzoic acid in Bramley's seedling apples infected by *Nectaria galligena*. *Trans. Brit. Mycol. Soc.* 64:505-7.

42. Uritani, I., and K. Oba. 1978. The tissue slice system as a model for studies of host-parasite relationships. In *Biochemistry of wounded tissues*, ed. G. Kahl, 287-308. New York: Walter de Gruyter & Co.

43. Uritani, I., K. Oba, M. Kojima, W. Kim, I. Ohuni, and H. Suzuki. 1977. Primary and secondary defense actions of sweet potato in response to infection by *Ceratocystis fimbriata* strains. In *Biochemistry and cytology of plant-parasite infection*, eds. K. Tomiyama, J. Daly, I. Uritani, H. Oku, and S. Ouchi, 239-52. Amsterdam: Elsevier.

44. Yang, S. F., and H. K. Pratt. 1978. The physiology of ethylene in wounded plant tissues. In *Biochemistry of wounded plant tissues*, ed. G. Kahl, 599-622. New York: Walter de Gruyter & Co.

16
Postharvest Treatments for Insect Control

F. Gordon Mitchell and Adel A. Kader

Introduction

Much has been learned about postharvest physiology, storage, and technology of horticultural crops but little is known about the effect of postharvest manipulations on the large number of insects that can be carried by fruits and vegetables during postharvest handling. Many of these insect species, especially the tephritid fruit flies, can seriously disrupt produce trade among countries and even among states within the U.S. Thus, effective insect control treatments, which are not harmful to the commodity, workers, or the consumer, are essential for allowing unrestricted movement of fresh horticultural crops in domestic or international commerce.

California has historically been in a position of potential rather than actual insect quarantine problems. Among fresh fruits and vegetables there has been a continuing need for small-scale quarantine treatment for exports to certain destinations in Canada and overseas. Among dried fruits, nuts, and grains there has been and remains a need for large-scale insect control treatments, but these are not normally against quarantined insects.

Among horticultural industries serious concern developed in California after World War II when the Oriental fruit fly was first identified in Hawaii and there was fear it might spread to California. The California legislature appropriated funds for an extensive study of quarantine treatments and of product tolerance to those treatments. Information gained from that study at the University of California, Davis and Riverside, provided the base for our knowledge of product tolerance. The emphasis of that study was on a variety of fumigants. Ethylene dibromide (EDB) and methyl bromide (MB) emerged as the most satisfactory treatments.

In 1980, with the medfly crisis in Central California, those early studies reappeared and produce industries prepared to use the information. Emergency programs to expand our knowledge of insect quarantine treatment and product tolerance were initiated, still primarily involving EDB and MB. During that crisis many large fumigation chambers were constructed and personnel certified as fumigation operators. Fortunately, the pest was eradicated as a result of a large-scale, expensive control program.

It now appears that the two commonly used fumigants (EDB and MB) might not be available in the future. EDB, which was on the RPAR (Rebutable Presumption Against Registration) list in 1980, recently had its registration withdrawn for most quarantine use effective 1 September 1984. MB is currently under attack based upon a recent report of carcinogenicity on laboratory animals, and under consideration are a lower residue tolerance or withdrawal of registration. Thus, alternate treatments must be developed if agriculture and the food supply are to be protected.

To avoid complacency about the dangers of future infestations or the need for quarantine treatments, let us explore the facts. There are many "fruit flies" around the world that are potential quarantine pests. In California during the early 1980s, several different fruit flies were found, some repeatedly, including the Mexican fruit fly, Caribbean fruit fly, Oriental fruit fly, Mediterranean fruit fly, Indonesian fruit fly, peach fly, and the apple maggot (a fruit fly). There are many other insects, codling moth is an example, that are quarantined by certain countries. Should any quarantined pest be found within a production area, immediate quarantine treatment would be needed. Unfortunately, honest quarantine problems are often intermingled with convenient quarantines that may be imposed by some countries as trade barriers.

Quarantine Treatment Research

Many quarantine treatment studies have been underway in recent years. The USDA-ARS laboratories in Hawaii, California, Washington, Florida, and Texas have been working extensively on these problems. In the fall of 1981, researchers at the University of California, Davis initiated a proposal for a regional research project to intensify the total activities in this area. The project, "Postharvest Biotechnology and Quarantine Treatments for Insect Control in Horticultural Crops," was initiated formally in October, 1982, with a 5-year duration. Researchers from the universities in California, Oregon, Hawaii, and Florida, along with the USDA-ARS labs, the Florida Department of Citrus, DSIR-New Zealand, and Agriculture Canada, are now participating. Responsibilities are being divided and resources and results pooled to maximize efforts on this broad project. It had been hoped that more time would be available to identify and to develop viable alternatives before the commonly used fumigants were removed from use.

Currently there are studies of many alternative treatments within this regional project. Some studies are to evaluate fumigants—ways of reducing residues, combination treatments to reduce concentrations, and alternate fumigants (including natural plant volatiles) as substitutes for EDB and MB. Other researchers are working on ways of avoiding infestation, such as harvest maturity (how mature must a fruit be before the flies will oviposit?).

Work is being done on heat, cold, very low oxygen levels, and very high carbon dioxide concentrations. Studies are underway on radiation including ultrasound, microwave, and the much publicized gamma radiation. Ultimately, combination treatments must also be considered.

At the moment there are very limited options available for insect quarantine treatment other than fumigation. As alternate treatments are developed their feasibility has to be studied from many aspects—worker safety, product safety and tolerance, problems of commercialization, engineering and economic considerations. Certainly no one treatment is expected to be universally applicable to all products. Each treatment has some inherent problems and limitations. Following is a brief discussion of the disinfestation (disinsection) methods in current use as well as those under investigation for possible future use.

Methyl Bromide Fumigation

Methyl bromide is still the most widely used insect quarantine treatment for horticultural commodities. The approved concentrations, durations, and temperatures depend upon the commodity. In certain cases, a lower concentration or shorter duration of fumigation combined with cold treatment is acceptable and may be less injurious to the commodity than the fumigation treatment alone.

All fumigation treatments must be done in an approved and certified fumigation chamber. Trained and certified fumigation operators must be on-site whenever fumigation is in progress.

The medfly fumigation treatments have required that the fruit pulp temperature be at 21°C (70°F) or above during fumigation, except that grapes could be at 18°C (65°F) or above. Current evidence may allow alternative schedules at lower temperatures. There is a potential problem of product injury if fumigated with free water on the surface (as might occur from sweating during rewarming of cold fruit). Injury may be more severe if the product has been treated with a water-soluble wax before fumigation. It may also be more difficult to purge the fumigant from many shipping containers than from field bins or lugs. Generally these problems provide a strong argument in favor of fumigation before cooling or packing, especially when energy efficiency is considered.

After the host material has received fumigation and/or cold treatment it must be protected from reinfestation. This means that any packing, storage, or loading facility handling the treated product must be isolated from the medfly. The County Agricultural Commissioner should be contacted to certify compliance. Isolation requires insect screening around the isolation area, use of plastic strip doors for access (the type used on cold storage rooms), with positive air flow out of the openings that cannot be adequately screened (cull fruit chutes, bin or box chutes, and so on). Use of traps both within and in the perimeter around the packing facility may be required in certifying isolation.

Federally acceptable insect quarantine treatments are established and administered by the U.S. Department of Agriculture, Animal and Plant Health Inspection Service (APHIS). In California, the Agricultural Commissioner in the county of production is the authority for currently approved treatments. The Agricultural Commissioner should, therefore, always be consulted before any treatment procedure is used.

Alternate Fumigants

Any alternate fumigant must be carefully screened for hazards to workers and consumers. One fumigant that is receiving attention is phosphine, which must be applied for 48 to 72 hours instead of the 2 hours necessary for EDB or MB. Perishable products would have to be under refrigeration during that long exposure time to avoid unacceptable deterioration. Extensive refrigerated fumigation facilities would be needed at a very high cost. Such a treatment would be more promising for the less perishable commodities (dried fruits, nuts, grains) where fumigation of bulk facilities may be possible. There is continuing interest in plant volatiles that may be promising as fumigants against certain surface-feeding insects.

Heat Treatment

The long standing, but seldom used, heat treatments in the quarantine manual require prolonged exposure (more than 8 hours plus heating time) to moist heat at temperatures as high as 112°F (44.4°C). This has been sufficiently injurious as to rule out its use on most products. Recently researchers at the USDA-ARS Laboratory in Hilo, Hawaii developed a 1-hour, 2-stage heating process using 108° and 120°F (42° and 49°C) to control fruit flies in papaya. If effective insect control can be shown in other products, and if those products will tolerate such treatment, this alternative might have a wide potential. The short treatment time would make this alternative especially appealing.

Cold Treatment

The quarantine manual currently allows cold treatment for control of some insects. The following cold treatments are allowed for fresh commodities from areas infested with the Mediterranean fruit fly:

10 days at 0°C (32°F) or below

11 days at 0.6°C (33°F) or below

12 days at 1.1°C (34°F) or below

14 days at 1.7°C (35°F) or below

16 days at 2.2°C (36°F) or below

There are strict requirements for temperature monitoring in cold storage facilities in order to certify compliance with these cold treatments. Attempts are underway to develop guidelines for monitoring temperatures which would allow application of the cold treatment during marine transport.

Some commodities are injured ("chilling injury") from exposure to such low temperatures. Recent studies have demonstrated that preconditioning at warm temperatures will cause some citrus fruits to tolerate the cold treatment. This procedure is now in the commercial test stage for treating grapefruit during transport from Florida to Japan.

Ten to sixteen days exceeds the potential market life for many perishable commodities. When the cold treatment must be completed before transport there is also a logistical problem with providing sufficient extra refrigerated storage capability in the midst of a heavy shipping period. Combination treatments, such as heat treatment followed by cold treatment, need study to determine if they would reduce the cold dwell time required for effective insect quarantine treatment.

High Carbon Dioxide and Low Oxygen Levels

The moderate modifications in these gases that are commercially used to improve the storage performance of some commodities (controlled-atmosphere storage) are inadequate to control insects. Some storage insects are controlled by exposure for 4 to 8 days to under 0.5 percent oxygen or above 80 percent carbon dioxide. Stored dried products should tolerate such conditions and may even show beneficial effects in terms of slower deterioration rate. Recent studies suggest that such treatment of bulk almonds may be more economical than the conventional MB fumigation.

The possibilities for quarantine treatment of fresh commodities are much less certain. Recent tests indicate that stone fruits may have a reasonable tolerance to very low oxygen for periods of up to 8 days. Tolerance to high carbon dioxide appears less promising. Studies are needed to evaluate the potential for effective control of the major quarantined insects under such atmospheres, alone and in combination.

If these modified atmosphere treatments are found effective, there is still a logistical problem in applying them to a large volume of fresh market produce, particularly if 4 to 8 days are required for treatment. The investment in specially constructed additional refrigerated controlled-atmosphere storage facilities would be too great to make it feasible as a "stand-by" treatment for potential infestations. The greatest hope is that modified atmospheres might be useful in combination with some other treatment to achieve more rapid insect control.

Ultrasound and Microwave

Insect control by ultrasound or microwave is in the very early stages of investigation. Use of these techniques shows preliminary indications of insect control, but much more study is needed to verify the effectiveness and to evaluate product tolerance and commercial potential. Some work has been done on the use of microwave heating combined with conventional heating to speed the overall warming of the product for heat treatment.

Radiation

By far the most publicized alternate quarantine treatment is the use of radiation. This involves exposing the product to a radiation source (isotopic source using Cobalt-60 or Cesium-137 or electrically-driven machine source) for a time period sufficient for it to absorb a required dose level of gamma or X rays. Successful use of this procedure is based on the determination that an undesirable organism (in this case the insect) will be "inactivated" at a dose level that is tolerated by the host commodity.

Current interest in radiation is based on a recent proposal by the Food and Drug Administration (FDA) to allow treatments up to 1 kGy (100 Krad) on foods (*Federal Register* 49[31]:5714-22, 14 Feb. 1984). This decision is based upon data from many sources over many years that have satisfied FDA of the safety of such usage.

Gamma radiation was extensively studied on fresh fruits and vegetables at University of California, Davis, and many other locations during the 1960s. The main emphasis at that time was on postharvest disease control which generally requires dose levels greater than the projected 1 kGy limit. Generally the conclusions from these studies were that the dose levels required to control disease organisms were sufficiently phytotoxic to the host as to make the treatment unsatisfactory. Whenever the procedure appeared marginally acceptable, less expensive alternatives were available.

During that same period certain other potentially beneficial effects of radiation were studied including growth inhibition on certain vegetables, retardation of ripening of some tropical fruits, and insect control. These effects were found at levels generally below the projected 1 kGy limit. Subsequently many studies of insect control by radiation have been conducted, primarily on potential quarantine pests. Results indicate that most insects are sterilized by doses below 0.75 kGy.

While this information makes the use of radiation for insect quarantine treatment appear quite promising, there are a number of considerations that dictate extreme caution in projecting radiation as an alternative to fumigation for insect quarantine treatment. Some are listed here:

1. Many potential hosts are reported to suffer significant detrimental effects from dose levels below 1 kGy. Conflicting results have been reported with other potential hosts, which suggests that more data are needed.

2. The USDA Animal and Plant Health Inspection Service (APHIS) will have to change its basis for acceptance of a quarantine treatment if it is to authorize use of radiation. Dose levels below 1 kGy will sterilize but not kill most quarantined insects to the level currently required by APHIS.

3. Dosimetry data must be accumulated to determine the dose range between the outside and center of a mass of product. If palletized fresh produce must be disassembled for treatment, the logistics become more difficult and the cost increases substantially.

4. Radiation is an injurious stress to the host commodity.

Thus good handling procedures, including good temperature management, are vital to minimize deterioration. Refrigeration capability must be available throughout the handling of fresh commodities, including the irradiator plant. Refrigeration before, during, and after radiation substantially increases the cost of the facility.

5. In an intensive horticultural area such as California, the logistics of radiation are great. Many multimillion dollar facilities have to be constructed to handle the volume of fruits and fruit-type vegetables that leave horticultural production areas.

6. Considerable engineering work is required to develop logistically sound radiation facilities. Because of the high volume involved these would be of quite a different scale than is currently available. The need to incorporate refrigeration further complicates the design.

7. Current information indicates that radiation plants must be used to capacity, essentially year-round, to be economical. Fresh fruit and vegetable production is seasonal in California. Further, because quarantine treatment is still primarily a "potential" threat in California, such facilities would be too costly to build for stand-by use.

8. Some serious social and public policy issues must be addressed. Will local governments accept environmental impact statements and allow radiation facilities to be constructed in their areas? Will consumers be willing to purchase and consume irradiated foods?

No Single Solution

Based upon the background studies that have been conducted, it is apparent that there is not a single alternate to EDB or MB fumigation for insect quarantine treatment. The choice is going to be based on many interacting factors. Of primary concern is adequate insect control. For bulk dry commodities, the potential for low oxygen and/or elevated carbon dioxide appears quite promising. This quarantine procedure must be explored further with fresh produce. Some fresh commodities may tolerate the new, short duration heat treatment, and that may prove to be the most rapid, least costly treatment for them. When storage time is not a concern, and the commodity is or can be conditioned to be tolerant to low temperatures, cold treatment may be the most satisfactory solution. For certain commodities and pests, alternate fumigants may prove safe, effective and economical. Given the right conditions of host tolerance, year-round production, and endemic pests that are sufficiently radiation-sensitive, radiation may be useful.

Regardless of the system that is selected, it is obvious that California food producers must have available suitable postharvest insect treatments. With changes in pesticide regulations and the increasing threat of infestations of quarantined pests, it is important not to wait for a quarantine threat before becoming prepared. Thus insect quarantine treatment capabilities will continue to be another cost in marketing California grown food products, even though they may be available simply on a stand-by basis.

References

1. Burditt, Jr., A. K. 1982. Food irradiation as a quarantine treatment of fruits. *Food Technol.* 36(11):51-54, 58-60, 62.

2. Couey, H. M. 1983. Development of quarantine systems for host fruits of the medfly. *HortScience* 18:45-47.

3. Gaunce, A. P., C. V. G. Morgan, and M. Meheriuk. 1982. Control of tree fruit insects with modified atmospheres. In *Controlled atmospheres for storage and transport of perishable agricultural commodities*, 383-90. Beaverton, OR: Timber Press.

4. Kader, A. A., W. J. Lipton, H. J. Reitz, D. W. Smith, E. W. Tilton, and W. M. Urbain. 1984. *Irradiation of plant products*. Comments from CAST 1984-1, Council for Agric. Sci. Technol. Ames, IA. 6 pp.

5. Kader, A. A., and F. G. Mitchell. 1981. Postharvest treatments for insect control in horticultural crops—An indexed reference list. *Perishables Handling* 47:8-28.

6. Mitchell, F. G., A. A. Kader, G. U. Crisosto, and G. Mayer. 1984. *The tolerance of stone fruits to elevated CO_2 and low O_2 levels*. Special Report to Calif. Tree Fruit Agreement. 11 pp.

7. Monro, H. A. U. 1969. *Manual of fumigation for insect control*. FAO Agric. Studies No. 79. 381 pp.

8. USDA. 1971. *Plant protection and quarantine programs, plant quarantine treatment manual*. USDA, Animal and Plant Health Inspection Service.

17
Transportation of Horticultural Commodities

ROBERT F. KASMIRE

Introduction

Transportation methods used for moving fresh market horticultural crops from shipping points to destination markets include railroads, trucks, airplanes, ships, and combinations of these (e.g., trailer on flat car [TOFC], also called "Piggy-back"). All methods are used for fruits and vegetables with trucks carrying the major portion (over 80%) in North America. Flowers are transported by trucks and airplanes. Export shipments involve all methods but, of course, only ships and airplanes are used for transoceanic shipments. Loads of mixed commodities are shipped mostly in trucks. There are factors, problems, limitations, and requirements common to all methods. Each method also has its own specific problems, limitations and requirements which must be considered to successfully use the system. Adequate knowledge of these can help users and suppliers of perishables transportation equipment to use transportation more effectively.

Equipment

Each method requires certain essential equipment in addition to the obvious structural features (e.g., wheels, frame, siding, doors).

Temperature and atmosphere control—necessary components

Refrigeration systems. Mechanical refrigeration systems are generally used. Top-ice (crushed ice placed on top of the load) and/or ice placed in each individual box is used with or without mechanical refrigeration. Use of liquid nitrogen as a refrigeration source has been tried on a very limited scale. Cooler, outside air drawn through ventilated transportation vehicles is occasionally used to provide limited temperature control for some commodities during transit to nearby markets or for dry onions to even more distant markets, but is inadequate for long-distance shipments of most perishable commodities. Most mechanical refrigeration systems can provide heat to the storage compartment if the vehicle is traveling through subfreezing conditions. A properly designed system will also provide high relative humidities in the storage compartment.

Air circulation system. Air circulation is necessary to move cooling or heating air through and around loads in order to absorb heat from internal sources (products' sensible and vital heat) and to absorb external heat conducted across a vehicle's outer surfaces and from air (heat) infiltration. In virtually all presently used mechanically refrigerated rail cars and trucks, the air circulation system is designed to provide an envelope of cold air *around* frozen food loads (there is no vital heat from frozen foods) or to circulate cold air around and *through* fresh produce loads (where vital heat removal is important).

Present usage of trucks and rail cars often seriously restricts this capability and causes product deterioration and losses. Most truck refrigeration units have fans that cease to circulate air at about 4.4 cm (1.75 inches) of water static pressure. However, most newer units have blowers that circulate a considerable amount of air against much higher static pressures. In most presently used rail cars and truck trailers (including TOFC and marine containers vans), cold air is blown over the top of the load (top-air delivery). Some newer marine container vans, a few railcars, and an increasing small number of highway truck and TOFC trailers use bottom-air delivery systems, which, under specific conditions, provide better air circulation and more uniform temperatures. Other features that aid air circulation include ribbed inner walls and T-beam floors.

Temperature control system. This is used in mechanically refrigerated systems only. It includes a thermostat(s), automatically operated (but with manual override) cooling and heating cycle, defrost mechanisms, and air-circulating fan speed controls. Thermostats are generally placed in the "return air" channel, which causes some problems because the return air temperature does not accurately represent air temperature around the commodity. Many newer units have thermostats in the "discharge-air" channel, or in both the discharge and return air channels.

Insulated product storage area. Insulation restricts the amount of heat conducted across walls, floor, doors, and roof of transportation vehicles. The storage area is also tightly sealed to restrict unwanted air leakage. These features limit the amount of ambient heat entering a vehicle during hot weather and the amount of internal heat (mostly from the product) escaping to the outside (causing product chilling or freezing) during freezing weather. Most presently used insulation is foamed-in-place material that deteriorates slowly.

Rail cars have the highest levels of insulation, trucks and container vans less, and air cargo containers very limited, if any, insulation. Most new refrigerated truck trailers are rated for a U-factor which is indicated on a metal plate attached to the trailer. This U-factor specifies the total number of BTUs of heat gain for the whole trailer/hour/°F difference between inside and outside temperatures under specified test conditions. Insulation may be damaged extensively (and its value lessened) in rail cars and trucks during lift truck loading and unloading. Insulation wetted by melted ice or condensation is also less effective than dry insulation.

Air exchange system. This is used for limited ventilation of the load compartment with outside air to reduce undesirable concentrations of CO_2, C_2H_4, or offensive odors. New air exchange units designed for marine container vans shipments are being used.

Modified atmosphere accommodations. These are used in some rail cars and container vans for long haul domestic and marine export shipments. They include atmosphere injection ports or a constant nitrogen supply system, special seals around doors, sometimes an atmospheric pressure compensation apparatus, and a control system for maintaining a desired atmosphere composition.

Equipment differences—problems and practices

Refrigerated rail cars (fig. 17.1). Cars are primarily used for long haul (>3,220 km [>2,000 miles]) domestic and Canadian shipments of generally only one commodity per load. A small percentage of the loads contain two or three commodities, rarely more. Some main characteristics of these cars are noted below.

Large load space (>114 cubic meters [>4,000 cubic feet]) and weight capacity (>45,360 kg [>100,000 pounds]) are standard.

Mechanical refrigeration system, diesel powered, provides cooling (or heating) air with thermostatic control (generally in the return-air channel, but discharge air temperature control is gaining in use).

Vertical top-to-bottom cooling (or heating) air circulation is supplied by fans. Wall flues permit some air to go around the load area to absorb heat conducted across walls of cars. Adequate air circulation and refrigeration capacity may provide slow product cooling if loads are not too tight. A few railway companies are experimenting with bottom-air delivery rail cars. One company provides mechanically refrigerated cars modified with bottom-air delivery and sloping floor racks for bulk potato and onion shipments.

Floor racks have small distances between slats but have 4 inches to 6 inches air space beneath them to permit adequate air circulation beneath the load.

Heavy insulation and air-tightness are characteristics when cars are new; occasionally these cause increases in CO_2 levels or decreases in O_2 levels. High CO_2 levels (above about 2%) are damaging to some commodities. Cars, if in good condition, may be adapted for modified atmosphere shipments.

Load divider doors are available on some cars for maintaining load integrity during transit.

Air-change capability is lacking for ventilating with outside air.

Design purpose was originally for transporting frozen food; presently cars are used mostly for fresh produce transportation.

Transportation times range from 6 to 10 days on transcontinental shipments in the U.S.

Major heat sources include sensible heat of product and shipping containers and vital heat from the product.

Problems. Mechanically refrigerated cars, last made in 1972 in the U.S., are (1) inadequately serviced; (2) no longer able to maintain proper product temperatures throughout the car; (3) loaded solidly so that adequate vertical air channels are not provided and, consequently, freezing can occur in top layer(s) and product warming can occur in center of load, at the car end (B-end) opposite location of the refrigeration unit.

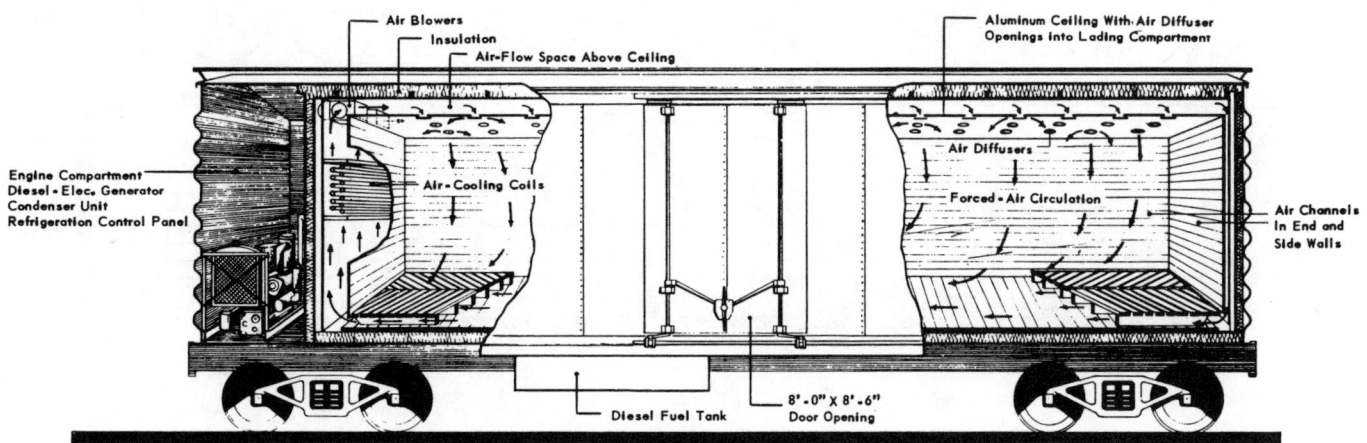

Fig. 17.1. Mechanically refrigerated rail car (from USDA Handbook No. 195).

Bulk-hopper rail cars. Very large, insulated, and refrigerated hopper cars (3-hoppers per car) have bottom-to-top vertical air flow in the hoppers, and bottom unloading. Cars were originally designed for transporting bulk fresh products for processing or for subsequent prepackaging at destination markets (e.g., onions, potatoes, oranges). They are rarely used now for fresh market products, partly because the quantity hauled per load (>68,039 kg [>150,000 pounds]) commonly exceeded the capacity of a receiver's facility or product supply needs.

Truck trailers. Included are over-the-road refrigerated truck trailers, trailer on flat car (TOFC or piggy-back, fig. 17.2), and marine container vans (fig. 17.3). Trucks are used for hauling loads of one commodity or mixed loads. Refrigerated truck trailer features are noted below.

Intermediate load space (57 to 99 cubic meters [2,000 to 3,500 cubic feet]) and weight load capacity (18,144 to 20,412 kg [40,000 to 45,000 pounds]) are standard. Some older marine container vans are much smaller (<40 cubic meters [<1,400 cubic feet]). Trucks are limited in gross weight by state highway load-limit regulations (36,287 kg = 80,000 pounds maximum gross weight). This requires careful loading for even-weight distribution to the axles. Trailers on flat cars are often more heavily loaded with resulting transit temperature management problems (poor air circulation due to excessive tightness of the load).

Refrigeration units are powered by diesel motors (truck trailers and TOFC), or by diesel motors or electricity on docks and ships for marine container vans, or by diesel-electric generator sets. Some foreign ships have cooling towers to which container vans are attached in the holds of the ships. Cold air from the ship's refrigeration system is circulated through the towers to the vans for maintaining product temperatures.

Cooling air-circulation pattern in most is presently lengthwise, i.e., front-to-rear, over the top of the load, down the sides and rear of the load, and rear-to-front through and/or under the load, and up the front to the return-air side of the refrigeration unit, providing there are adequate return-air channels. Air-circulation capacity is designed for maintaining product temperatures, not for cooling.

Air circulation is provided by a fan in the refrigeration units and is aided by air delivery chutes over the top of the load. This feature helps to deliver more air to the rear of the load and prevents, or reduces, freezing of products in the top layer(s) at the front of the load, a location which would otherwise be exposed to the coldest air from the refrigeration coil. Loads must be secured away from rear doors to provide adequate air circulation down over the rear of the load. Inner walls may be flat (as in most truck trailers) or with vertical channels (some TOFC and some marine container vans) which allow some cool air circulation between side walls and the load.

Some newer marine container vans have vertical, bottom-to-top air flow through the load compartment, making them more effective for transporting fresh produce. These trailers are able to provide better and more uniform product

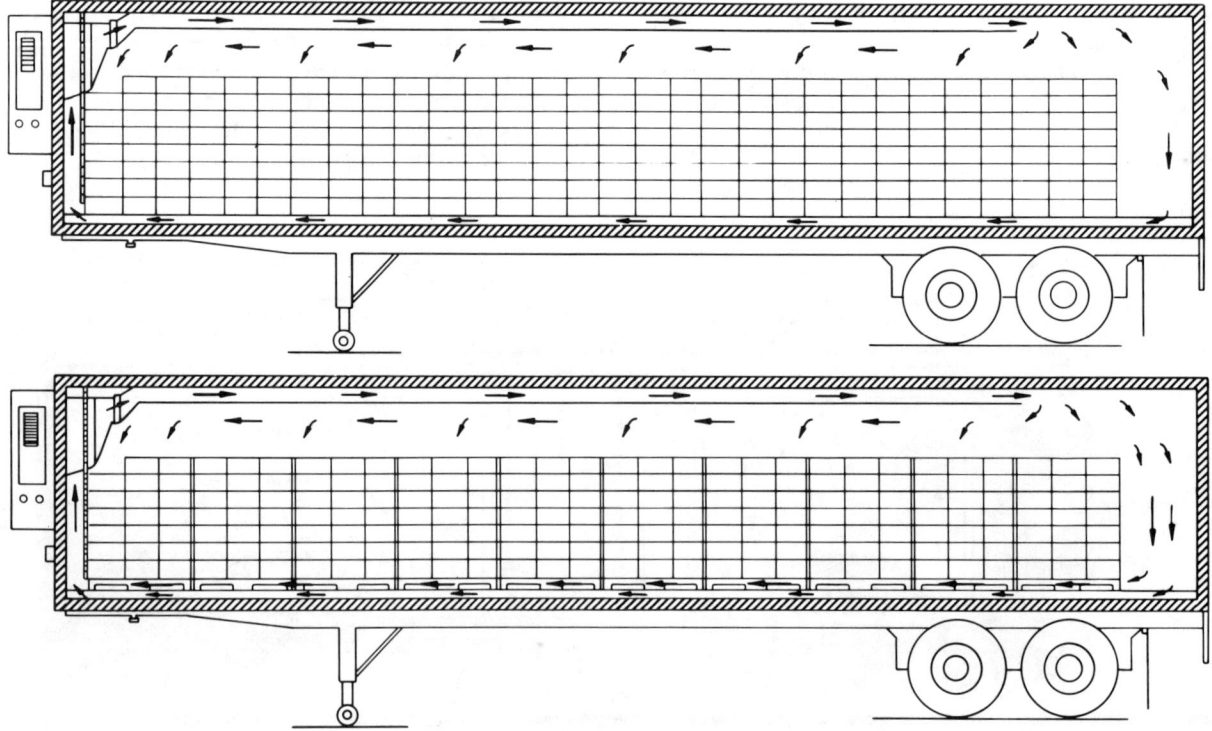

Fig. 17.2. *Top:* **Air circulation in and around a solid load directly on the trailer floor.** *Bottom:* **Air circulation in and around a palletized load.**

transit temperatures providing load pattern requirements are correctly met. To date, only a small percentage of long-haul truck trailers and TOFC trailers have incorporated this feature. The deep 'T' beam floors, solid front bulkheads, and higher capacity blowers used are more costly and reduce the net pay load that can be carried.

A solid return-air bulkhead is necessary at the front of the load to facilitate air circulation and to keep the load away from the trailer's front wall.

Ductboard floors, shallow and lengthwise in most truck trailers, provide inadequate air circulation under loads to prevent heating (in summer) or freezing (in winter) of the

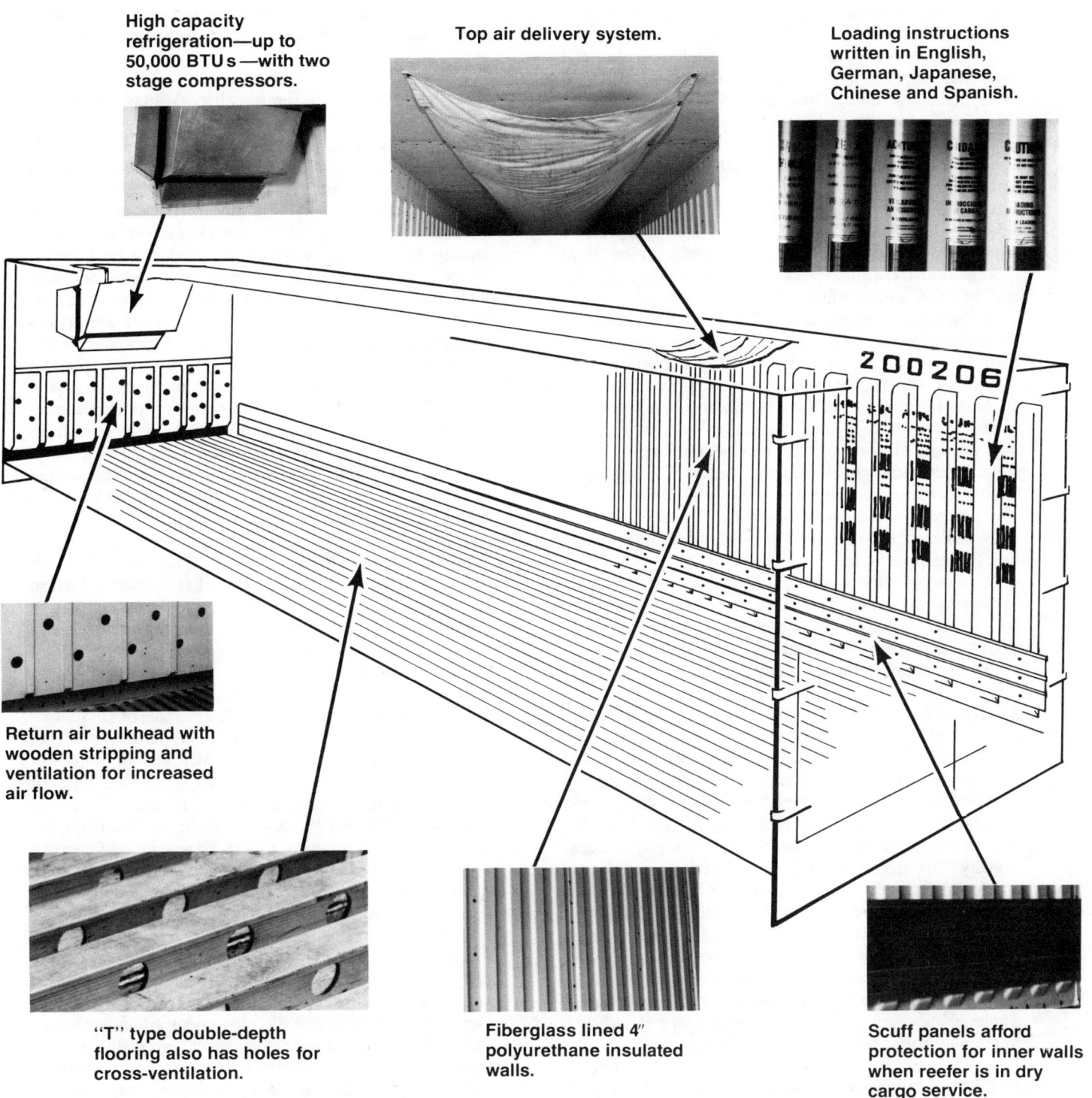

Fig. 17.3. Components of a marine transport container. Courtesy of Sea Land Corp.

bottom layer of a load. This is not a problem in marine container vans because they have deeper floor racks, are not as tightly loaded, and have less ambient heat conduction, especially through the floor.

Tight loads without adequate vertical or lengthwise air channels prevent circulation of cooling air and enhance product warming. However, air circulation under loads is much greater and product transit temperatures are more uniform in palletized loads or those on wood racks than in on-the-floor loads (fig. 17.2).

Insulation is considerably less in trailers than in rail cars, allowing for more heat conduction across walls, roof, floor, and doors. Insulation is often damaged by lift-trucks in loading and unloading. Tightness of trailers (especially doors) decreases rapidly with abuse.

Rear and side doors provide loading, unloading, and inspection openings. Because they are constantly used and are hinged, doors are easily and often damaged. Considerable heat leakage occurs around doors, especially those with damaged seals, hinges, or locking pins. Heat leakage is also considerable around nose-mounted refrigeration units, through floor drains, and seams.

Air exchange capability is available in many truck trailers and newer marine container vans for brief venting with ambient air to prevent potentially harmful carbon dioxide or ethylene levels from occurring in trailers.

Maintenance of modified atmospheres is not possible in most over-the-road truck trailers because they are not tight enough. However, newer marine container vans and some TOFC vans can be made tight enough for modified atmosphere requirements.

Transcontinental travel time in the U.S. is 3 to 6 days for trucks and 5 to 7 days for TOFC loads. Marine shipment times range from 5 to 25 days.

Equipment maintenance problems, such as damaged insulation, walls and doors, air delivery chutes, and floors with debris in grooves are common problems. Responsibility and payment for damage to trailers and refrigeration systems are continuing and controversial matters.

Air transport. Air shipment is mainly used for transporting highly perishable and valuable commodities (e.g., ornamentals, berries, tropical fruits) to distant domestic and overseas markets or to supply markets with limited supplies during periods of high prices and very strong demand. Products are transported in closed container units or in net covered pallet loads, in air freighters, or occasionally in the freight compartments of passenger airplanes.

Air travel time is often about 6 to 18 hours but waiting time at origin and destination terminals may be 1 to 2 days, often under nonrefrigerated storage. Most air transport containers lack any refrigeration system for product temperature maintenance. A few lightly refrigerated container units are used. No humidity control is available. Some airlines use cold storage rooms at origin and destination airports, but not on a regular basis.

Breakbulk marine transport. This type of transportation is used to a limited extent for shipping perishable commodities to and from ports lacking container-van loading facilities, or where the only ships used are older, smaller ships having common cold storage rooms in which products are stored during shipment. The breakbulk designation refers to the older system in which individual packages are rehandled each time the cargo is transferred from one mode of transport to another. This method can be very costly because of slow loading and unloading, rough handling, high labor costs, and inadequate temperature maintenance capability. These problems are minimized in newer ships that have the capability of handling palletized loads and better refrigeration systems.

Product Transit Temperature Management

Refrigeration and air circulation system. To be effective a system needs forced-air delivery and air-return channel(s) large enough to enable fans to operate at near-peak performance. Anything that interferes with this air circulation reduces air volume output of the blowers (fans), thereby reducing the amount of circulating cold air for temperature maintenance. A thick layer (8 inches to 16 inches) of top-ice over a load prevents cold air from penetrating the ice and cooling the load. In trailers with frame-front bulkheads, this results in only the top product layer, in contact with the ice, being kept cool, while product warming occurs in lower layers in tight loads. Applying top-ice in a windrowed pattern prevents this from occurring and provides faster cooling. In trailers with solid-front return air bulkheads, top-ice causes a false, "already cold" signal to be transmitted to the thermostat. This causes the refrigeration unit to go to the heating cycle, which melts the top-ice and causes product warming.

Transportation vehicle features. Flat sidewalls, shallow, ribbed floors (or absence of racks), or lack of a return air plenum (bulkhead) can impede proper air circulation and product temperature management in truck trailers. Large side and end-wall flues in rail cars permit much of the cooling air to bypass the load compartment. Sharply declining fan output, as resistance to air circulation increases, is a problem. Fans must be capable of circulating air against considerable resistance. Some newer model trailers have refrigeration units which include fans with greatly improved air circulation capacity, even against considerable load resistance. Direct mechanical linkage of the refrigeration compressor and the air circulation fan results in the fan shifting to low speed when the compressor does. The fan should be capable of operating at high speed at all times. This feature is incorporated into some newer model refrigeration units.

Shipping packages. Shipping container (package) design, construction, and use are variable. Inadequately vented packages (primarily cartons and large polyethylene bags) prevent sufficient air movement around products for effective temperature management, especially in solid, tight loads. Partial collapse of weak cartons results in formation of a solid load mass which completely prevents circulating air from permeating through the load in both rail and truck shipments. Weak packages are also more easily damaged by rough handling than strong ones.

Load patterns and load sizes. Product transit temperature maintenance is affected by load size and pattern. Loads must have open air channels through and around them that are either vertical (for rail shipments, bottom-air delivery marine container vans, TOFC, or highway trailers), or horizontal, lengthwise (for most truck trailers, TOFC, and some marine container vans). Provisions must be made in assembling load patterns to assure that their integrity will be maintained in transit. This is generally achieved by unitizing load units on pallets or slip sheets. Various types of load securing gates, braces, and/or locking bars are used to maintain the integrity of loads.

Use of incentive freight rates (in which per-package-freight cost decreases as load weight increases) has created serious transit temperature maintenance problems in recent years. Increases in load size, weight, and tightness have occurred, making adequate transit temperature maintenance difficult, especially with products that are not properly cooled before loading. Overloading TOFC vans (because there is no need to comply with highway load tolerances) has structurally weakened many to the extent that they cannot be used for modified atmosphere shipments. Overloading also blocks air circulation in the vans so that product temperatures increase during transit. Properly constructed palletized loads, with sufficient vertical air channels, assure desirable transit temperatures in heavy loads. However, in bottom-air delivery vans all floor space not covered by the load must be completely covered to ensure that the circulating air will be forced up through the load. In hot or very cold weather, loads should be palletized or on wood racks and loaded away from the side walls to prevent excessive warming or freezing of product in wall rows or bottom layers.

Mixed loads. Maintaining optimum product temperatures in mixed loads is difficult, especially in loads containing several (e.g., five or more) commodities. Commodities are generally packed in different sizes, shapes, and numbers of packages that are then loaded in different load patterns in various parts of trailers. With current types of shipping containers these variations often result in very tight loads that restrict air circulation in a trailer. When products with different optimum storage temperatures are shipped together, compromise transit temperature settings are often used that are designed to protect the most perishable, or the most valuable, commodity in a load. Some products in mixed loads may not be precooled or may be inadequately precooled before loading. Frequent opening of doors during loading of mixed loads causes warming or cooling of already-loaded products.

Condition of transit vehicle. The maintenance of desired transit temperatures is affected by the physical condition of a transit vehicle. Intact sidewalls and insulation, clean floors and drains, refrigeration units properly serviced and maintained (including calibration of the thermostat), intact air delivery chute(s), and tight, undamaged doors and seals are essential to proper temperature maintenance. The carrier owner/operator is primarily responsible for the equipment's condition. However, users (shippers, buyers, brokers, or receivers) or their representatives are responsible for assessing the equipment's condition prior to loading their product(s) and for damage resulting during loading or unloading operations by their workers. Loading good products into faulty transportation equipment is wasteful, increases the rate of product quality loss, and can increase marketing losses.

Recording thermometers. Many shippers place a recording thermometer in each loaded transit vehicle. In truck trailers the thermometer is generally secured high on a sidewall at about three-quarters the trailer length or on top of the load toward the rear, and high on the sidewall just inside a door in rail cars. In these positions thermometers measure and record only discharge air temperature at that specific location(s) and provide a performance record(s) of the refrigeration unit used. They do not measure or record product temperatures in loads.

Thermostats. Accurate temperature control is provided only if thermostats are accurately calibrated and are in an air stream that is representative of the air temperature in the load. Thermostats should be calibrated periodically. If they are not, operators are most worried about freezing a load and will set the thermostat several degrees high to protect themselves. While this prevents freezing it may result in product temperatures that are well above those desired. If the air return to the refrigeration unit is blocked (usually by boxes), or the stacking pattern does not allow air to flow through the boxes, air returning past the thermostat in the return air stream is cold because it has absorbed very little heat from the load. In such cases the thermostat signals the refrigeration unit to shift to low speed refrigeration, or in extreme cases, to a heating cycle. This may result in part of the load (top layer and rear stack) being frozen and the rest being warmed, or result in extensive cycling (heating and cooling) of the refrigeration unit. In trailers, air that circulates over the top of a load and returns directly to the refrigeration unit side or bottom parts, does not pass by or influence the thermostat. It is important to make sure the air-return passage is not blocked and that adequate air can circulate through or around the load. The temperature sensors for the thermostats in some new models of refrigeration units are in the supply air stream. This feature provides better control of the circulating air temperature and greatly reduces the risk of top layer product freezing. An increasing number of older trailers are being modified to change the thermostat from return to supply air monitoring.

Modified atmospheres—accidental or planned. Modified atmospheres can result from air passages (e.g., floor drains) becoming plugged with debris or ice, resulting in depletion of the available oxygen supply by the contained product's respiration. This is a common problem in winter rail shipments of sprouting broccoli, rappini, and brussels sprouts, all of which are shipped with package ice and under top-ice. This problem is overcome by providing open-air channels from inside rail cars to the ambient air. Use of planned modified atmospheres is discussed below and in chapter 11.

Product Compatability

In mixed loads certain product compatibility factors must be considered. These include the following:

Temperature compatibility. Differences in temperatures that are needed for various products in a load must be considered. For example, strawberries that must be kept near 32°F (0°C) should not be shipped with summer squash, cucumbers, or tomatoes, all of which are sensitive to chilling injury below about 55°F (12.5°C).

Ethylene production and sensitivity compatibility. Care must be taken not to ship commodities that produce large amounts of ethylene (e.g., apples, pears, avocados, and certain muskmelons) with commodities that are very sensitive to ethylene (broccoli, carrots, lettuce, kiwifruit, and most ornamentals). The incidence of russet spotting on lettuce (caused by exposure to ethylene) is about three times greater in mixed loads than in straight loads in truck shipments.

Product odor(s) compatibility. Some products produce odors (e.g., onions, garlic) which can be absorbed by other products, causing the latter to have an objectionable odor, and less market appeal.

Moisture compatibility. Some products benefit from package-ice or a high relative humidity in the ambient atmosphere (e.g. leafy vegetables, sweet corn, berries) while other commodities benefit from intermediate humidity levels (e.g., garlic, dry onions). Humidity control is especially important during long transit periods.

Modified Atmospheres

Some commodities benefit from maintenance of modified atmospheres in transit vehicles, while others do not. Successful use of modified atmospheres is largely dependent upon the tightness of the transit vehicles used. In general, some rail cars and newer marine container vans can be used for maintenance of modified atmospheres. Older or damaged rail cars are no longer tight enough to use for this purpose. Modified atmospheres are applied to marine containers at ports of embarkation. Over-the-road trucks and many TOFC trailers are generally not tight enough or do not remain tight (e.g., around rear doors) to maintain modified atmospheres. In these vehicles, modified atmospheres can be established and maintained within pallet covers (polyethylene bags) secured to the pallet bases. Some newer trailers can maintain desired atmospheres by controlled purging with nitrogen from liquid nitrogen tanks that are carried in a special compartment or underneath the trailer.

Conclusions

Successful use of various means of transporting horticultural products to markets depends upon products being cooled to, and loaded at, their desired transit temperatures; users and carriers being well-informed about the capabilities and limitations of each type of equipment; condition of transportation equipment; types of packages, optimum load patterns and loading methods; ability to maintain modified atmospheres; and the compatibilities of various commodities shipped in mixed loads. Failure to consider each of the factors can cause marketing losses.

References

1. Ashby, B. H. 1970. *Protecting perishable foods during transport by motor truck*. USDA Agric. Handb. 105.

2. ASHRAE. 1982. *ASHRAE handbook and product directory, applications volume*, sect. IV, chaps. 44-47. Atlanta, GA: Am. Soc. Heating, Refrigeration and Air Conditioning Engineers.

3. Brooks, E. E., and R. J. Byrne. 1979. *Piggybacking fresh vegetables—California to Midwest and Northeast*. USDA ESCS Farmer Coop. Res. Rep. No. 10.

4. Hinsch, R. T., R. Rij, and R. F. Kasmire. 1981. *Transit temperatures of California Iceberg lettuce shipped by truck during the hot summer months*. USDA Mark. Res. Rep. No. 1117.

5. Kasmire, R. F., and R. T. Hinsch. 1982. *Factors affecting transit temperatures in truck shipments of fresh produce*. Univ. Calif. Perishables Handling, Transportation Bull. No. 1.

6. Kasmire, R. F., R. T. Hinsch, and R. Rij. 1980. *Truck inspection poster*. Univ. of Calif. Coop. Ext.

7. Lipton, W. J., and J. M. Harvey. 1977. *Compatibility of fruits and vegetables during transport in mixed loads*. USDA Mark. Res. Rep. No. 1070.

8. Redit, W. H. 1969. *Protection of rail shipments of fruits and vegetables*. USDA Agric. Handb. 195.

9. Ryall, A. L., and W. J. Lipton. 1979. *Handling, transportation, and storage of fruits and vegetables*. Vol. 1. *Vegetables and Melons*, 244-93.Westport, CT: AVI Publishing Co.

18

Handling of Horticultural Crops at Destination Markets

ROBERT F. KASMIRE

Handlers of fresh produce in destination markets are an integral link between shippers and consumers—they handle approximately 90 percent of the fresh market fruits and vegetables in the U.S. Presently, about 70 percent to 75 percent of the produce is shipped directly to distribution centers of chain food stores, and 5 percent to 10 percent directly to institutional receivers and fast-food chains. Most wholesale produce terminal market growth has been in the area of food service.

Roles of Wholesaling and Retailing

Wholesale operations

1. Buying and accumulating products for selling to retailers, jobbers, purveyors, and institution outlets
2. Distribution to, and servicing of, retailers and institutions
3. Preparing and shipping mixed loads to other markets, by terminal market operators called "mixers"
4. Supplying small-volume items (specialty commodities) to chain store and other retailers, institutions, and purveyors
5. Ripening (sometimes), regrading, and prepackaging into consumer units

Retail operations

1. Accumulating, preparing (trimming, sorting, and consumer packaging), and presenting products for sale to consumers
2. Activities related to promoting various produce items

Types of Wholesalers and Retailers

Wholesalers

1. Chain store distribution centers servicing own stores
2. Service wholesalers supply produce to independent and/or chain retail stores
3. Carlot receivers divide and sell large quantities to retailers, brokers, jobbers, purveyors, and institutions, and may service retail stores
4. Commission merchants may be carlot receivers who sell consigned shipments on a fixed percent commission and may perform the same functions as carlot receivers, including service of retail stores
5. Jobbers handle products from carlot receivers to small, independent retailers
6. Mixers buy from other wholesalers, generally carlot receivers, and make up mixed loads of various commodities for shipping on order to distant markets
7. Purveyors service restaurants, institutions, and/or carriers and may also be processors of prepared foods
8. Wholesale auctions sell certain commodities on a price-bid basis

Retailers

1. Chain stores may belong to corporations, individuals (families), or to cooperatives—they may be large or small (e.g., "7-11")
2. Independent stores and other outlets include neighborhood supermarkets, small ("Mom & Pop") retail stores, greengrocer stores that sell only produce and produce-related items, and produce carts
3. Direct marketing outlets include farmers' markets, roadside stands, "pick-your-own" operations, and "rent-a-tree" operations

Marketing channels for fresh fruits and vegetables are illustrated in figure 18.1.

Product Handling

Considerable variation exists in product handling practices. Good product-temperature management practices and facilities are essential to proper handling and to providing consumers with the best possible quality produce. Unfortunately, some practices and facilities in use result in product warming or chilling, either of which will cause marketing losses. Rough handling of products is common. This can be attributed to labor-related problems in those operations, to the use of outdated, poorly maintained, and undercapacity facilities, and to poor product-handling practices. All these factors are especially important for products subsequently shipped in mixed loads to other markets.

Sanitation procedures are necessary in both wholesale and retail operations. Proper discarding of decayed produce, and cleaning and sanitizing of storage facilities, preparation areas, and display bins, help to maintain product quality and to reduce marketing losses.

Wholesale receivers

Most receivers handle a large number of commodities. They may have a "wet" cold room set at about 1.7° to 4.4°C (35° to 40°F) for leafy and root vegetables and a "dry" cold room set at about 0°C (32°F) for temperate fruits and other cool-season vegetables. Some have a room at a compromise temperature of 7.2° to 10°C (45° to 50°F), often too cold, in which chilling-sensitive commodities are stored. In other facilities, chilling-sensitive commodities are stored in a warehouse area that may be refrigerated at about 10° to 12.8°C (50° to 55°F) or not refrigerated at all. Modern chain store distribution centers and service wholesalers tend to have better-designed and better-maintained cold rooms than those found at terminal market operations. This is partly because many terminal markets are old, neglected, and have inadequate space for expansion. Products are often displayed for several hours at ambient temperatures (sometimes very cold or warm) in terminal markets, causing loss of shelf-life.

Retail handlers

Retail handling varies as much as wholesale handling. Retailers must handle a large number of produce items and there is generally only one cold room in a store. The room is usually set at about 4.4°C (40°F), is too small, and often is not well-maintained, especially in smaller stores. Product temperature management is minimal in many roadside markets. A major problem associated with poor retail handling is the lack of product handling education programs in the industry, although more emphasis is being given to this area.

Ethylene damage

Damaging concentrations of ethylene in ambient storage atmospheres at the wholesale and the retail levels is another important, yet unsolved, problem. Because of limited cold storage space, ethylene-generating and ethylene-sensitive products are commonly stored together in the same room. This problem is compounded when products are stored for more than 24 hours under such conditions. A common example of this problem is the storing of iceberg lettuce in cold rooms with apples, pears, stone fruits, and muskmelons. The ethylene from these fruits and melons causes russet spotting, a physiological disorder,

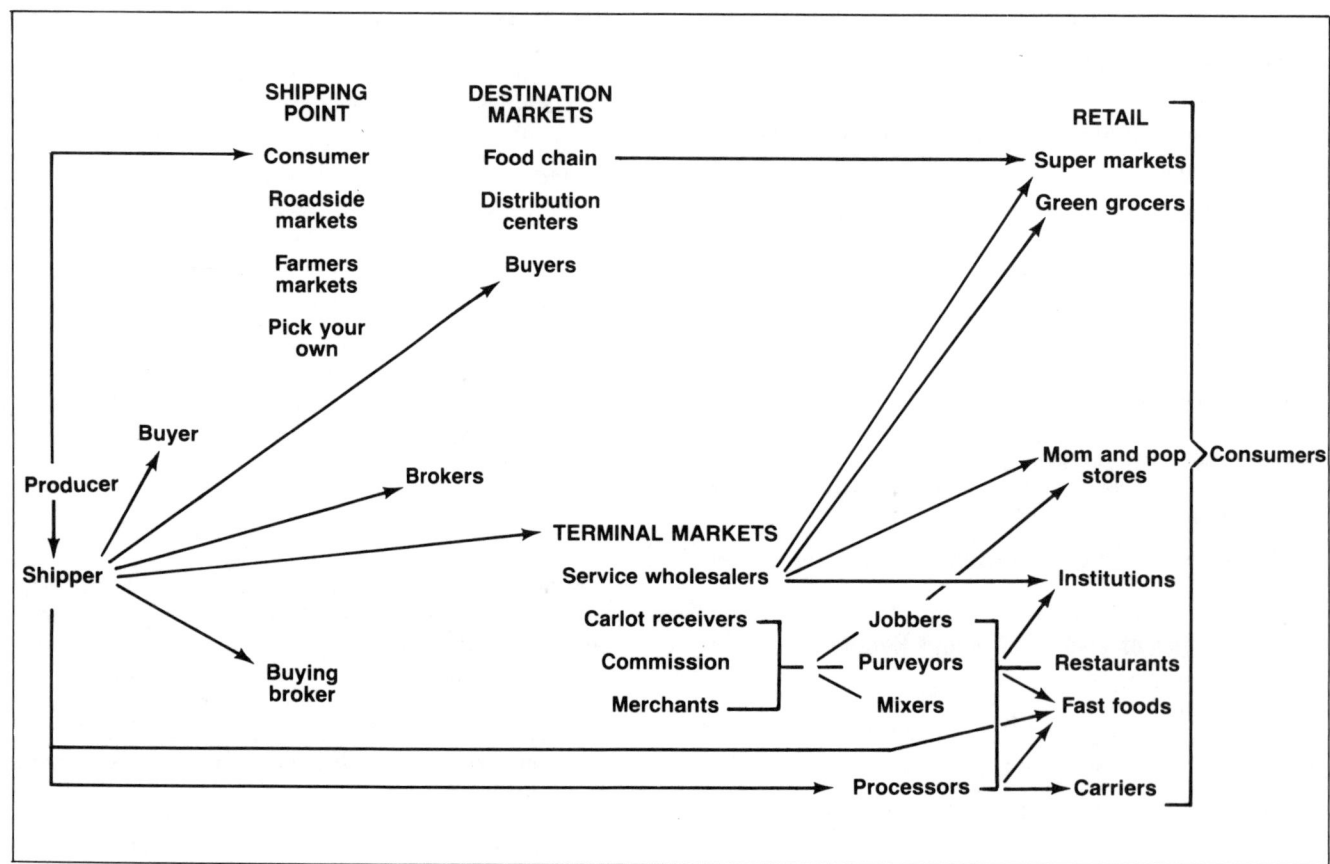

Fig. 18.1. Marketing channels for fresh fruits and vegetables.

on the lettuce. Ethylene-producing and ethylene-sensitive products are best stored in separate rooms. Propane-powered lift trucks used in destination-market produce warehouses are also major sources of ethylene and cause excessive heat loads on cold room refrigeration systems. Another important source of ethylene in some distribution facilities is the release of residual ethylene during the unloading of unvented ripening rooms into nearby storage areas that are used for produce.

Container handling

Stacking containers of varying sizes, shapes, constructions, and designs for loading in trucks destined for retail markets causes another major problem. There are more than 500 different produce shipping container sizes, shapes, and designs used in the U.S. This wide variety causes stacking and unitizing problems for handlers and distributors in destination marketing channels. An industry program is underway to reduce this large number of container sizes, shapes, etc., to a limited number (e.g., 12 to 24) of modular shapes and sizes that can be handled in unitized loads during distribution. Better, less costly handling, and reduced marketing losses would result from this change.

Produce shipped to distribution centers and service wholesalers on pallets other than hardwood 100 cm by 120 cm (40 inches by 48 inches) require additional, sometimes rough, handling. The disposition of nonstandard-size pallets is a major and costly problem for destination market handlers.

There is no continuing education program in postharvest handling available for destination market handlers, although a few trade associations conduct programs addressing the subject. Only a few states conduct extension programs with fruit, vegetable, and ornamental plant shippers and none of these programs are formally extended into distribution markets. Such programs must be made available to destination market handlers.

Problems of Wholesalers and Retailers

Wholesalers' problems include:

1. Produce warehouse management personnel need more training in horticultural product-handling requirements.
2. Products are not of uniform quality.
3. Purchased or received commodities are often immature, overripe, or of mixed maturities and require extra handling, space, and time; this causes marketing losses.
4. Quality control at both shipping points and at the wholesale level needs to be improved. There is a need for better, more objective communication of product quality.
5. Physical facilities used in many operations are inadequate for proper product handling, especially with respect to temperature maintenance, sanitation, and ethylene concentrations in storage atmospheres.
6. Transportation problems, including product temperature management and physical damage to shipping containers and products during transit, must be solved.
7. Education programs in product handling must be made available to wholesale handlers. Wholesaler handling practices can either compound or alleviate previous mishandling, maturity, or temperature-management problems.
8. Extra handling is required for many products that are received on nonstandard-size pallets.

Retailers' problems include:

1. It is difficult to obtain uniform-quality supplies of each commodity.
2. The lack of uniform maturity and ripeness is a problem with various commodities, and is common among muskmelons, tomatoes, and some temperate fruits.
3. Quality control at the shipping point and at wholesale levels is inadequate.
4. Knowledgeable personnel must be recruited and retained.
5. Inadequate physical facilities exist for temporary holding and display of produce under optimum conditions.
6. Product-handling education programs for retail produce personnel are lacking.

Current Trends and Developments

Some current developments will aid in better handling of produce in destination markets. These include:

1. Better product temperature management during transit from shipping point to destination markets, increased use of refrigerated trucks in distribution from wholesale warehouses or terminal markets to retail stores, increased and improved use of refrigerated storage facilities at wholesale and retail levels, and increased use of rapid cooling methods (hydrocooling, forced-air cooling, and package-icing) to cool local produce received at distribution centers.
2. Improved unitization of containers in distribution from wholesale to retail. Unitization can be improved by unitizing containers on standard-size 100 cm by 120 cm (40 inches by 48 inches) pallets or slip sheets, using the layer concept of ordering (i.e., ordering enough packages of a given commodity to make one or more layers of containers on a pallet or slip-sheet unit), and developing a minimum number of standard-size, metric-measured, modular shipping containers, which would facilitate mixing several sizes of modular container on a pallet or slip sheet.
3. Increased use of quality control personnel at wholesale (mostly by service wholesalers and food chains). The FOB buyer also serves as a quality control person.
4. Increased use of product handling educational programs

at wholesale and retail. Such programs may be provided or conducted by the company's own staff, produce consultants, produce trade associations, commodity group representatives (e.g., California Fresh Market Tomato Advisory Board, California Iceberg Lettuce Commission, California Avocado Commission) or marketing organizations (e.g., Sunkist, Blue Anchor, United Brands).

5. Expanded efforts by national produce associations in promoting the nutritional value of fresh produce and in supporting studies on the effects of handling practices on maintenance of nutritional quality.

6. Improved communications among handlers in the various handling steps, primarily through trade associations' programs. This includes educational activities aimed at food service produce handlers.

References

1. Kretchman, D. W. 1973. *Care and handling of fresh fruits and vegetables in retail markets*. Produce Marketing Association Yearb. 96, 98.

2. Lewis, W. E. 1957. Maintaining produce quality in retail stores. USDA Agric. Handb. 117, 30 pp.

3. UFFVA (no date). *Basic produce department operation*. Alexandria, VA: United Fresh Fruit and Veg. Assn. 81 pp.

4. UFFVA (various dates). *Fruit and vegetables facts and pointers*. A series of reports on each of 81 commodities. Alexandria, VA: United Fresh Fruit and Veg. Assn.

5. Volz, M. D., and J. J. Karitas. 1973. *Handling and space costs for selected food wholesalers in urban food distribution centers*. USDA Mark. Res. Rep. 992. 24 pp.

19
Energy Use in Postharvest Technology Procedures

JAMES F. THOMPSON

Energy Use in the Total Food System

Production, processing, distribution, and preparation of food requires about 12 percent to 16 percent of the total annual energy use in the U.S. About one-third of this energy (i.e., 5% to 7% of the total U.S. energy budget) is utilized in postharvest procedures associated with processing and transporting food products. With increasing energy costs and the potential for uncertain availability, how should postharvest practices be modified? The answer lies in the present use of energy in various postharvest handling operations and the potential for energy conservation technology in these operations.

It is important to consider how a handling system affects the total energy use from production to consumption. Whittlesey and Lee (1976) catalogued the direct energy use of various methods of handling potatoes and apples. Tables 19.1 and 19.2 summarize the results. Energy use for the manufacture of equipment and facilities is not included.

Total energy use varies significantly among handling systems for each commodity. Because handling fresh commodities requires no energy (or at least very little) for processing, one might assume this would result in less total energy use than systems that require more processing. This assumption holds true of potatoes but fresh apples require roughly 50 percent more energy than canned apple juice or dried apples. Processing often reduces the weight of the commodity, resulting in lower transportation costs. In the case of canned apple juice and dried apples, the reduced use of transportation energy more than offsets the energy used in processing. It is apparent that selecting the least energy-intensive method of handling a perishable commodity is not a simple choice.

In the future, conservation practices should reduce energy consumption in each of the six categories of energy use, production, processing, transportation, trade, home storage, and home preparation. The categories that use large quantities of heat, i.e., processing and home preparation, hold the greatest potential for reduction. Authorities think energy use in food processing could be reduced by 50 percent. If this level of reduction were attained, systems that involve processing would, in many cases, be less energy intensive than fresh handling. Reduction of energy use in home preparation would shift the advantage back toward fresh, but only a small portion of domestic energy consumption is used for cooking. The processor is more apt to make changes for the sake of reduced utility costs than the homeowner.

Transportation energy costs constitute a significant portion of total energy costs for commodities shipped long distances. This cost can be reduced by utilizing more efficient transportation systems or by growing commodities closer to the consumer. Table 19.3 lists the energy efficiencies of shipping methods for perishable commodities.

About three-quarters of all fresh market commodity shipments originating from the west are transported by truck. A shift to railroad transportation could reduce energy

Table 19.1. Energy use in potato handling systems

Handling system	Energy use in 10^6 BTU/ton						
	Production	Processing	Transportation	Trade	Home storage	Home preparation	Total
Fresh	.819	0	1.97	.082	0	3.19	6.06
Frozen	.819	4.86	1.27	1.14	.423	1.04	9.55
Dehydrated	.819	6.84	.36	.150	0	.915	9.09

Table 19.2. Energy use in apple handling systems

Handling system	Energy use in 10^6 BTU/ton						
	Production	Processing	Transportation	Trade	Home storage	Home preparation	Total
Fresh	1.89	0	4.08	.314	.984	*	7.23
Canned juice	1.89	.587	1.43	.197	.641	0	4.76
Canned sauce	1.89	2.95	3.13	.516	0	0	8.46
Dried	1.89	1.80	.611	.117	0	0	4.38

*If fresh apples are used to make apple crisp, home preparation energy is 5.07×10^6 BTU/ton.

use by 70 percent. However, rail transit usually requires more time to reach eastern markets than truck shipment, making it less desirable. Rail service needs to be improved before it can be used as a low-energy method of transportation for fresh perishables. At present rail shipment is more suitable for transporting processed commodities.

Growing produce closer to its final market may reduce transportation energy use but it is often impractical. The major U.S. markets are concentrated east of the Mississippi River, but many of the best growing areas with long harvest seasons are in the West. Local production should be encouraged wherever possible, but the consumers' demand for a wide variety of fresh commodities year-round cannot be satisfied by local production. Greenhouse production of vegetables under eastern winter weather conditions probably would require more energy than transporting field-grown western produce to the East.

Some have maintained that local production should be carried one step further, claiming energy use can be further reduced by home food production. While the potential exists for home production to contribute to energy savings, the homeowner must be very careful to make correct energy-use decisions or this potential advantage may not be realized. For example, a 5-mile trip to haul 10 pounds of tomatoes to a friend makes the same per ton energy use as a truck hauling fresh tomatoes across the country. Other poor energy-use choices are buying a rototiller for 10 hours of use per year, making a trip to the local chicken farm to get a good deal on organic fertilizer, home canning, and so on. Consumption of home-grown fresh produce has the potential of low energy use, but it is subject to seemingly insignificant practices which can greatly increase energy use.

Table 19.3. Energy use of various methods of transportation

Transportation method	Energy use (BTU/ton-mile)
Water	490
Railroad	750
Motor freight	2,460
Air	42,700
Auto hauling 10 lb of fruit	1,650,000

Energy Use in Postharvest Handling of Fresh Produce

The data presented in tables 19.1 and 19.2 indicate there is no energy use in processing fresh produce. This is not exactly the case. Fresh commodities are often sorted, packed, and cooled at central handling facilities and are sometimes stored. Data for a wide variety of fruits and vegetables indicate this processing requires 0.1 to 0.2 $\times$ 10^6 Btu/ton of fresh produce. This is less than 3 percent of the total energy use figures listed in the tables for fresh handling, but there are ways to reduce this energy use.

Storage facilities

Energy use in a cold storage facility is affected by the amount of heat the refrigeration equipment must remove and the efficiency of the equipment. The main sources of heat in the facility are heat transmitted through walls, fans on the evaporator coils, lights, air leakage, and respiration of the stored commodity.

Heat entering a cold storage facility through walls can be minimized by increasing the wall insulation and by painting the exterior walls a light color. Doubling the wall insulation (as measured by R value) will reduce transmitted heat by one-half. Newer facilities in the U.S. are using insulation levels as high as R-30 in walls and ceilings. In general, it is advisable to build with more insulation than utility costs presently warrant, because it is cheaper to install insulation during construction than after construction is completed.

Sun shining on walls and roof can dramatically increase the effective outside temperature, increasing heat flow into a storage facility. Table 19.4 shows the relationship between wall or roof orientation, color, and effective outside air temperature. (Effective outside air temperature is the normal air temperature plus a factor to account for the sun shining on a surface.) A dark, flat roof can be almost 44°C (80°F) warmer than the outside air temperature. Painting a south-facing wall a light color can reduce the effective wall temperature by 11°C (20°F) compared to a dark wall. Walls and roof of a cold storage facility should be painted a light color or shaded from direct sun.

Fans are necessary in cold storage facilities to move air past the evaporator coils for cooling and to circulate cooled air uniformly around the stored commodity. During the initial filling of a storage facility, constant air movement is needed to remove field heat from the product. However, after the commodity has reached the desired storage temperature, air movement can be reduced. In winter, less air movement is needed to intercept heat entering through the walls.

Lights inside the cold storage room should be turned off when not needed. Warm outside air leaking into the cold storage room increases energy use required of the refrigeration equipment. Air leakage should be minimized—this is vital in long-term controlled atmosphere storage where gas concentrations are controlled. Heat produced by respiration of the stored commodity can be minimized by keeping the commodity cool and using controlled atmospheric storage if possible.

Table 19.4. Effect of surface color and orientation on effective outside temperature*

Color and orientation	Effective temperature	
	°C	°F
Outside air	35	95
Light-colored southwall	46	115
Dark-colored southwall	57	135
Dark-colored flat roof	77	170

*Data are maximum temperatures for a clear July 21 in Fresno, CA.

The most significant factor affecting the refrigeration system as a whole is the temperature of the refrigerant fluid after it is cooled in the condenser. For example, in a cold-storage facility maintaining 0°C (32°F) a condensing temperature of 52°C (125°F) requires 50 percent more power than a condensing temperature of 35°C (95°F) while providing the same amount of refrigeration. In warm areas, well-water cooled or evaporatively cooled condensers should be selected over air-cooled units. Compressors operate efficiently only in a certain range of refrigerant flow. Select a compressor that operates efficiently for your facility even though it may be larger and cost more than a slightly smaller unit which will not operate as efficiently. A large compressor operating at low refrigerant flow is also inefficient. Energy costs can be reduced if several small units are used, and turned on and off independently as needed, rather than throttling one large compressor up and down.

Some storage facilities may reduce energy costs by using evaporative cooling, nighttime cooling, or other alternate methods of cooling listed in chapter 9.

Forced-air cooling facilities

Forced-air cooling facilities can utilize all the energy conservation techniques that are applicable to cold storage plants plus several more. Additional energy saving possibilities exist in reducing field heat of the commodity and heat from lift trucks. Table 19.5 shows the magnitude of heat sources in a strawberry cooling facility in California. Almost 75 percent of the heat input is from warm fruit and propane-powered forklifts.

Energy use for product cooling can be minimized by bringing fruit to the cooler at the lowest possible temperature. Night or early morning harvest results in much lower fruit temperatures than harvesting in the heat of the day. Produce should be removed from the field as quickly as possible and shaded from the sun at all times. Fruit that will be culled eventually should not be cooled.

Where possible, propane forklifts operated in storage and cooling facilities should be replaced by electric lifts. An internal combustion engine has an efficiency of about 25 percent, meaning that for every 4 gallons of propane burned only 1 gallon's worth of mechanical work is accomplished. The 3 gallons produce heat that is rejected in the exhaust gases, radiator, and so on. (Even the mechanical work ends up as heat which must be removed.) An electric forklift, when operating, produces very little wasted heat and when not in use can be shut off for brief periods rather than left to idle, unlike a propane forklift. The change to electric lifts decreases the total heat input by about 90 percent. Additional electricity required for charging electric lift batteries nearly equals the decreased electricity used by the refrigeration system. However, the cost of purchasing and handling propane is completely eliminated. Propane-powered forklifts should also be kept out of the cold room because the ethylene gas they produce may damage some perishable commodities.

Design the facility for smooth traffic flow. This will minimize forklift operation and reduce both forklift and refrigeration energy use.

Hydrocooling facilities

Hydrocoolers vary tremendously in their energy use per ton of commodity cooled. Some coolers use less than one-half the energy of others. Low energy use can be achieved through a number of straightforward techniques. Prevent unnecessary outside heat from entering the unit by shading from the sun, painting with a light color, and insulating exterior surfaces. It may also be possible to enclose the cooler in an insulated building. Reduce the distance between the top of the commodity and the water distribution pan. Provide flaps to enclose the space between the top of the commodity and the bottom of the water distribution pan. These procedures will decrease the amount of outside air flowing in and out of the unit. Minimize cooled water leakage. Minimize water holding capacity so that large amounts of refrigeration are not required to start the system. The latter is a particular problem if the cooler is cycled on and off frequently; the pumps that move water in the unit require a large amount of energy. Energy use per ton of commodity can be minimized by ensuring that the cooler operates at maximum capacity and is shut off when no product is moving through it. Do not cool cull produce.

Vacuum cooling facilities

Little research has been done on energy conservation in vacuum coolers. However, some obvious suggestions are to insulate refrigerant lines and receivers, ensure the unit is always fully loaded, and don't cool cull commodities. Some operators believe energy use for refrigeration could be reduced by increasing coil surface area and designing coils in the chamber to condense the water vapor more efficiently. This allows suction pressures to be increased, reducing energy use of the refrigeration compressor.

Table 19.5. Distribution of heat input to a strawberry cooling facility

Heat source	Heat input (%)
Fruit	40
Propane forklifts	33
Walls	10
Fans and motors	10
Lights	4
Air leakage	3

References

1. Cervinka, V., et al. 1974. *Energy requirements for agriculture in California*. Joint study by Calif. Dept. of Food and Agric. and the Agric. Eng. Dept., Univ. of Calif., Davis.

2. Whittlesey, N. K., and C. Lee. 1976. *Impacts of energy price changes on food costs*. Washington State Univ. College of Agriculture Research Center Bull. 882.

20

Quality Factors: Definition and Evaluation for Fresh Horticultural Crops

ADEL A. KADER

Table. Quality components of fresh fruits and vegetables

Main factors	Components
Appearance (visual)	Size: dimensions, weight, volume
	Shape and form: diameter/depth ratio, smoothness, compactness, uniformity
	Color: uniformity, intensity
	Gloss: nature of surface wax
	Defects: external, internal
	Morphological
	Physical and mechanical
	Physiological
	Pathological
	Entomological
Texture (feel)	Firmness, hardness, softness
	Crispness
	Succulence, juiciness
	Mealiness, grittiness
	Toughness, fibrousness
Flavor (taste & smell)	Sweetness
	Sourness (acidity)
	Astringency
	Bitterness
	Aroma (volatile compounds)
	Off-flavors and off-odors
Nutritive value	Carbohydrates (including dietary fiber)
	Proteins
	Lipids
	Vitamins
	Minerals
Safety	Naturally-occurring toxicants
	Contaminants (chemical residues, heavy metals)
	Mycotoxins
	Microbial contamination

Quality is defined as *any of the features that make something what it is* or *the degree of excellence or superiority*. The word *quality* is used in various ways in reference to fresh fruits and vegetables such as *market* quality, *edible* quality, *dessert* quality, *shipping* quality, *table* quality, *nutritional* quality, *internal* quality, and *appearance* quality.

Quality of fresh horticultural commodities is a combination of characteristics, attributes, and properties that give the commodity value to humans for food (fruits and vegetables) and enjoyment (ornamentals). Producers are concerned that their commodities have good appearance and few visual defects, but for them a useful cultivar must score high on yield, disease resistance, ease of harvest, and shipping quality. To receivers and market distributors, appearance quality is most important; they are also keenly interested in firmness and long storage life. Consumers consider good quality fruits and vegetables to be those that look good, are firm, and offer good flavor and nutritive value. Although consumers buy on the basis of appearance and feel, their satisfaction and repeat purchases are dependent upon good edible quality.

Quality Components

The various components of quality listed in the table are used to evaluate commodities in relation to specifications for grades and standards, selection in breeding programs, and evaluation of responses to various environmental factors and postharvest treatments. The relative importance of each quality factor depends upon the commodity and its intended use (fresh or processed). Appearance factors are the most important quality attributes of ornamental crops.

Many defects can influence the appearance quality of horticultural crops. Morphological defects include sprouting of potatoes, onions, and garlic, rooting of onions, elongation of asparagus, curvature of asparagus and cut flowers, seed germination inside fruits such as tomatoes and peppers, presence of seedstems in cabbage and lettuce, doubles in cherries, floret opening in broccoli, and so on. Physical defects include shrivelling and wilting of all commodities; internal drying of some fruits; mechanical damage such as punctures, cuts and deep scratches, splitting and crushing, skin abrasions and scuffing, deformation (compression), and bruising; growth cracks (radial, concentric); and so on. Temperature-related disorders (freezing, chilling, sunburn, sunscald), puffiness of tomatoes, blossom-end rot of tomatoes, tipburn of lettuce, internal breakdown of stone fruits, water core of apples, and black heart of potatoes are examples of physiological defects. Pathological defects include decay caused by fungi or bacteria and virus-related blemishes, irregular ripening, and other disorders. Other defects result from damage caused by insects, birds, and hail; chemical injuries; and scars, scabs, and various blemishes (e.g., russeting, rind staining).

The texture of horticultural crops is important for eating and cooking quality and is a factor in withstanding shipping stresses. Soft fruits cannot be shipped long distances without extensive losses due to physical injuries. In many cases, this necessitates harvesting fruits at less than ideal maturity for flavor quality.

Evaluating flavor quality involves perception of tastes and aromas of many compounds. Objective analytical determination of critical components must be coupled with subjective evaluations by a taste panel to yield useful, meaningful information about fresh fruit and vegetable flavor quality. This approach can be used to establish a minimum acceptable level. To learn consumer flavor preferences for a given commodity, large-scale testing by a representative sample of consumers is required.

Fresh fruits and vegetables play a significant role in human nutrition, especially as sources of vitamins (C, A, B_6, thiamin, niacin), minerals, and dietary fiber. Postharvest losses in nutritional quality, particularly vitamin C content, can be substantial and increase with physical damage, extended storage, high temperatures, low relative humidity, and chilling injury.

Safety factors include levels of naturally-occurring toxicants in certain crops (such as glycoalkaloids in potatoes) which vary according to genotype and are routinely monitored by plant breeders so they do not exceed safe levels. Contaminants such as chemical residues and heavy metals on fresh fruits and vegetables are also monitored by various agencies to assure compliance with established maximum tolerance levels. Sanitation procedures throughout harvesting and postharvest handling operations are essential to minimize microbial contamination. Proper preharvest and postharvest handling procedures that reduce the potential for growth and development of mycotoxin-producing fungi must be used.

Interrelationships among Quality Components

It is important to define the interrelationships among each commodity's quality components and to correlate subjective and objective methods of quality evaluation. This information is essential for selection of new cultivars, choice of optimum production practices, definition of optimum harvest maturity, and identification of optimum postharvest handling procedures. The point of all this effort is to provide high-quality fruits and vegetables for the consumer.

Quality Criteria in Standards for Fresh Fruits and Vegetables

See tables 21.1 to 21.5 in chapter 21 for a summary of quality criteria used in the U.S. grading standards and the California Agricultural Code. Note the emphasis on appearance quality factors in most commodities. In many cases, good appearance does not necessarily mean good flavor and nutritional quality. A fruit or vegetable that is misshapen or has external blemishes may be just as tasty and nutritious as one that is perfect in appearance. For this reason, it is important to include quality criteria other than appearance that more accurately reflects consumer preferences. Such quality indices must be relatively easy to evaluate, and objective methods for evaluation should be developed.

Factors Influencing Quality

Many pre- and postharvest factors influence the composition and quality of fresh horticultural crops. These are:

1. **Genetic factors**

    Selection of cultivars, rootstocks

2. **Preharvest environmental factors**

    Climatic—temperature, light, wind, rainfall, pollutants

    Cultural conditions—soil type, nutrient and water supply, mulching, pruning, thinning, agricultural chemicals, time and method of harvest

3. **Harvesting**

    Stage—maturity, ripeness, physiological age

4. **Postharvest treatments**

    Environmental factors (e.g., temperature, relative humidity, atmospheric composition), handling methods, duration between harvesting and consumption

5. **Interaction among the various factors listed above**

Methods of Quality Evaluation

Quality evaluation methods can be destructive or nondestructive. They include both objective (based on instrument readings) and subjective (based on human judgment, using hedonic scales) methods.

Appearance quality (visual)

1. **Size**

    Dimensions—measured with sizing rings, calipers

    Weight—correlation is generally good between size and weight; size can also be expressed as number of units of commodity per unit of weight

Volume—determined by water displacement or by calculation from measured dimensions

2. Shape

Ratio of dimensions—such as diameter/depth ratio; used as indices of shape in fruits

Diagrams and models of shape—some commodity models are used as visual aids for quality inspectors

3. Color

Uniformity and intensity—important appearance qualities

Visual matching—color charts, guides, and dictionaries to match and describe colors of fruits and vegetables

Light reflectance meter—measures color on basis of amount of light reflected from surface of the commodity; examples include Gardner and Hunter Color Difference Meters (tristimulus colorimeters) and Agtron E5W spectrophotometer

Light transmission meter—measures the amount of light transmitted through the commodity; may be used to determine internal color and various disorders, such as water core of apples and black heart of potatoes

Determination of pigment content—evaluates the color of horticultural crops by pigment content, i.e., chlorophylls, carotenoids (carotene, lycopene, xanthophylls), and flavonoids (anthocyanins)

4. Gloss (bloom, finish)

Wax platelets—amount, structure, and arrangement on the fruit surface affect the gloss quality; measured using a Gloss-meter or by visual evaluation.

5. Presence of defects (external and internal)

The incidence and severity of defects are evaluated using a scoring system of 1 to 5 (1 = no symptoms, 2 = slight, 3 = moderate, 4 = severe, and 5 = extreme) which may be expanded to a 1 to 7 or 1 to 9 hedonic scale if more categories are needed. To reduce variability among evaluators, detailed descriptions and photographs may be used as guides in scoring for a given defect.

Textural quality

1. Yielding quality (firmness, softness)

Hand-held testers—determine penetration force using testers such as the Magness-Taylor Pressure Tester and the Effegi penetrometer

Stand-mounted testers—determine penetration force using testers with a more consistent speed of punch such as the UC Fruit Firmness Tester and the Effegi penetrometer mounted on a drill stand

Laboratory testing—fruit firmness can be determined by measuring penetration force using an Instron Universal Testing machine or a Texture Testing system; or by measuring fruit deformation using a Deformation Tester

2. Fibrousness and toughness

Measured on basis of shear force determinations—use an Instron or a Texture Testing system

Resistance to cutting—determined by using a Fibrometer

Chemical analysis—fiber content or lignin content

3. Succulence and juiciness

Measurement of water content—an indicator of succulence or turgidity

Measurement of extractable juice—an indicator of juiciness

4. Sensory textural qualities

Sensory evaluation procedures—evaluate grittiness, crispness, mealiness, chewiness, and oiliness

Flavor quality

1. Sweetness

Sugar content—determined by chemical analysis procedures for total and reducing sugars or for individual sugars; indicator papers for quick measurement of glucose in certain commodities, such as potatoes

Total soluble solids content—measured using refractometers or hydrometers; can be used as indicator of sweetness because sugars are major component of soluble solids

2. Sourness (acidity)

pH (hydrogen ion concentration) of extracted juice—determined using a pH meter or pH indicator paper

Total titratable acidity—determined by titrating a specific volume of the extracted juice; 0.1 N NaOH to pH 8.1, then calculate titratable acidity as citric, malic, or tartaric acid (depending on which organic acid is predominant in the particular commodity)

3. Saltiness

Fresh vegetables and fruits—usually not applicable

4. Astringency

Determined by taste testing or by measuring tannin content, solubility, and degree of polymerization

5. Bitterness

Determined by taste testing or measurement of the alkaloids or glucosides responsible for the bitter taste

6. Aroma (odor)

Determined by sensory panels in combination with identification of volatile components responsible for specific aroma of a commodity (using gas chromatographic methods)

7. **Sensory evaluation**

    Human subjects—judge and measure combined sensory characteristics (sweetness, sourness, astringency, bitterness, overall flavor intensity) of a commodity

    Laboratory panels—detect and describe differences among samples; determine which volatile compounds are organoleptically important in a commodity

    Consumer panels—indicate quality preferences

Nutritional value

Various analytical methods are available for determination of total carbohydrates, dietary fiber, proteins and individual amino acids, lipids and individual fatty acids, vitamins, and minerals in fruits and vegetables. Several companies have a continuing developmental effort to automate these analytical procedures for use in situations where nutritional labeling is required and large numbers of samples have to be analyzed routinely.

Safety factors

Analytical procedures, using thin-layer chromatography, gas chromatography, and high-pressure liquid chromatography, are available for determining minute quantities of the following toxic substances:

1. Naturally-occurring toxicants—such as cyanogenic glucosides in lima beans and cassava, nitrates and nitrites in leafy vegetables, oxalates in rhubarb and spinach, thioglucosides in cruciferous vegetables, and glycoalkaloids (solanine) in potatoes
2. Natural contaminants—such as fungal toxins (mycotoxins), bacterial toxins and heavy metals (Hg, Cd, Pb)
3. Synthetic toxicants—such as environmental contaminants and pollutants, and residues of agricultural chemicals

References

1. Amerine, M. A., R. M. Pangborn, and E. B. Roessler. 1965. *Principles of sensory evaluation of food.* New York: Academic Press. 602 pp.
2. Arthey, V. D. 1975. *Quality of horticultural products.* New York: Halstead Press, John Wiley and Sons. 228 pp.
3. Bourne, M. C. 1980. Texture evaluation of horticultural crops. *HortScience* 15:51-57.
4. Dull, G. G., G. S. Birth, and J. B. Magee. 1980. Nondestructive evaluation of internal quality. *HortScience* 15:60-63.
5. Eskin, N. A. M. 1979. *Plant pigments, flavors and textures: The chemistry and biochemistry of selected compounds.* New York: Academic Press. 219 pp.
6. Finney, E. E., Jr. 1970. *Measurement techniques for quality control of agricultural products.* St. Joseph, MI.: Am. Soc. Agric. Eng. Spec. Publ., ASAE. 53 pp.
7. Francis, F. J. 1980. Color quality evaluation of horticultural crops. *HortScience* 15:58-59.
8. Gaffney, J. J., comp. 1976. *Quality detection in foods.* St. Joseph, MI.: Am. Soc. Agric. Eng. ASAE Publ. 1-76. 240 pp.
9. Goddard, M. S., and R. H. Matthews. 1979. Contribution of fruits and vegetables to human nutrition. *HortScience* 14:245-47.
10. Gould, W. A. 1977. *Food quality assurance.* Westport, CT: AVI Publ. Co. 314 pp.
11. Heintz, C. M., and A. A. Kader. 1983. Procedures for the sensory evaluation of horticultural crops. *HortScience* 18:18-22.
12. Kader, A. A. 1983. Postharvest quality maintenance of fruits and vegetables in developing countries. In *Postharvest Physiology and Crop Preservation*, ed. M. Lieberman, 455-70. New York: Plenum.
13. Lipton, W. J. 1980. Interpretation of quality evaluations of horticultural crops. *HortScience* 15:64-66.
14. Mohsenin, N. N. 1970. *Physical properties of plant and animal materials: Structure, physical characteristics, and mechanical properties.* New York: Gordon and Breach Sci. Publ. 742 pp.
15. Stevens, M. A., and M. Albright. 1980. An approach to sensory evaluation of horticultural commodities. *HortScience* 15:48-50.
16. USDA. 1983. *Composition of foods: raw, processed, prepared.* In USDA Agric. Handb. 8 (revised), sec. 9 (Fruits and fruit products) and sec. 10 (Vegetables and vegetable products).
17. Watada, A. E. 1980. Quality evaluation of horticultural crops—The problem. *HortScience* 15:47.
18. Williams, A. A. 1979. The evaluation of flavour quality in fruits and fruit products. In *Progress in Flavour Research*, eds. D. G. Land and H. E. Nursten, 287-305. Essex: Appl. Sci. Publ.

21

Standardization and Inspection of Fresh Fruits and Vegetables

ADEL A. KADER

Introduction

Grade standards identify the degrees of quality in a given commodity that provide the basis for its usability and value. Such standards are important tools in fresh fruit and vegetable marketing because they (1) provide a common language for trade among growers, handlers, processors, and receivers at terminal markets; (2) help producers and handlers do better jobs of preparing and labeling fresh horticultural commodities for market; (3) provide a basis for incentive payments rewarding better quality; (4) serve as the basis for market reporting—prices and supplies quoted by the Federal-State Market News Service in the different markets can only be meaningful if they are based on products of comparable quality; and (5) help settle damage claims and disputes between buyers and sellers.

U.S. Grade Standards

U.S. standards for fresh fruit and vegetable grades are voluntary, except when they are required by state and local regulations, by industry marketing orders (federal or state), or for export marketing. They are also used by many private and government procurement agencies when purchasing fresh fruits and vegetables. The U.S. Department of Agriculture (USDA), Food Safety and Quality Service (FSQS) is responsible for developing, amending, and implementing grade standards.

The first U.S. grade standards were developed for potatoes in 1917. Currently there are more than 150 standards covering 80 different commodities. The quality factors used in these standards for fresh fruits, vegetables, and tree nuts are summarized in tables 21.1 to 21.5 at the end of this chapter.

The number of grades and grade names included in the U.S. standards for a given commodity vary with the number of distinct quality gradations that the industry normally recognizes and with the established usage of grade names. Currently, grades include three or more of the following: *U.S. Fancy, U.S. No. 1, U.S. No. 2, U.S. No. 3, U.S. Extra No. 1, U.S. Extra Fancy, U.S. Combination, U.S. Commercial*, and so on. The FSQS is gradually phasing in the first four grades as uniform grades for all fresh fruits and vegetables, to represent available levels of quality.

Steps to establish or change U.S. standards

1. Demonstrate need, interest, and support from the industry.
2. Study physical characteristics, quality factors, and their normal ranges for the commodity in the main production areas.
3. Consult all interested parties as part of data collection.
4. Develop a proposal which must be practical to use.
5. Publish the proposal in the *Federal Register,* and publicize it through various means with an invitation for comments. Public hearings may be held for the same purpose.
6. Amend the proposal on the basis of comments received from interested parties.
7. Publish the standards in their final form in the *Federal Register* with a specified date on which they become effective (at least 30 days after publication date).

Applying the standards

USDA inspectors are located at most shipping points and at terminal markets. In many cases cooperative agreements between the USDA and the states are in place to allow federal-state grading by USDA-licensed state inspectors. Some inspectors are full-time employees, while others are seasonal employees hired during the peak production season in a given location.

Methods of inspection include:

1. *Continuous inspection*. One or more inspectors are assigned to a packinghouse. They make frequent quality checks on the commodity along the packing lines and examine samples of the packed product to determine whether it meets the U.S. grade specifications for which it is being packed. The inspector gives oral and/or written reports to management so that they can correct problems.

2. *Inspection on a sample basis*. Representative samples of a prescribed number of boxes out of a given lot are randomly selected and inspected to determine the

quality and condition of the commodity according to grade specifications. Automatic sampling systems are used for some commodities that are handled in bulk bins or trailers, such as tomatoes, grapes, and cling peaches destined for processing.

When inspection is completed, certificates are issued by the inspector on the basis of the official standards that are applicable to the situation. USDA inspectors can also inspect quality or condition based on a state grade or other specifications agreed upon by the parties involved. The cost of inspection is paid by the party requesting the service.

An *allowance* is made within each grade for a percentage of individual units within a lot that do not meet the standard. This is done because of practical limitations in sorting perishable products accurately into grades within a limited time. *Tolerances,* or the number of defective units or types of defects allowed, are more restrictive in U.S. No. 1 grade than in U.S. No. 2. The penalty for noncompliance with the U.S. grade specified on a given container may be rejection, resorting, and repacking, or reclassification to a lower grade.

To insure uniformity of inspection, (1) inspectors are trained to apply the standards, (2) visual aids (color charts, models, diagrams, photographs, and the like) are used whenever possible, (3) objective methods for determining quality and maturity are used whenever feasible and practical, and (4) good working environments with proper lighting are provided.

California Standards

California is one of the few states in the U.S. having quality standards for horticultural crops produced within the state. The standards for fresh fruits and vegetables in the California Agricultural Code are *mandatory minimum standards*. The quality factors provided for various fresh fruits and vegetables by the California Agricultural Code are summarized in tables 21.1 to 21.5. Enforcement of these standards is carried out by the California Department of Food and Agriculture (CDFA), Division of Inspection Services, Fruit and Vegetable Quality Control through each county agricultural commissioner's staff. The cost of this inspection is paid by taxpayers. Noncompliance results in destruction of the commodity or its resorting and repacking to meet minimum requirements.

Steps for establishing new standards or revising existing ones are similar to those mentioned above for U.S. grade standards, except that they are carried out at the state level. Establishment, modification, and application of these standards are responsibilities of the CDFA, Division of Inspection Services, Fruit and Vegetable Quality Control.

Uniformity of inspection is insured by methods similar to those mentioned above for U.S. grade standards.

Industry Standards

Some industries establish their own quality standards or specifications for a given commodity. Examples include apricots, clingstone peaches, processing tomatoes, and walnuts. These standards are established by agreement between producers and processors who pay application costs. Inspection is performed by such independent agencies as the California Dried Fruit Association and the Federal-State Inspection Service.

Some companies, cooperatives, and other organizations have quality grades that are applied by their quality-control personnel. Examples include quality grades for bananas, papayas, pineapples, and pistachios.

International Standards

International standards for fruits and vegetables were defined by the Economic Commission for Europe (EEC), Geneva, in 1954. Many standards have since been introduced, mainly under the Organization for Economic Co-operation and Development (OECD) scheme drawn up for this purpose. The first European International Standards were promulgated in 1961 for apples and pears. Now, there are standards for 37 commodities. Each includes three quality classes with appropriate tolerances: *Extra class* = superior quality; *Class I* = good quality; and *Class II* = marketable quality. *Class I* covers the bulk of produce entering into international trade. These standards or their equivalents are mandatory in EEC countries for imported and exported fresh fruits and vegetables. Inspection and certification is done by exporting and/or importing EEC countries.

References

1. Anon. 1973. *USDA standards for food and farm products*. USDA Agric. Handb. 341.

2. Anon. 1983. *Fruit and vegetable quality control standardization*. Extracts from the Administrative Code of California. Sacramento: Dept. Food Agric., 154 pp.

3. Anon. Various dates. *U.S. standards for grades of fresh fruits and vegetables*. USDA, Food Safety and Quality Service.

4. Anon. Various dates. *International standardization of fruits and vegetables*. Paris: Organization for Economic Cooperation and Development.

Table 21.1 Quality factors for fresh fruits in the U.S. standards for grades (US) and the California Food and Agricultural code (CA)

Fruit	Standard (date*)	Quality factors
Apple	US (1976)	Maturity, color (color charts) related to grade, firmness, shape, and size, and freedom from decay, internal browning, internal breakdown, scald, scab, bitter pit, Jonathan spot, freezing injury, water core, bruises, russeting, scars, insect damage, and other defects.
	CA (1983)	Maturity (as determined by soluble solids content [SSC] and firmness tests) <table><tr><td>*Cultivar*</td><td>*SSC (%)*</td><td>*Firmness (lb)*</td></tr><tr><td>Red Delicious</td><td>11.0</td><td>18</td></tr><tr><td>Golden Delicious</td><td>12.0</td><td>18</td></tr><tr><td>Jonathan</td><td>12.0</td><td>19</td></tr><tr><td>Rome</td><td>12.5</td><td>21</td></tr><tr><td>Newtown Pippin</td><td>11.0</td><td>23</td></tr><tr><td>McIntosh</td><td>11.5</td><td>19</td></tr><tr><td>Gravenstein</td><td>10.5</td><td>—</td></tr></table> Size, color, flesh condition, freedom from defect (such as scald, spot, internal breakdown, watercore, bruises, sunburn, russeting), and decay.
Apricot	US (1928)	Maturity, size, shape, and freedom from defect and decay.
	CA (1983)	Maturity (>3/4 of external surface area has attained a color equal to No. 3 yellowish green of the CDFA standard color chart or at least 1/2 has attained No. 4 yellow) and freedom from insect injury, decay, and mechanical damage.
Avocado	US (1957)	*For Florida avocados:* Maturity, shape, texture, skin and flesh color, and freedom from decay, anthracnose, freezing injury, bruises, russeting, scars, sunburn, mechanical damage, and other defects.
	CA (1983)	Maturity (17% to 20.5% dry weight of the flesh depending on cultivar), size, and freedom from defect, insect damage, freezing injury, rancidity, and decay.
Blueberry	US (1966)	Maturity, color, size, and freedom from defect and decay.
Cherry, sweet	US (1971)	Maturity, color, size, shape, and freedom from cracks, hail damage, russeting, scars, insect damage, and decay.
	CA (1983)	Maturity (entire surface with at least a solid light red color and/or 14% to 16% soluble solids depending on the cultivar), and freedom from bird pecks, insect injury, shriveling, growth cracks, other defects, and decay.
Citrus *Grapefruit*	US (1950)	*California and Arizona:* Maturity, color, firmness, size, shape, skin thickness, smoothness, and freedom from defect and decay.
	US (1980)	*Florida:* Maturity, color (color charts), firmness, size, smoothness, shape, and freedom from discoloration, defect, and decay.
	US (1969)	*Texas and other states:* Maturity, color, firmness, size, shape, smoothness, and freedom from discoloration, defect, and decay.
	CA (1983)	Maturity (minimum soluble solids/acid ratio of 5.5 or 6 [desert areas] and >2/3 of fruit surface showing yellow color—0.9 GY 6.40/5.7 Munsell color) and freedom from decay, freezing damage, scars, pitting, rind staining, and insect damage.
Lemon	US (1964)	Maturity (28% or 30% minimum juice content by volume depending on grade), firmness, shape, color, size, smoothness, and freedom from discoloration, defect, and decay.
	CA (1983)	Maturity (30% or more juice by volume), size uniformity, and freedom from decay, freezing damage, drying, mechanical damage, rind stains, red blotch, shriveling, and other defects.
Lime	US (1958)	Color, shape, firmness, smoothness, and freedom from stylar end breakdown, bruises, dryness, other defects, and decay.
	CA (1983)	Maturity, and freedom from defect (freezing injury, drying, mechanical damage) and decay.
Orange	US (1957)	*California and Arizona:* Maturity, color, firmness, smoothness, size, and freedom from defect and decay.
	US (1980)	*Florida:* Maturity, color (color charts), firmness, size, shape, and freedom from discoloration, defect, and decay (used also for tangelos).
	US (1969)	*Texas and other states:* Maturity, color, firmness, shape, size, and freedom from discoloration, defect, and decay.
	CA (1983)	Maturity (soluble solids/acid ratio of 8 or higher and orange color on 25% of the fruit—7.5 Y 6/6 Munsell color—or soluble solids/acid ratio of 10 or higher and orange color on 25% of fruit—2.5 GY 5/6 Munsell color), size uniformity, and freedom from defect and decay.
Tangerine and mandarine	US (1948)	*States other than Florida:* Maturity, firmness, color, size, and freedom from defect and decay.
	US (1980)	*Florida:* Maturity, color (color charts), firmness, size, shape, and freedom from defect and decay.
	CA (1983)	Maturity (yellow, orange, or red color on 75% of fruit surface and soluble solids/acid ratio of 6.5 or higher), size uniformity, and freedom from defect and decay.
Cranberry	US (1971)	Maturity, firmness, color, and freedom from bruises, freezing injury, scars, sunscald, insect damage, and decay.

*Date when standard was issued or revised.

Continued on next page

Table 21.1—Continued

Fruit	Standard (date*)	Quality factors
Date	CA (1983)	Freedom from insect damage, decay, black scald, fermentation, and other defects.
Dewberry, blackberry	US (1928)	Maturity, color, and freedom from calyxes, decay, shriveling, mechanical damage, insect damage, and other defects.
	CA (1983)	Maturity and freedom from decay and damage due to frost, bruising, insects, or other causes.
Grape, table *European Vinifera type*	US (1983)	Maturity (as determined by percent soluble solids as set forth by the producing states), color, uniformity, firmness, berry size, and freedom from shriveling, shattering, sunburn, waterberry, shot berries, dried berries, other defects, and decay. Bunches: fairly well-filled but not excessively tight. Stems: not dry and brittle, and at least yellowish green in color. For states other than California and Arizona, and countries exporting to U.S.: Cultivar — Minimum SSC (%) Muscat — 17.5 Cardinal, Ribier, Olivette, Blanche, Emperor, Perlette, Rish Baba, Red Malaga — 15.5 All other cultivars — 16.5
	CA (1983)	Maturity (minimum percent soluble solids of 14 to 17.5, depending on cultivar and production area, or soluble solids/acid ratio of 20 or higher, or a combination of a minimum soluble solids/acid ratio and percent soluble solids), and freedom from decay, freezing injury, sunburned or dried berries, and insect damage. Same for Arizona.
American bunch type	US (1983)	Maturity (juiciness, ease of separation of type skin from pulp), color, firmness, compactness, and freedom from defect and decay.
Kiwifruit	US (1982)	Maturity (more than 6.5% soluble solids), firmness, and freedom from growth cracks, insect injury, broken skin, bruises, sunscald, freezing injury, internal breakdown, and decay.
Nectarine	US (1966)	Maturity, color depending on variety, shape, and size, and freedom from growth cracks, insect damage, scars, bruises, russeting, split pits, other defects, and decay.
	CA (1983)	Maturity (surface ground color, fruit shape), and freedom from insect injury, split pits, mechanical damage, and decay.
Olive	CA (1983)	Freedom from insect injury, especially scale.
Peach	US (1952)	Maturity (shape, size, ground color), and freedom from decay and defect (split pit, hail injury, insect damage, growth cracks).
	CA (1983)	Maturity (skin and flesh color, and fullness of shoulders and suture), and freedom from defect and decay.
Pear *Winter*	US (1955)	Maturity (color, firmness), size, and freedom from internal breakdown, black end, russeting, other defects, and decay.
Summer and fall	US (1955)	Maturity (color, firmness), shape, size, and freedom from defect and decay.
	CA (1983)	Maturity (*Bartlett:* Average firmness test of <23 lb., and/or soluble solids content 13%, and/or yellowish green color . . . CDFA color chart), and freedom from insect damage, mechanical damage, decay, and other defects.
Persimmon	CA (1983)	Maturity as indicated by surface color: *Hachiya:* Blossom end's color is orange or reddish color equal to or darker than Munsell color 6.7 YR 5.93/12.7 on at least 1/3 of the fruit's length with the remaining 2/3 a green color equal to or lighter than Munsell color 2.5 GY 5/6. Other cultivars: A yellowish green color equal to or lighter than Munsell color 10 Y 6/6. Freedom from growth cracks, mechanical damage, decay, and other defects.
Pineapple	US (1953)	Maturity, firmness, uniformity of size and shape, and freedom from decay, sunscald, bruising, insect damage, and cracks. Tops: color, length, and straightness.
Plum and fresh prune	US (1969)	Maturity, color, shape, size, and freedom from decay, sunscald, split pits, hail damage, mechanical damage, scars, russeting, and other defects.
	CA (1983)	Maturity as indicated by surface color (minimum color requirements are described for 56 cultivars), and freedom from decay, insect damage, bruises, sunburn, hail damage, gum spot, growth cracks, and other defects.
Pomegranate	CA (1983)	Maturity (<1.85% acid content in juice and red juice color equal to or darker than Munsell color 5 R 5/12), freedom from sunburn, growth cracks, cuts or bruises, and decay.
Quince	CA (1983)	Maturity, and freedom from insect damage, mechanical damage, and decay.
Raspberry	US (1931)	Maturity, color, shape, and freedom from defect and decay.
	CA (1983)	Maturity, and freedom from decay and damage due to insects, sun, frost, bruising, or other causes.
Strawberry	US (1965)	Maturity (>1/2 or >3/4 of surface showing red or pink color, depending on grade), firmness, attached calyx, size, and freedom from defect and decay.
	CA (1983)	Maturity (>2/3 of fruit surface showing a pink or red color), and freedom from defect and decay.

*Date when standard was issued or revised.

Table 21.2 Quality factors for fresh vegetables in the U.S. standards for grades (US) and the California Food and Agricultural code (CA)

Fruit	Standard (date*)	Quality factors
Anise, sweet	US (1973)	Firmness, tenderness, trimming, blanching, and freedom from decay and damage caused by growth cracks, pithy branches, wilting, freezing, seedstems, insects, and mechanical means.
Artichoke	US (1969)	Stem length, shape, overmaturity, uniformity of size, compactness, and freedom from defect and decay.
	CA (1983)	Freedom from decay, insect damage, and freezing injury.
Asparagus	US (1966)	Freshness (turgidity), trimming, straightness, freedom from damage and decay, stalk diameter, percent green color.
	CA (1983)	Turgidity, straightness, percent showing white color, stalk diameter, and freedom from decay, mechanical damage, and insect injury.
Bean, lima	US (1938)	Uniformity, maturity, freshness, shape, and freedom from damage (defect) and decay.
Bean, snap	US (1936)	Uniformity, size, maturity, freshness (firmness), and freedom from defect and decay.
Beet, bunched or topped	US (1955)	Root shape, trimming of rootlets, firmness (turgidity), smoothness, cleanness, minimum size (diameter), and freedom from defect.
Beet greens	US (1959)	Freshness, cleanness, tenderness, and freedom from decay, other kinds of leaves, discoloration, insects, mechanical injury, and freezing injury.
Broccoli	US (1943)	Color, maturity, stalk diameter and length, compactness, base cut, and freedom from defects and decay.
	CA (1983)	Freedom from decay and damage due to overmaturity, insects, or other causes.
Brussels sprouts	US (1954)	Color, maturity (firmness), no seed stems, size (diameter and length), and freedom from defect and decay.
	CA (1983)	Freedom from decay, from burst, soft, or spongy heads, and from insect damage.
Cabbage	US (1945)	Uniformity, solidity (maturity or firmness), no seed stems, trimming, color, and freedom from defect and decay.
	CA (1983)	Conform to U.S. commercial grade or better.
Cantaloupe	US (1968)	Soluble solids (>9%), uniformity of size, shape, ground color and netting; maturity and turgidity; and freedom from "wet slip," sunscald, and other defects.
	CA (1983)	Maturity (soluble solids >8%), and freedom from insect injury, bruises, sunburn, growth cracks, and decay.
Carrot, bunched	US (1954)	Shape, color, cleanness, smoothness, freedom from defect, freshness, length of tops, and root diameter.
	CA (1983)	Number, size, and weight per bunch, freshness, and freedom from defect and decay (tops).
Carrot, topped	US (1965)	Uniformity, turgidity, color, shape, size, cleanness, smoothness, and freedom from defect (growth cracks, pithiness, woodiness, internal discoloration).
	CA (1983)	Freedom from defect (growth cracks, doubles, mechanical injury, green discoloration, objectionable flavor or odor) and decay.
Carrots with short trimmed tops	US (1954)	Roots: Firmness, color, smoothness, and freedom from defect (sunburn, pithiness, woodiness, internal discoloration, and insect and mechanical injuries) and decay. Leaves: (Cut to <4 inches). Freedom from yellowing or other discoloration, disease, insects, and seed stems.
Cauliflower	US (1968)	Curd cleanness, compactness, white color, size (diameter), freshness and trimming of jacket leaves, and freedom from defect and decay.
	CA (1983)	Freedom from insect injury, decay, freezing injury, and sunburn.
Celery	US (1959)	Stalk form, compactness, color, trimming, length of stalk and midribs, width and thickness of midribs, no seed stems, and freedom from defect and decay.
	CA (1983)	Freedom from pink rot and other decay, blackheart, seed stems, pithy condition, and insect damage.
Collard greens and broccoli greens	US (1953)	Freshness, tenderness, cleanness, and freedom from seed stems, discoloration, freezing injury, insects, and diseases.
Corn, green	US (1954)	Uniformity of color and size, freshness, milky kernels, cob length, freedom from defect, coverage with fresh husks.
	CA (1983)	Milky, plump, well-developed kernels, and freedom from insect injury, mechanical damage, and decay.
Cucumber	US (1958)	Color, shape, turgidity, maturity, size (diameter and length), and freedom from defect and decay.
Cucumber, greenhouse	US (1934)	Freshness, shape, firmness, color, size (>5.5 inches), and freedom from decay, cuts, scars, and other defects.
Dandelion greens	US (1955)	Freshness, cleanness, tenderness, and freedom from damage caused by seed stems, discoloration, freezing, diseases, insects, and mechanical injury.
Eggplant	US (1953)	Color, turgidity, shape, size, and freedom from defect and decay.
Endive, escarole, or chicory	US (1964)	Freshness, trimming, color (blanching), no seed stems, and freedom from defect and decay.
Garlic	US (1944)	Maturity, curing, compactness, well-filled cloves, bulb size, and freedom from defect.
	CA (1983)	Size (bulb diameter).
Honeydew and honey ball melons	US (1967)	Maturity, firmness, shape, and freedom from decay and defect (sunburn, bruising, hail spots, and mechanical injuries).

*Date when standard was issued or revised.

Continued on next page

Table 21.2 Continued

Fruit	Standard (date*)	Quality factors
Honeydew (Cont.)	CA (1983)	Maturity, soluble solids (>10%), and freedom from decay, sunscald, bruises, and growth cracks. Honey ball melons should be netted and should have pink flesh.
Horseradish roots	US (1936)	Uniformity of shape and size, firmness, smoothness, and freedom from hollow heart, other defects, and decay.
Kale	US (1934)	Uniformity of growth and color, trimming, freshness, and freedom from defect and decay.
Lettuce, crisp-head	US (1975)	Turgidity, color, maturity (firmness), trimming (number of wrapper leaves), and freedom from tipburn, other physiological disorders, mechanical damage, seed stems, other defects, and decay.
	CA (1983)	Freedom from insect damage, decay, seed stems, tipburn, freezing injury, broken midribs, and bursting. For sectioned, chopped, or shredded lettuce: Same as intact heads plus freedom from discoloration and excessive moisture.
Lettuce, greenhouse leaf	US (1964)	Well-developed, well-trimmed, and freedom from coarse stems, bleached or discolored leaves, wilting, freezing, insects, and decay.
Lettuce, romaine	US (1960)	Freshness, trimming, and freedom from decay and damage caused by seed stems, broken, bruised, or discolored leaves, tipburn, and wilting.
Melon, casaba and Persian	CA (1983)	Maturity, and freedom from growth cracks, decay, mechanical injury, and sunburn.
Mushroom	US (1966)	Maturity, shape, trimming, size, and freedom from open veils, disease, spots, insect injury, and decay.
	CA (1983)	Freedom from insect injury.
Mustard greens and turnip greens	US (1953)	Freshness, tenderness, cleanness, and freedom from damage caused by seed stems, discoloration, freezing, disease, insects, or mechanical means. Roots (if attached): firmness and freedom from damage.
Okra	US (1928)	Freshness, uniformity of shape and color, and freedom from defect and decay.
Onion, dry Creole	US (1943)	Maturity, firmness, shape, size (diameter), and freedom from decay, wet sunscald, doubles, bottlenecks, sprouting, and other defects.
Bermuda-Granex-Grano	US (1962)	
Other cultivars	US (1971)	
	CA (1983)	Freedom from insect injury, decay, sunscald, freezing injury, sprouting, and other defects.
Onion, green	US (1947)	Turgidity, color, form, cleanness, bulb trimming, no seed stems, and freedom from defect and decay.
Onion sets	US (1940)	Maturity, firmness, size, and freedom from decay and damage caused by tops, sprouting, freezing, mold, moisture, dirt, disease, insects, or mechanical means.
Parsley	US (1930)	Freshness, green color, and freedom from defects, seed stems, and decay.
Parsnip	US (1945)	Turgidity, trimming, cleanness, smoothness, shape, freedom from defect and decay, and size (diameter).
Pea, fresh	US (1942)	Maturity, size, shape, freshness, and freedom from defect and decay.
	CA (1983)	Maturity, and freedom from mechanical damage, insect damage, decay, yellowing, and shriveling.
Pea, Southern (cowpea)	US (1956)	Maturity, pod shape, and freedom from discoloration and other defects.
Pepper, sweet	US (1963)	Maturity, color, shape, size, and freedom from defects (sunscald, freezing injury, hail, scars, insects, mechanical damage) and decay.
	CA (1983)	Freedom from insect damage and decay.
Potato	US (1972)	Uniformity, maturity, firmness, cleanness, shape, size, and freedom from sprouts, blackheart, greening, and other defects.
	CA (1983)	A minimum equivalent of U.S. No. 2 grade. Maturity is described in terms of extent of skin missing or feathered.
Radish (topped)	US (1968)	Tenderness, cleanness, smoothness, shape, size, and freedom from pithiness and other defects.
Rhubarb	US (1966)	Color, freshness, straightness, trimming, cleanness, stalk diameter and length, and freedom from defect.
Shallot, bunched	US (1946)	Firmness, form, tenderness, trimming, cleanness, and freedom from decay and damage caused by seed stems, disease, insects, mechanical and other means. Tops: freshness, green color, and no mechanical damage.
Spinach leaves	US (1946)	Color, turgidity, cleanness, trimming, and freedom from seed stems, coarse stalks, and other defects.
Spinach plants	US (1956)	Freshness, cleanness, trimming, and freedom from decay and damage caused by coarse stalks or seed stems, discoloration, insects, and mechanical means.
Squash, summer	US (1984)	Immaturity, tenderness, shape, firmness, and freedom from decay, cuts, bruises, scars, and other defects.
Squash, winter and pumpkin	US (1983)	Maturity, firmness, freedom from discoloration, cracking, dry rot, insect damage, and other defects; uniformity of size.
Sweet potato	US (1963)	Firmness, smoothness, cleanness, shape, size, and freedom from mechanical damage, growth cracks, internal breakdown, insect damage, other defects, and decay.
	CA (1983)	Freedom from decay, mechanical damage, insect injury, growth cracks, and freezing injury.

*Date when standard was issued or revised.

Continued on next page

Table 21.2—Continued

Fruit	Standard (date*)	Quality factors
Tomato	US (1976)	Maturity and ripeness (color chart), firmness, shape, size, and freedom from defect (puffiness, freezing injury, sunscald, scars, catfaces, growth cracks, insect injury, and other defects) and decay.
	CA (1983)	Freedom from insect and freezing damage, sunburn, mechanical damage, blossom-end rot, catfaces, growth cracks, and other defects.
Tomato, greenhouse	US (1966)	Maturity, firmness, shape, size, and freedom from decay, sunscald, freezing injury, bruises, cuts, shriveling, puffiness, catfaces, growth cracks, scars, disease, and insects.
Turnip and rutabaga	US (1955)	Uniformity of root color, size, and shape, trimming, freshness, and freedom from defects (cuts, growth cracks, pithiness, woodiness, water core, dry rot).
Watermelon	US (1978)	Maturity and ripeness (optional internal quality criteria: soluble solids content = >10% very good, >8% good), shape, uniformity of size (weight), and freedom from anthracnose, decay, sunscald, and whiteheart.
	CA (1983)	Maturity (arils around the seeds have been absorbed and flesh color is >75% red), and freedom from decay, sunburn, flesh discoloration, and mechanical damage.

Table 21.3 Quality factors for processing fruits in the U.S. standards for grades (US) and the California Food and Agricultural code (CA)

Fruit	Standard (date*)	Quality factors
Apple	US (1961)	Ripeness (not overripe, mealy, or soft), and freedom from decay, worm holes, freezing injury, internal breakdown, and other defects that would cause a loss of >5% (U.S. No. 1) or >12% (U.S. No. 2) by weight.
Berries	US (1947)	Color, and freedom from caps (calyxes), decay, and defect (dried, undeveloped and immature berries, crushing, shriveling, sunscald, insect damage, and mechanical injury).
Blueberry	US (1950)	Freedom from other kinds of berries, clusters, large stems, leaves and other foreign material, and freedom from damage caused by decay, shriveling, dirt, overmaturity, or other means.
Cherry, red sour	US (1941)	Color uniformity, and freedom from decay, pulled pits, attached stems, hail marks, windwhips, scars, sunscald, shriveling, disease, and insect damage.
Cherry, sweet for canning or freezing	US (1946)	Maturity, shape, freedom from decay, worms, pulled pits, doubles, insect and bird damage, and mechanical injury, and freedom from damage caused by freezing, softness, shriveling, cracks and skin breaks, scars, and sunscald. Tolerance is 7% (U.S. No. 1) or 12% (U.S. No. 2) by count.
Cherry, sweet for sulfur brining	US (1940)	Maturity (ease of pit separation), firmness, shape, and freedom from decay and defect (bruises, bird and insect damage, skin breaks, russeting, shriveling, scars, sunscald, and limbrubs).
Cranberry, red sour	US (1957)	Maturity, color, firmness, size, and freedom from defect (insect damage, bruises, scars, sunscald, freezing injury, and mechanical injury) and decay.
Currant	US (1952)	Color, stem attached, and freedom from decay and damage caused by crushing, drying, shriveling, insects, and mechanical means.
Grape, American type for processing and freezing	US (1975)	Maturity (>15.5% soluble solids), color, freedom from shattered, split, crushed, or wet berries, and freedom from decay and from damage caused by freezing, heat, sunburn, disease, insects, or other means.
Grape, juice (European or vinifera type)	US (1939)	Maturity (>16% to 18% soluble solids depending on cultivar), freedom from crushed, split, wet, waterberry and redberry, and freedom from defect (insect, disease, mechanical injury, sunburn, and freezing damage).
	CA (1983)	Maturity (minimum soluble solids content of 14% to 17.5% depending on cultivar *or* soluble solids/acid ratio of 20 or higher), and freedom from decay, freezing injury, waterberry, redberry, and other defects.
Grape for processing and freezing	US (1977)	Maturity (>15.5% soluble solids content), and freedom from decay and defect (dried berries, discoloration, sunburn, insect damage, and immature berries).
Peach, freestone for canning, freezing, or pulping	US (1966)	Maturity, color (not greener than yellowish green), shape, firmness, and freedom from decay, worms and worm holes, split pits, scab, bacterial spot, insects, and bruises. Grade is based on the severity of defects with 10% tolerance.
Pear for processing	US (1970)	Maturity, color (less than yellowish green), shape, firmness, and freedom from scald, hard end, black end, internal breakdown, decay, worms and worm holes, scars, sunburn, bruises, and other defects. Grade is based on the severity of defects with 10% tolerance.
Raspberry	US (1952)	Color, and freedom from decay and defect (dried berries, crushing, shriveling, sunscald, scars, bird and insect damage, discoloration, or mechanical injury).
Strawberry, growers' stock for manufacture	US (1935)	Color, freedom from decay and defect (crushed, split, dried or undeveloped berries, sunscald, and bird or insect damage), size, and cap removal.
Strawberry, washed and sorted for freezing	US (1935)	Color, cleanness, size, cap removal, and freedom from decay and defect (crushed, split, dried or undeveloped berries, bird and insect damage, mechanical injury).

*Date when standard was issued or revised.

Table 21.4 Quality factors for processing vegetables in the U.S. standards for grades (US)

Fruit	Standard (date*)	Quality factors
Asparagus, green	US (1972)	Freshness, shape, green color, size (spear length), and freedom from defect (freezing damage, dirt, disease, insect injury, and mechanical injuries) and decay.
Bean, shelled lima	US (1953)	Tenderness, green color, and freedom from decay and from injury caused by discoloration, shriveling, sunscald, freezing, heating, disease, insects, or other means.
Bean, snap	US (1959)	Freshness, tenderness, shape, size, and freedom from decay and from damage caused by scars, rust, disease, insects, bruises, punctures, broken ends, or other means.
Beet	US (1945)	Firmness, tenderness, shape, size, and freedom from soft rot, cull material, growth cracks, internal discoloration, white zoning, rodent damage, disease, insects, and mechanical injury.
Broccoli	US (1959)	Freshness, tenderness, green color, compactness, trimming, and freedom from decay and damage caused by discoloration, freezing, pithiness, scars, dirt, or mechanical means.
Cabbage	US (1944)	Firmness, trimming, and freedom from soft rot, seed stems, and from damage caused by bursting, discoloration, freezing, disease, birds, insects, or mechanical or other means.
Carrot	US (1984)	Firmness, color, shape, size (root length), smoothness, not woody, and freedom from soft rot, cull material, and from damage caused by growth cracks, sunburn, green core, pithy core, water core, internal discoloration, disease, or mechanical means.
Cauliflower	US (1959)	Freshness, compactness, color, and freedom from jacket leaves, stalks, and other cull material, decay, and damage caused by discoloration, bruising, fuzziness, enlarged bracts, dirt, freezing, hail, or mechanical means.
Corn, sweet	US (1962)	Maturity, freshness, and freedom from damage by freezing, insects, birds, disease, cross-pollination, or fermentation.
Cucumber, pickling	US (1936)	Color, shape, freshness, firmness, maturity, and freedom from decay and from damage caused by dirt, freezing, sunburn, disease, insects, or mechanical or other means.
Mushroom	US (1964)	Freshness, firmness, shape, and freedom from decay, disease spots, and insects, and from damage caused by insects, bruising, discoloration, or feathering.
Okra	US (1965)	Freshness, tenderness, color, shape, and freedom from decay and insects, and from damage caused by scars, bruises, cuts, punctures, discoloration, dirt, or other means.
Onion	US (1944)	Maturity, firmness, and freedom from decay, sprouts, bottlenecks, scallions, seed stems, sunscald, roots, insects, and mechanical injury.
Pea, fresh shelled for canning/freezing	US (1946)	Tenderness, succulence, color, and freedom from decay, scald, rust, shriveling, heating, disease, and insects.
Pea, Southern	US (1965)	Pods: Maturity, freshness, and freedom from decay. Seeds: Freedom from scars, insects, decay, discoloration, splits, cracked skin, and other defects.
Pepper, sweet	US (1948)	Firmness, color, shape, and freedom from decay, insects, and damage by any means that results in 5% to 20% trimming (by weight) depending on grade.
Potato	US (1963)	Shape, smoothness, freedom from decay and defect (freezing injury, blackheart, sprouts), size, specific gravity, glucose content, and fry color.
Potato for chipping	US (1978)	Firmness, cleanness, shape, freedom from defect (freezing, blackheart, decay, insect injury, and mechanical injury), size; optional tests for specific gravity and fry color are included.
Spinach	US (1956)	Freshness, freedom from decay, grass weeds, and other foreign material, and freedom from damage caused by seed stems, discoloration, coarse stalks, insects, dirt, or mechanical means.
Sweet potato for canning/freezing	US (1959)	Firmness, shape, color, size, and freedom from decay and defect (freezing injury, scald, cork, internal discoloration, bruises, cuts, growth cracks, pithiness, stringiness, and insect injury).
Sweet potato for dicing/pulping	US (1951)	Firmness, shape, size, and freedom from decay and defect (scald, freezing injury, cork, internal discoloration, pithiness, growth cracks, insect damage, and stringiness).
Tomato	US (1983)	Firmness, ripeness (color as determined by a photoelectric instrument), and freedom from insect damage, freezing, mechanical damage, decay, growth cracks, sunscald, gray wall, and blossom-end rot.
Tomato, green	US (1950)	Firmness, color (green), and freedom from decay and defect (growth cracks, scars, catfaces, sunscald, disease, insects, or mechanical damage).
Tomato, Italian type for canning	US (1957)	Firmness, color uniformity, and freedom from decay and defect (growth cracks, sunscald, freezing, disease, insects, or mechanical injury).

*Date when standard was issued or revised.

Table 21.5 Quality factors for tree nuts in the U.S. standards for grades (US) and the California Food and Agricultural code (CA)

Fruit	Standard (date*)	Quality factors
Almond, shelled	US (1960)	Similar varietal characteristics (shape, appearance), size (count per ounce), degree of dryness, cleanness (freedom from dust, particles, and foreign materials), and freedom from decay and defect (rancidity, insect injury, doubles, split or broken kernels, shriveling, brown spot, or gumminess).
Almond, in-shell	US (1964)	Shell: Similar varietal characteristics (shape, hardness), cleanness (freedom from loose extraneous and foreign materials), size (thickness), brightness and uniformity of color, and freedom from discoloration, insect infestation, adhering hulls, and broken shells. Kernel: Degree of dryness, and freedom from decay and defect (rancidity, insect damage, shriveling, brown spot, gumminess, and skin discoloration).
Brazil nut, in-shell	US (1966)	Shell: Degree of dryness, cleanness (freedom from dirt, extraneous, and adhering foreign materials), size (diameter), and freedom from damage caused by splits, breaks, punctures, oil stains, and mold. Kernel: Degree of development (must fill more than 50% of the shell capacity), freedom from decay and defect (rancidity, insect damage, and discoloration).
Filbert, in-shell	US (1970)	Shell: Shape, size (diameter), cleanness, brightness, and freedom from defect (blanks, broken or split shells, stains, and adhering husk). Kernel: Degree of dryness (less than 10% moisture content), development (must fill more than 50% of the shell capacity), shape, and freedom from decay and defect (insect injury, shriveling, rancidity, and discoloration).
Mixed nuts, in-shell	US (1970)	Each species of nut must conform to a minimum size and grade (same quality criteria used for that species). Grade of the mix is also determined by percent allowable for each component (almonds, brazils, filberts, pecans, walnuts).
Pecan, shelled	US (1969)	Degree of dryness, degree of development (amount of meat in proportion to width and length), color (plastic models for color standards are available), color uniformity, size (number of halves per pound or diameter of pieces), freedom from decay and defect (shriveling, insect damage, internal discoloration, dark spots, skin discoloration, and rancidity), and cleanness (freedom from dust, dirt, and adhering material).
Pecan, in-shell	US (1976)	Shell: Color uniformity, size (number of nuts per pound), cleanness, and freedom from decay and defect (insect damage, dark stains, split or cracked shells, and broken shells). Kernel: Same as for shelled pecans (*above*).
Walnut, shelled	US (1968)	Color (USDA color chart), degree of dryness, cleanness (freedom from shells, dirt, dust, and foreign material), freedom from decay and defect (insect injury, rancidity, shriveling, and meat discoloration), and size (diameter of halves or pieces).
Walnut, in-shell	US (1976)	Shell: Dryness, cleanness, brightness, freedom from decay and defect (splits, discoloration, broken shells, perforated shells, and adhering hulls), and size (diameter). Kernel: Same as for shelled walnuts (*above*).
	CA (1983)	Shell: Dryness, size, and freedom from blanks, decay, and defect (insect damage, adhering hulls, and perforations affecting more than 1/8 of the surface). Kernel: Size, and freedom from decay and defect (insect damage, shriveling, and rancidity).

*Date when standard was issued or revised.

22

Postharvest Handling Systems: Leafy, Root, and Stem Vegetables

ROBERT F. KASMIRE

Nonfruit vegetables can be divided into two subgroups: leafy and succulent vegetables, and bulky vegetative organs (underground structures). This chapter reviews the general characteristics of each group.

Leafy Succulent Tissues

The leafy and succulent tissues group includes the following vegetables:

- *Leafy vegetables*—lettuce, cabbage, Chinese cabbage, brussels sprouts, celery, rhubarb, spinach, chard, kale, endive, escarole, parsley, green onion
- *Stem vegetables*—asparagus, kohlrabi, fennel
- *Floral vegetables*—artichoke, broccoli, cauliflower

Most of these vegetables are available on the market throughout the year from various California production areas. For this reason, no long-term storage is required. Long-term storage is not possible for the more perishable commodities.

Harvest and Postharvest Handling Procedures

Harvesting

Leafy vegetables. Virtually all leafy vegetables are cut by hand, but harvesting aids are used with some (brussels sprouts, celery, and parsley). Mechanical harvesting systems have been developed for crisp-head lettuce, celery, cabbage, brussels sprouts, and cauliflower, but they are not used commercially. Celery is commonly top-mowed mechanically to a uniform height (length) before harvesting. The lettuce harvester uses gamma rays or X-rays to determine head density, an indication of maturity. Either method is more accurate than subjective determination by hand. Maturity determination varies with commodities. The various solidity (maturity) classes used for crisp-head lettuce are shown in table 22.1. Maturity of other leafy vegetables is judged mainly on size and compactness.

Stem vegetables. Most stem vegetables are hand harvested. A limited amount of asparagus may be experimentally machine harvested. Asparagus is generally hand cut when spears have grown at least 23 cm (9 inches) above the soil surface.

Floral vegetables. All floral vegetables are hand harvested, but harvest aids are sometimes used with broccoli.

Field packing versus packinghouse packing

Field packing (fig. 22.1) is used for all leafy vegetables, except brussels sprouts and green onions in most operations. The products are selected for maturity and quality, and then cut, trimmed, packed in cartons or crates, transported to coolers, cooled, put into temporary cold storage prior to loading or loaded directly, and transported to market (fig. 22.2). Field packing generally provides greater marketable yields because of reduced damage from handling. Wrapped and unwrapped lettuce, celery, cauliflower, broccoli, and spinach are mostly field packed, though the latter three are still packed in packinghouses by a few shippers.

Small celery stalks are generally field packed in bulk containers after harvesting and transported to packinghouses for trimming, sorting, prepackaging, and packing as celery hearts. Wrapped lettuce and cauliflower are hand selected, cut, and trimmed, and then placed on mobile field units where they are wrapped and packed into cartons. Next they go to vacuum coolers or other cooling facilities for cooling and subsequent handling. Rough handling in field packing operations is a major cause of lettuce and cauliflower marketing losses. Cleanliness of the commodity is a problem in field packing operations, particularly when fields are muddy.

Table 22.1. Solidity (maturity) classes of crisp-head lettuce

Solidity class		Postharvest considerations
(1)	Soft, no head formation	More susceptible to physical damage; has a higher respiration rate than more mature lettuce; unacceptable for market
(2)	Fairly firm, slight head formation	Higher respiration rate
(3)	Firm, good head formation; optimum density (maturity)	Maximum storage life
(4)	Hard, maximum density, but no split ribs	More susceptible to russet spotting, pink rib, and other physiological disorders; decreased storage life
(5)	Extra hard; split mid-ribs common; extreme internal pressure	Has minimum storage and shelf-life remaining because of advanced maturity; most difficult to vacuum cool

Fig. 22.1. A crew field packing crisp-head lettuce.

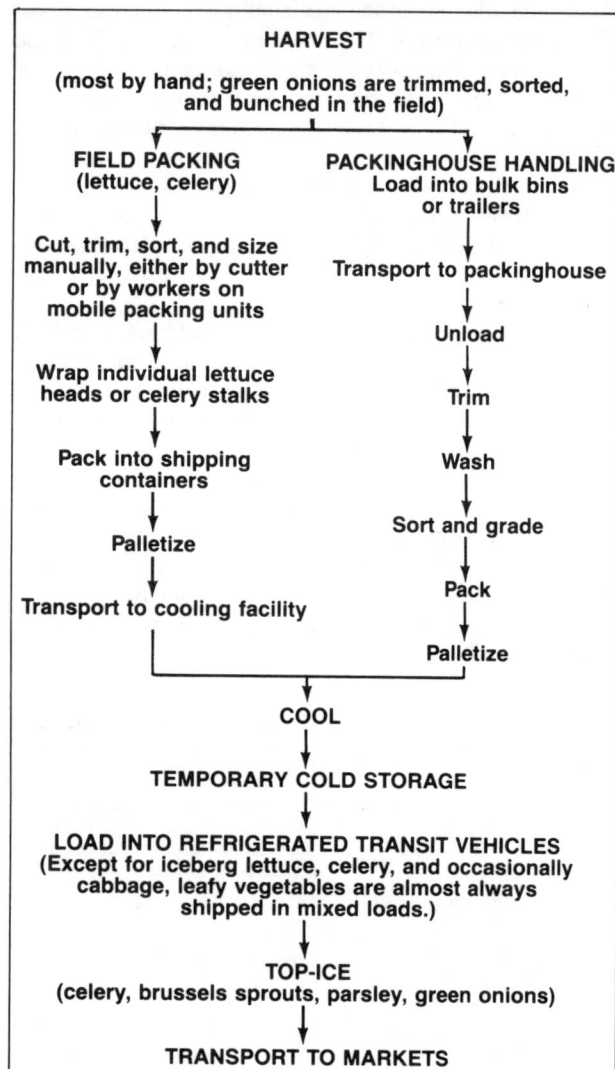

Fig. 22.2. Harvest and postharvest operations for leafy vegetables.

Packinghouse operations

All stem vegetables and floral vegetables not packed in the field are selected, cut, placed in bulk containers, transported to packinghouses, trimmed, sorted, sized, packed into cartons or crates, cooled, placed temporarily in cold storage for subsequent loading or loaded directly into precooled refrigerated transport vehicles, and transported to market (figs. 22.3, 22.4). Compared to field packing, the packinghouse method requires more energy, more product handling, and causes more product damage, reducing marketable yield. All of these factors make the packinghouse method the more expensive of the two.

Hard heads of crisp-head lettuce are hand cut, closely trimmed, loaded in bulk bins, transported to vacuum coolers, and cooled for subsequent processing at the shipping point or shipped to destination market processors. The processing includes coring, chopping or shredding, washing in cold water (near 0°C = 32°F), sorting to eliminate defective products, spin drying, packaging in polyethylene bags, applying of modified atmospheres (where used), packing into cartons, placing in temporary cold storage, and shipping to markets in refrigerated transport vehicles.

Packinghouse operations used in preparation for market include the following:

- trimming and cleaning with chlorinated water (desirable concentration is about 200 ppm of chlorine)
- sorting to eliminate defective products
- sizing, in some cases (all sizing is subjective and done by hand)
- wrapping individual units (cauliflower), or in some cases, prepackaging (brussels sprouts, broccoli, cauliflower florets)
- packing in wax-impregnated fiberboard shipping containers or wood crates to withstand hydrocooling, in-package ice, or top-ice during transit

Cooling

Cooling methods vary with commodity, as follows:

- vacuum cooling for crisp-head lettuce, leaf lettuce, spinach, cauliflower, Chinese cabbage, bok choy, cabbage, and other leafy vegetables
- Hydro-Vac cooling (a modification of vacuum cooling) for celery and other leafy vegetables
- hydrocooling for leaf lettuce, celery, spinach, some green onions, leek, artichoke, and other leafy vegetables
- package-icing, including liquid-icing, for broccoli, spinach, parsley, green onions, and brussels sprouts
- room cooling, primarily for artichoke and cabbage, and for other vegetables in some operations
- forced-air cooling, primarily for cauliflower and to a limited extent for some other leafy and stem vegetables

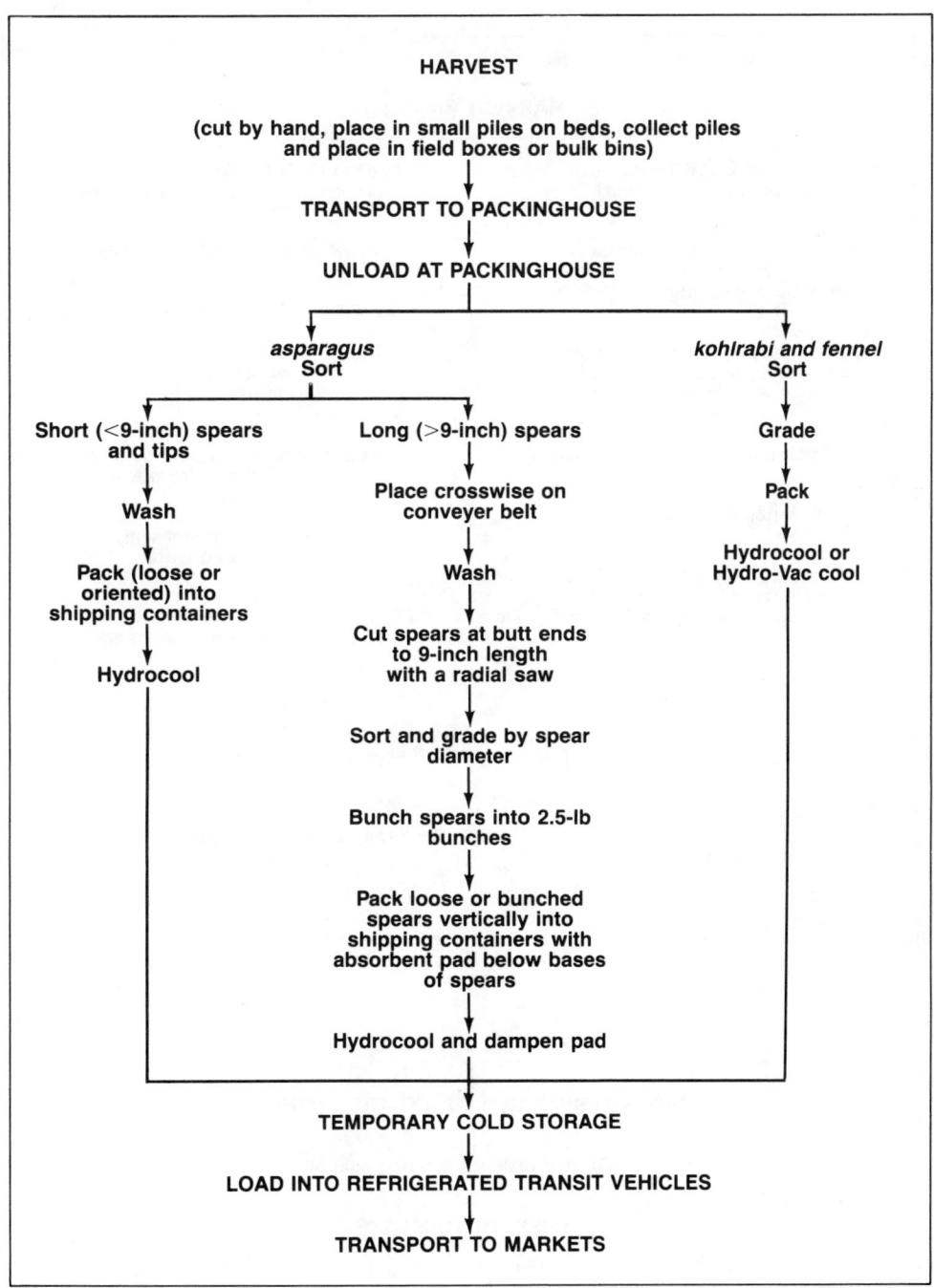

Fig. 22.3. Harvest and postharvest operations for stem vegetables (asparagus, kohlrabi, and fennel).

Recommended storage conditions

Temperature management procedures:

- avoid delays between harvesting and cooling, especially during warm weather
- cool to 1°C (±1°C) [34°F (±2°F)]; avoid freezing
- immediately after cooling, transfer into refrigerated transit vehicles or cold-storage rooms
- maintain a temperature of 1°C (±1°C) and a relative humidity of 90 percent to 95 percent during transit and temporary storage

Long-term storage is *not* recommended, except with cabbage, Chinese cabbage, and celery. In storage, keep air movement at the minimum required for proper temperature control, removal of CO_2, and maintenance of adequate O_2 levels. Avoid exposure to ethylene throughout the handling system.

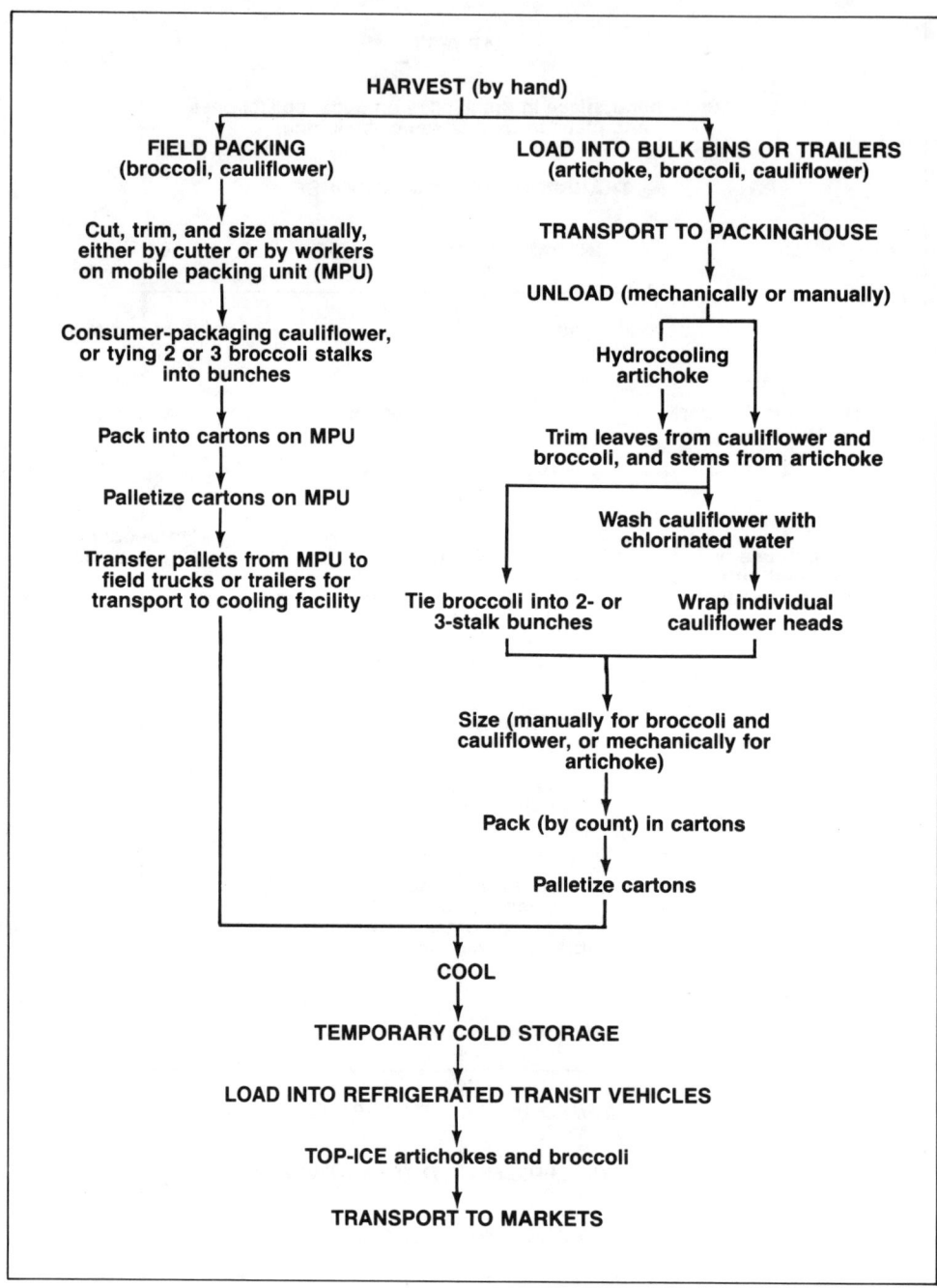

Fig. 22.4. Harvest and postharvest operations for floral vegetables (artichoke, broccoli, cauliflower).

Bulky Vegetative Organs

The edible portion of the bulky vegetative organs group develops mostly underground. This group of vegetables includes several botanical structures:

- *Roots*—beet, carrot, celeriac, radish, horseradish, parsnip, turnip, sweet potato, cassava
- *Tubers*—potato, Jerusalem artichoke, yam (*Dioscorea* spp.)
- *Bulbs*—onion, garlic, shallot
- *Rhizomes*—ginger
- *Corms*—taro (dasheen)

Vegetables within the group can also be divided into two subgroups based on their postharvest requirements:

- *Temperate zone "root" vegetables*—beet, carrot, celeriac, radish, horseradish, parsnip, turnip, potato, onion, garlic, shallot
- *Subtropical and tropical "root" vegetables*—sweet potato, yam, cassava, ginger, taro

Harvesting and Postharvest Handling Procedures

Harvesting (fig. 22.5)

Maturity (harvest) indices vary with commodity. These are the criteria used for a few crops:

- *Carrot*—size of root

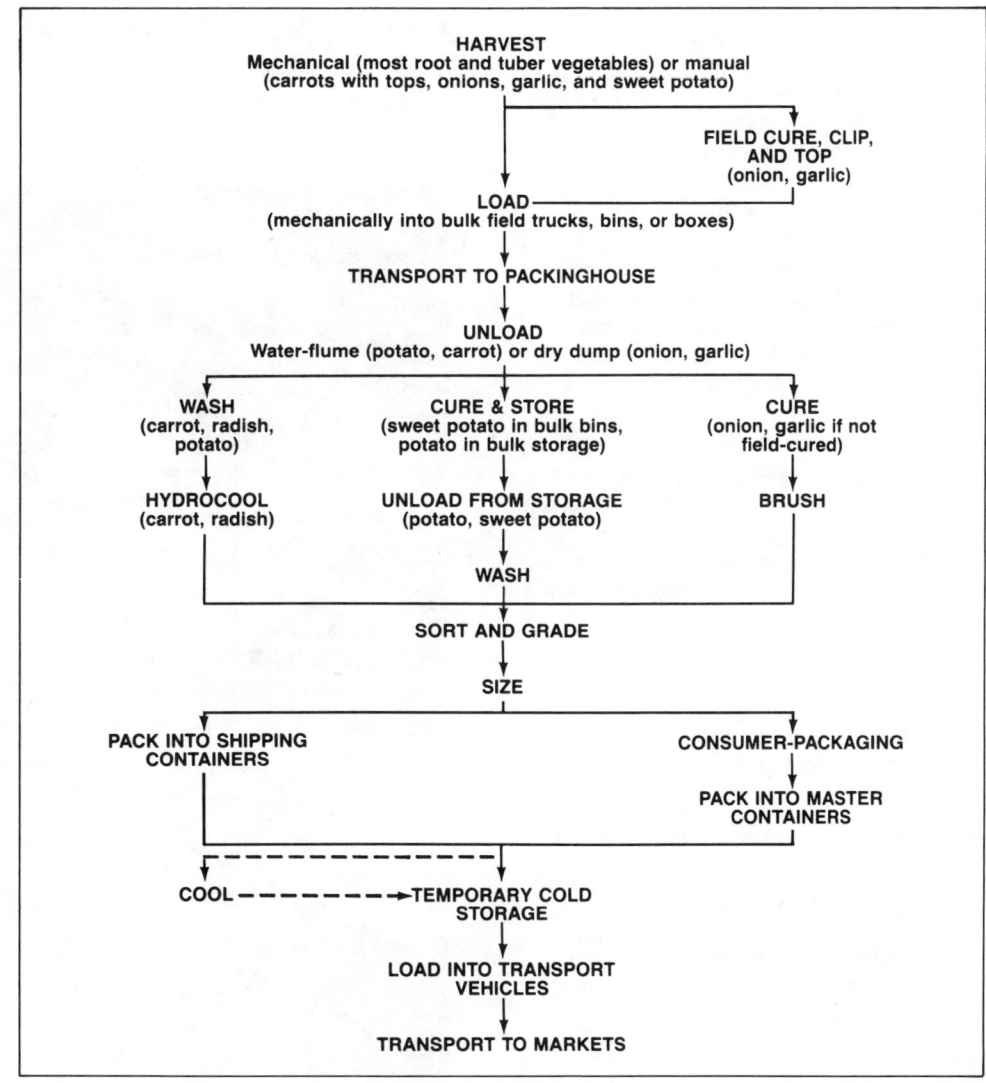

Fig. 22.5. Harvest and postharvest operations for root, tuber, and bulb vegetables.

- *Radish*—days from planting
- *Potato*—drying of foliage, setting of skins
- *Taro*—drying of foliage begins
- *Garlic and onion*—drying of tops, neck tissues begin to soften
- *Sweet potato*—senescence of vines

Both mechanical and hand harvesting are used on this group of vegetables. Most root vegetables are harvested mechanically and handled in bulk to packinghouses or processing plants. Garlic and onion for fresh market are hand harvested following lifting, cured in the field, and transported in bulk to packinghouses. Sweet potato harvesting is still mostly by hand; heavy vines and very tender, easily damaged roots limit mechanization. Vine-killing chemicals are used on potatoes before mechanical harvesting. Physical damage during harvesting operations can be excessive and is a major cause of postharvest deterioration.

Curing

One of the most simple and effective ways to reduce water loss and decay during postharvest storage of onion, garlic, potato, sweet potato, and other tropical root vegetables is curing. Injured or bruised surfaces are allowed to heal, and wound periderm thickens on potato, sweet potato, and similar commodities. Some water loss takes place during curing. The removal of decaying products prior to curing and storage insures a greater percentage of usable product after storage. The conditions required for curing vary according to commodity, as shown in table 22.2.

When onions and garlic are cured in the field, they are undercut, hand pulled, and allowed to dry for 5 to 10 days (depending on ambient temperatures) before topping. Curing may be done in windrows with tops covering bulbs to prevent sunburn of bulbs, or, with the roots and tops clipped, the bulbs may be placed in burlap bags for 3 to 14 days of field curing. Curing is sometimes done at or near the packinghouse in special facilities that use warm forced air.

Storage

Onions, garlic, Irish potatoes, and sweet potatoes are often stored after curing and before preparation for market (grading, sizing, and packing). These products may be stored for 3 to 10 months.

Preparation for market and marketing

The following steps bring bulky vegetative organs to market in salable condition:

- Cleaning: Dry brush or wash and partially dry root and tuber vegetables to remove excess moisture from their surfaces.
- Sorting: Eliminate defective products and plant debris (fig. 22.6).
- Postharvest fungicide treatment SOPP and Botran are used on sweet potatoes.
- Sizing: Mechanical sizers or by hand (subjective). Mechanical sizers are generally diverging rollers or weight sizers. Modified volumetric sizers are used for potatoes.
- Grading: Separate into quality grades (fig. 22.7).
- Packing: Pack into shipping or bulk containers. Some products (potato, garlic, root vegetables) are commonly

Table 22.2 Optimum conditions for curing root, tuber, and bulb vegetables

Commodity	Temperature Celsius (degrees)	Temperature Fahrenheit (degrees)	Relative humidity (%)	Duration (days)
Potato	15-20	59-68	85-90	5-10
Sweet potato	30-32	85-90	85-90	4-7
Yam	32-40	90-104	90-100	1-4
Cassava	30-40	86-104	90-95	2-5
Onion & garlic	35-45	95-113	60-75	0.5-1*

*With warm forced air.

Fig. 22.6. Workers grading potatoes in a packinghouse.

packed in consumer units that are then packed into master shipping containers.

- Loading into transit vehicles: Bulk transport to processing plants is sometimes used for onions, potatoes, and radishes. Most products, however, are loaded in packed shipping containers.
- Storage: Refrigerated or ventilated storage facilities keep commodities for extended periods to regulate product flow to markets.
- Destination handling: Consumer packaging at destination may be done by special handlers known as prepackers, by service wholesalers, or by chain food stores.
- Retail handling: This stage varies with the type of retail food store, but always includes temporary storage, preparation of products for display to consumers, and maintenance of displays.

Cooling

All temperate-zone root crops except potatoes, onions, and garlic can be hydrocooled, and a few shippers hydrocool potatoes during extremely hot weather. Tropical root crops, Irish potatoes, onions, and garlic are occasionally room-cooled before shipment to market. Cooling before shipment is growing in use for potatoes shipped in hot weather. Irish potatoes and onions destined for storage are cooled during the early part of the storage period with cool air forced through storage piles or bins. Cooling may be done with ambient air, using temperature and relative humidity controls, or with air cooled by mechanical refrigeration.

Special treatments

For storage, onions and Irish potatoes are generally sprayed with MH (maleic hydrozide) a few weeks before harvest to inhibit sprouting during storage. Aerosol applications of CIPC (3-chloro-isopropyl-N-phenyl carbamate) are often circulated around stored potatoes to further inhibit sprouting. Rodent control (mice and rats) is also necessary in potato and sweet potato storage facilities.

Nonrefrigerated storage methods

Some growers occasionally store mature Irish potatoes in the ground for several weeks prior to harvest. Pits, trenches, and clamps are used for storage of tropical roots and tubers. Pits are occasionally used for short-term, small-scale storage of Irish potatoes in some areas, but this is much less effective than a cold-storage facility for long-term storage.

Ventilated storage in cellars and aboveground warehouses is used for potatoes, sweet potatoes, garlic, and onions. Newer facilities have temperature and relative humidity controls, and provide forced-air circulation through bulk piles of Irish potatoes or onions, or through and around stacks of bulk bins containing onions, garlic, or sweet potatoes.

Recommended Storage Conditions

Temperate-zone root vegetables

In California, temperate-zone root vegetables are not usually stored. When they are stored, the following conditions should be maintained: 0°C (32°F) temperature, 95 percent to 98 percent relative humidity (RH), and adequate air circulation to remove vital heat from the product and prevent CO_2 accumulation.

Irish potatoes can be stored for 10 to 12 months under proper storage conditions. Most long-term potato storage facilities are in the northern states. In California, potatoes are stored in the Klamath Basin near the Oregon border. For fresh market, potatoes should be stored under the following conditions: 4° to 7°C (39.2° to 44.6°F) temperature, 95 percent to 98 percent RH, enough air circulation to prevent CO_2 accumulation (about 0.8 cubic feet [.02 cubic meter] per minute (cfm) per 100 pounds [45 kg] of potatoes), and exclusion of light to avoid greening. For processing (e.g., chipping), the proper conditions are: 8° to 12°C (46.4° to 53.6°F), 95 percent to 98 percent RH, adequate ventilation, and exclusion of

Fig. 22.7. A USDA inspector grades potatoes for quality and condition.

light. Potatoes for seed are best kept at 0° to 2°C (32° to 36°F), 95 percent to 98 percent RH, and adequate ventilation.

Garlic should be kept at 0°C (32°F) for long-term storage (6 to 7 months); 28° to 30°C (82.4° to 86°F) can be used for up to 1 month's storage. Seventy percent RH, and ventilation of about 1 cubic meter air per minute per cubic meter of garlic are effective.

Onions vary in their storage capability. The hot types (high solids content) store longer, and mild types (low solids content) are rarely stored for more than one month. Storage temperatures should be either 0° to 5°C (32° to 41°F) or 28° to 30°C (82.4° to 86°F); intermediate temperatures favor sprouting. Relative humidity should be maintained at 65 percent to 70 percent, and ventilation should be from 0.5 to 1.0 cubic meters of air per minute for each cubic meter of onions. Avoid light exposure to prevent greening. The storage potential of onions depends on the cultivar.

Tropical-zone root vegetables

Optimum storage conditions for tropical-zone root vegetables are summarized in table 22.3.

Table 22.3. Optimum storage conditions for tropical "root" vegetables

Commodity	Temperature Celsius (degrees)	Fahrenheit (degrees)	Relative humidity (%)	Storage life
Cassava	5- 8	41.0-46.4	80-90	2-4 weeks
Ginger	12-14	53.6-57.2	65-75	≤6 months
Sweet potato	12-14	53.6-57.2	85-90	≤6 months
Taro	13-15	55.4-59.0	85-90	≤4 months
Yam	13-15	55.4-59.0	near 100	≤6 months
		or		
	27-30	80.6-86.0	60-70	3-5 weeks

References

Leafy and Succulent Tissues

1. Isenberg, F. M. R., ed. 1977. Symposium on vegetable storage. *Acta Hortic.* 62. Int. Soc. Hortic. Sci. 361 pp.

2. Kader, A. A., J. M. Lyons, and L. L. Morris. 1974. Quality and postharvest responses of vegetables to preharvest field temperatures. *HortScience* 9:523-27.

3. Lipton, W. J., J. K. Stewart, and T. W. Whitaker. 1972. *An illustrated guide to the identification of some market disorders of head lettuce.* USDA Mark. Res. Rep. 950. 7 pp. and 19 plates.

4. Morris, L. L., A. A. Kader, and J. A. Klaustermeyer. 1974. Postharvest handling of lettuce. *ASHRAE Trans.* 80:341-49.

5. Morris, L. L., J. A. Klaustermeyer, and A. A. Kader. 1974. Postharvest requirements of lettuce to control physiological disorders. In *Proc. 26th ann. conf., handling perishable agricultural commodities,* 22-29. Mich. State Univ.

6. Ramsey, G. B., B. A. Friedman, and M. A. Smith. 1959. *Market diseases of beets, chicory, endive, escarole, globle artichokes, lettuce, rhubarb, spinach, and sweet potatoes.* USDA Agric. Handb. 155. 42 pp.

7. Ramsey, G. B., and M. A. Smith. 1961. *Market diseases of cabbage, cauliflower, turnips, cucumbers, melons, and related crops.* USDA Agric. Handb. 184. 49 pp.

8. Ryall, A. L., and W. J. Lipton. 1979. *Handling, transportation and storage of fruits and vegetables.* Vol. 1, *Vegetable and melons.* 2d ed., Chap. 6. Westport, CT: AVI Publ. Co.

Bulky Vegetative Organs

1. Booth, R. H. 1974. Postharvest deterioration of tropical root crops: Losses and their control. *Trop. Sci.* 16(2):49-63.

2. Booth, R. H., and R. L. Shaw. 1981. *Principles of potato storage.* Lima: International Potato Center (CIP). 105 pp.

3. Buffington, D. E., S. K. Sastry, J. C. Gustashaw, Jr., and D. S. Burgis. 1976. Artificial curing of onions - Progress report. Paper no. 76-3553, presented at the winter meeting of the American Society of Agricultural Engineers, Chicago, IL.

4. Dewey, D. D. 1980. *For commercial growers: Using temperature and humidity as guides to curing and storing onions.* Mich. State Univ. Ext. Bull. E-1409. File 26.17.

5. Edmond, J. B., and G. R. Ammerman. 1971. *Sweet potatoes: Production, processing, marketing.* Westport, CT: AVI Publ. Co.

6. Gunkel, W. W., J. W. Lorbeer, R. F. Bensin, and H. A. Smith, Jr. 1976. Application of the artificial heating method to control botrytis neck rot in pallet box stored onions. 33rd. Annual Progress Report to the New York Farm Electrification Council.

7. Kushman, L. J., and F. S. Wright. 1969. *Sweet-potato storage.* USDA Agric. Handb. 358. 35 pp.

8. Orr, P. H. 1971. *Handling potatoes from storage to packing line.* USDA Mark. Res. Rep. 890. 52 pp.

9. Paterson, W. D. 1979. *How onions are marketed.* U.S. Dept. Agric. Mark. Bull. 65. 22 pp.

10. Ryall, A. L., and W. J. Lipton. 1979. *Handling, transportation, and storage of fruits and vegetables.* Vol. 1, *Vegetables and melons.* 2d ed., Chap. 9. Westport, CT: AVI Publ. Co.

11. Smith, M. A., L. P. McColloch, and B. A. Friedman. 1966. *Market diseases of asparagus, onions, beans, peas, carrots, celery, and related vegetables.* USDA Agric. Handb. 303. 65 pp.

12. Smith, Ora. 1977. *Potatoes: Production, storing, processing.* 2d. ed. Westport, CT: AVI Publ. Co.

13. Smith, Jr., W. L., and J. B. Wilson 1978. *Market diseases of potatoes.* USDA Agric. Handb. 479. 99 pp.

14. Ting-gau, Huang, and W. W. Gunkel. 1974. Theoretical and experimental studies of the heating front in a deep bed hygroscopic product. *Am. Soc. Agric. Eng.* 17(2):346-54.

15. Williams, L. G., and D. L. Franklin. 1971. *Harvesting, handling, and storing yellow sweet Spanish onions.* Univ. Idaho Agric. Exp. Stn. Bull. 526. 31 pp.

23
Postharvest Handling Systems: Fruit Vegetables

ROBERT F. KASMIRE

Introduction

With the exceptions of peas and broad beans, all fruit vegetables are warm-season crops, and with the exception of sweet corn, are subject to chilling injury. Fruit vegetables are not generally adaptable to long-term storage. Exceptions are hard-rind (winter) squash, pumpkins, and dry legumes. A useful classification for postharvest treatment of fruit vegetables is based on the vegetable's stage of maturity at harvest:

- *Immature fruit vegetables*—legumes (lima bean, snap bean, pea, cowpea), cucurbits (cucumber, soft-rind squash), eggplant, pepper, okra, sweet corn

- *Mature fruit vegetables*—muskmelon, watermelon, pumpkin, hard-rind squashes, tomato

This chapter presents the general postharvest requirements of this group of commodities in relation to handling systems. A flow diagram of the tomato-handling system is shown in figure 23.1.

Field Operations

Harvesting

Most fruit vegetables that are harvested mature for fresh marketing are harvested by hand. Some harvesting aids may be used, including melon pickup machines, conveyors, and the like. Mature-green tomatoes are mostly hand harvested, but are machine harvested by a few shippers in California. Crenshaw melons are very tender, require special care, and are loaded only one melon deep in field vehicles for transport to packinghouses.

Fruit vegetables that are harvested immature, except for sweet corn, have very tender skins that are easily damaged in handling. Special care must be taken in all handling operations to prevent product damage and subsequent decay. Sweet corn and snap beans are harvested both mechanically and by hand.

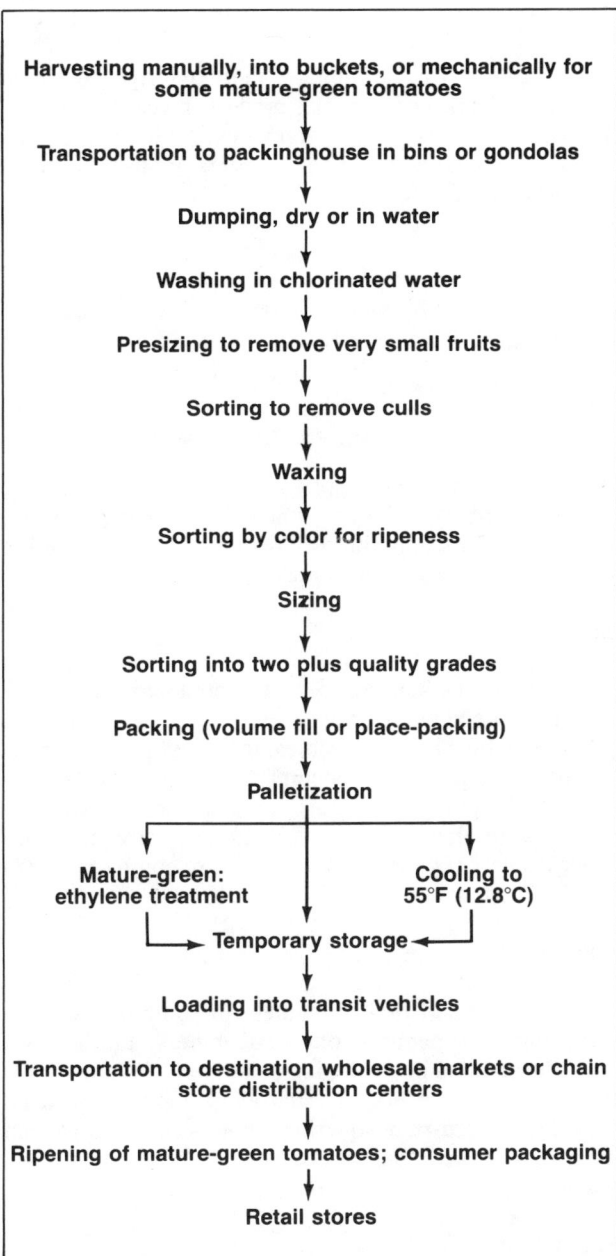

Fig. 23.1. Handling system for fresh market tomatoes.

Peas and cucumbers are harvested by hand and mechanically. Lima beans are mechanically harvested. Chinese peas (snow peas) are hand harvested.

Harvesting at night, when products are coolest, is common for sweet corn and is gaining in use for cantaloupe. This practice reduces cooling needs and costs, results in better and more uniform cooling, and helps maintain product quality.

Field Packing

There is an increasing trend toward commercial field grading, sorting, sizing, and packing of cantaloupes and summer squashes in California because this system reduces handling costs and damage, and offers a potential for higher marketable yields. This system is also used to a lesser extent for field-packing watermelons in cartons.

Hauling to Packinghouse

Packinghouses for crenshaw melons are generally very close to fields of production, because this melon is so tender that it is easily damaged in all handling operations. Mobile packing facilities are commonly towed through the fields for sweet corn, peppers, and eggplant in Florida, and are used by some shippers in California for cantaloupes, crenshaw melons, honeydew melons, and summer squashes. Less product damage should occur when these units are used than when products are hauled to separate packinghouses for subsequent handling operations. Immature fruit vegetables are commonly grown in fields close or adjacent to their packinghouses. This is especially true of products grown and packed by small-scale farmers (e.g., okra, cucumber, chili pepper, soft-rind squash, eggplant). These products are almost always harvested into the field boxes in which they are hauled to packinghouses.

Mature fruit vegetables are generally hauled to packinghouses, storage, or loading facilities in bulk bins (hard-rind squashes, peppers, pink tomatoes), gondolas (mature-green tomatoes), or bulk field trailers or trucks (muskmelons, watermelons, hard-rind squashes, pumpkins).

Packinghouse Operations

Receiving

Loaded field vehicles should be parked in shaded areas to prevent product warming and sunburning prior to unloading. Products may be hand unloaded (soft-rind squashes, eggplant, some muskmelons, cucumbers, watermelons), dry-dumped onto sloping, padded ramps (cantaloupes, honeydew melons, pink tomatoes, sweet peppers) or moving conveyor belts, or wet-dumped into tanks of moving water to reduce bruising (honeydew melons, mature-green tomatoes, sweet peppers, and sweet corn). Considerable mechanical damage occurs in dry-dumping operations. Bruising, scratching, abrading, and splitting are common examples. The water temperature in wet-dump tanks should be slightly warmer than the product temperature to prevent uptake of water and decay-causing organisms into the products.

Presizing

For many commodities, fruits below a certain size are eliminated manually or mechanically by a presizing belt or chain. The undersize fruits are diverted to a cull conveyor or used for processing.

Sorting

The sorting process eliminates cull, overripe, and misshapen products, and separates products by color, maturity, and ripeness classes (e.g., tomatoes, cantaloupes, other muskmelons). Electronic color sorters are used by some tomato shippers.

Grading

Fruits are sorted by quality into two or more grades according to U.S. standards for grades, California grade standards, or a shipper's own grade standards.

Waxing

Food-grade waxes are applied to cucumbers, eggplant, sweet peppers, cantaloupes, and tomatoes to inhibit water loss and enhance appearance. Waxing may be done before or after sizing. Fungicides may be added to waxes.

Sizing

Fruit-type vegetables may be segregated into several size categories depending on the commodity. Sizing is subjective for some fruit vegetables, including legumes, soft- and hard-rind squashes, cucumbers, eggplant, chili peppers, okra, pumpkins, muskmelons, and watermelons (although the latter may be mechanically sized). Cantaloupes, sweet peppers, and tomatoes are commonly sized by diameter, volume, or weight.

Packing

Mature-green and pink tomatoes (in some operations), sweet and chili peppers, okra, cucumbers, and legumes are volume-filled into shipping containers. All other fruit-type vegetables are place-packed into shipping containers, bulk bins (hard-rind squashes, pumpkins, casaba and honeydew melons, and watermelons) or bulk trucks (e.g., watermelons).

Palletizing

Packed shipping containers of most fruit vegetables in large-volume operations are palletized (stacked on pallets) for shipment. This is a common practice with cantaloupes, muskmelons, sweet peppers, and tomatoes. Except sweet corn, immature fruits are commonly handled in low-volume operations where palletizing is not commonly used, and most of these commodities are shipped in mixed loads with many other commodities. Palletizing for these commodities is done after hydrocooling or package-ice cooling, but before forced-air cooling.

Cooling

Various methods are used for cooling fruit vegetables. The most common methods are:

- *Forced-air cooling*—For cantaloupes, muskmelons, peas, peppers, soft-rind squashes, and tomatoes. Forced-air evaporative cooling is used to a limited extent for soft-rind squashes, peppers, eggplant, and cherry tomatoes.

- *Hydrocooling*—Before grading, sizing, and packing, for cantaloupes and sweet corn. Sorting of defective products is done both before and after cooling. Hydrocooling cycles are rarely long enough during hot weather. To avoid labor problems, shippers are more concerned with keeping packers adequately supplied with melons than with cooling their melons thoroughly. Labor problems can occur when packers run out of melons, because the workers are paid on the basis of the number of boxes packed.

- *Package icing and liquid icing*—Used to a limited extent for cooling cantaloupes and as supplementary cooling for packed sweet corn. Inadequate icing is a common problem.

Temporary cold storage

In large-volume operations, most fruit vegetables are placed in cold storage rooms after cooling and before shipment. Cold rooms are seldom used in small farm operations. Some small farmers, however, use cooperatively owned coolers and cold storage facilities.

Loading for transport

Some tomatoes, cantaloupes, and muskmelons are shipped in refrigerated railcars, but most fruit vegetables are shipped in refrigerated trucks—over-the-road, piggy back, or marine container vans. Possibly 80 percent to 85 percent are shipped in over-the-road trucks. Except for major volume commodities such as cantaloupes and tomatoes, most are shipped in mixed loads, sometimes with ethylene-sensitive commodities. Transit times to domestic markets vary from a few hours up to 7 days. To avoid yellowing of cucumbers and calyx abscission on eggplants, neither should be shipped with ethylene-producing commodities. Exposure of watermelons to ethylene can result in tissue maceration and rind separation.

Destination Handling

Handling practices vary with types of receiving operations. Products go to chain store distribution centers, service wholesalers, and various other terminal market receivers. Legumes and ripened fruit-type vegetables are prepacked in consumer units by some terminal-market operators. Careful handling, proper temperature management, and expedited marketing are essential for success of a destination-market handling operation.

Special Treatments

Ripening

For faster and more uniform ripening, ethylene treatments may be used at shipping points or at destination for mature-green tomatoes and honeydew, casaba, and crenshaw melons. Satisfactory ripening occurs within the temperature range of 12.8° to 25°C (55° to 77°F); the higher the temperature, the faster the vegetable will ripen.

Modified atmospheres

Modified atmospheres (MA) are rarely used with these commodities, but there is some use of MA (3% to 5% O_2 + 5% to 10% CO) in mature-green tomatoes for export.

Recommended Storage Conditions

Fruits harvested mature (consumed fully ripe)

Most mature-harvested fruit vegetables are sensitive to chilling injury at temperatures below about 12.5°C (54.5°F). Chilling injury is cumulative and its severity depends upon the product temperature and the duration of exposure to chilling temperatures. Figure 23.2 illustrates the effects of temperature on the quality and ripening of tomato fruits.

Optimum temperatures for short-term storage and transport are:

- Mature-green tomatoes, pumpkins, hard-rind squash: 12.8° to 15.6°C (55° to 60°F)

- Partially ripe tomatoes, muskmelons (except cantaloupes): 10° to 12°C (50° to 53.6°F). Honeydew melons whose ripening has been initiated with ethylene are best shipped at 5° to 7.2°C (41° to 45°F)

- Fully ripe tomatoes and watermelons: 7° to 10°C (44.6° to 50°F)

- Ripe cantaloupes: 4° to 6°C (39.2° to 42.8°F)

The optimum relative humidity range is 90 percent to 95 percent, except for pumpkins and hard-rind squashes (60% to 70%).

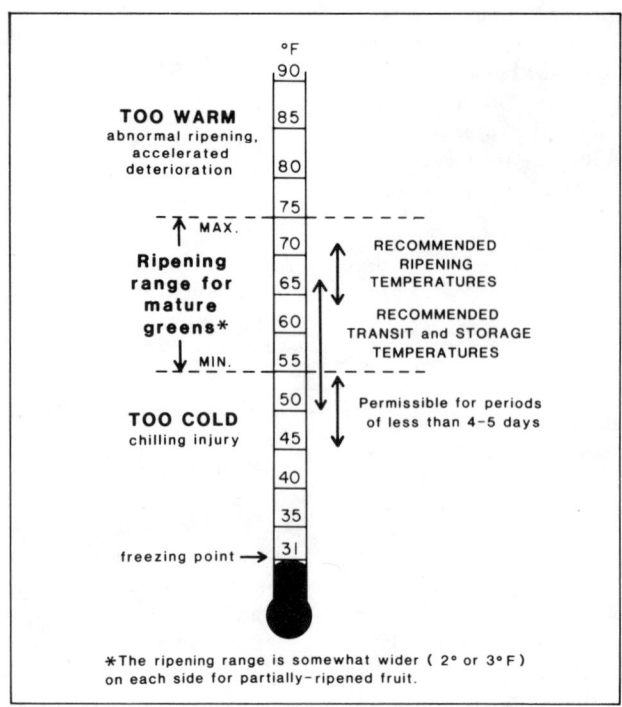

Fig. 23.2. Effect of temperature on quality and ripening of tomato fruits.

Fruits harvested immature

All immature-harvested fruit vegetables are sensitive to chilling injury, except broad beans, peas, and sweet corn, which are best kept at 0°C and 95 percent relative humidity. Optimum temperatures for short-term storage and transport are:

- Eggplant, cucumber, soft-rind squash, okra: 10° to 13°C (50° to 55.4°F)
- Peppers: 5° to 7°C (41° to 45°F)
- Lima beans, snap beans, cowpeas: 5° to 8°C (41° to 46.4°F)

Relative humidity should be maintained at 90 percent to 95 percent.

References

1. Davies, J. N., and G. E. Hobson. 1981. The constituents of tomato fruit—The influence of environment, nutrition and genotype. *CRC Crit. Rev. Food Sci. Nutr.* 15(3):205-80.

2. Fahy, J. V. 1976. *How fresh tomatoes are marketed*. USDA Mark. Bull. 59. 31 pp.

3. Gould, W. A. 1974. *Tomato production, processing and quality evaluation*. Westport, CT: AVI Publ. Co. 445 pp.

4. Hobson, G. E., and J. N. Davies. 1971. The tomato. In *The biochemistry of fruits and their products*, ed. A. C. Hulme, Vol. 2, 437-82. New York: Academic Press.

5. Isenberg, F. M. R. 1979. Controlled atmosphere storage of vegetables. *Hortic. Rev.* 1:337-94.

6. Kader, A. A., L. L. Morris, M. A. Stevens, and M. Albright-Holton. 1978. Composition and flavor quality of fresh market tomatoes as influenced by some postharvest handling procedures. *J. Am. Soc. Hortic. Sci.* 103:6-13.

7. Kasmire, R. F. 1973. Precooling, refrigeration and postharvest handling of tomatoes and cantaloupes. *ASHRAE Symp.* LO-37-7. pp. 19-20.

8. Kasmire, R. F. 1977. *Precooling tomatoes*. Calif. Fresh Mark Tomato Adv. Board Inf. Bull. 17. 6 pp.

9. Kasmire, R. F., and A. A. Kader. 1978. Handling tomatoes at wholesale and retail: A guide for better quality and greater profits. *UFFVA Outlook* 5(3):5-12.

10. Kasmire, R. F., et al. 1981. *Muskmelon production in California*. Univ. Calif. Div. Agric. Sci. Leaf. 2671. pp. 16-23.

11. McColloch, L. P., H. T. Cook, and W. R. Wright. 1968. *Market diseases of tomatoes, peppers, and eggplants*. USDA Agric. Handb. 28. 74 pp.

12. Morris, L. L., and A. A. Kader. 1978. *Postharvest physiology of tomatoes as related to transit and marketing problems*. Calif. Fresh Mark. Tomato Adv. Board Inf. Bull. 19. 4 pp.

13. Pratt, H. K. 1971. Melons. In *The biochemistry of fruits and their products,* ed. A. C. Hulme, Vol. 2, 207-32. New York: Academic Press.

14. Ryall, A. L., and W. J. Lipton. 1979. *Handling, transportation and storage of fruits and vegetables*. Vol. 1, *Vegetables and melons*. 2d ed. Westport, CT: AVI Publ. Co.

15. Sherman, M., R. K. Showalter, J. R. Bartz, and G. W. Simone. 1981. *Tomato packinghouse dump tank sanitation*. Univ. Florida Vegetable Crops Fact Sheet, VC-31, 4 pp.

24
Postharvest Handling Systems: Temperate Fruits

F. GORDON MITCHELL

Introduction

A wide range of handling systems is used for temperate fruit crops depending on whether they are marketed as fresh or processed, relative susceptibility to various injuries, cooling and temperature management requirements, and local customs. The handling systems described here for pome fruits (primarily Bartlett pears) and for strawberries illustrate very different systems in use in California. Even for strawberries that are harvested, graded, and packed right in the field by the picker, the total handling system has many steps.

Pome Fruits

Pear- and apple-handling practices are highly mechanized. Considerable investment in cooling and storage facilities now allows most varieties to be stored for relatively long periods. This ability to store has also led to extensive export marketing of these commodities.

Postharvest Diseases and Physiological Disorders

Pome fruits are subject to storage loss from Botrytis rot, which can develop in wounds or at either the stem or calyx end of the fruit. Calyx-end Botrytis rot has been a problem in some California Bartlett pears in recent years, possibly because of the increasing use of water dumps. Frequent cleaning and chlorine treatment of water dumps can reduce, but will not eliminate, the problem. In some instances, rapid forced-air cooling after packing seems to have eliminated or greatly reduced the problem.

Other common rot problems, especially after long-term storage of apples, include Alternaria rot in wounds and sunburned areas of the fruit, and Penicillium rot in wounds. Fungicidal treatments often precede long-term storage.

Storage scald

Physiological disorders of pome fruits can also cause serious losses, especially after long-term storage. Storage scald, which can occur on apples and pears, is most severe on low-maturity fruit. This disorder appears as brown discolored patches on the fruit surface, and may not develop until the fruit has been warmed following storage. Harvesting at optimum maturity helps reduce the problem, as does a consistently low temperature during storage. Often, diphenylamine (DPA) or ethoxyquin dips or drenches are used before storage to reduce the incidence and severity of scald.

Water core

Water core can be a problem of apples. This disorder, which involves flooding (water soaking) of the intercellular spaces, is often associated with low calcium in the fruit, and is most severe in fruit from young, vigorous, lightly cropped trees. Fruit of advanced maturity is generally more severely affected. There may be some disappearance of water core symptoms in fruits during storage, although such fruits should not be stored for long periods.

Bitter pit

Bitter pit involves the development of dry, pithy spots or "pits" near or below the fruit surface. This is another important cause of apple loss. Tree conditions associated with bitter pit are similar to those associated with water core. The problem may be visible at harvest, but often develops during storage. Tests indicate that fast cooling and high relative humidity can reduce symptom development during storage. In some growing areas calcium sprays are being used before harvest and calcium dips after harvest to reduce the problem.

Senescent breakdown

Senescent breakdown can occur, especially with pears, as a result of late harvest, slow cooling, high storage temperatures and extended storage periods. Symptoms usually start with a soft brown breakdown of the core area, later progressing through the flesh.

Watery breakdown

Watery breakdown is associated especially with Bartlett pear, and involves soft, watery breakdown in portions of the fruit, usually without brown discoloration. This enzymatic softening can affect any part of the fruit, and is probably the result of severe physiological stress to the fruit. Fast cooling and low storage temperatures have been effective at minimizing the problem.

Temperature Management

Pome fruits respond best to fast cooling and storage at as low a temperature as is possible without danger of freezing. For Bartlett pears we recommend a temperature of 0° to 0.5°C (32° to 33°F) for up to one month of storage, −1°C (30°F) for storage beyond one month. The lowest safe storage temperature is related to the extent of temperature fluctuation in the room and the soluble solids content (SSC) of the fruit. Freezing injury in pears and apples usually starts in the core area, because this is the area of lowest SSC. Therefore, the SSC of the core area must be known in order to predict the lowest safe temperature.

Cooling should be as quick as possible. Cooling delays are associated with increased physiological and pathological disorders and flesh softening. For Bartlett pears in California, we now recommend cooling to near storage temperature within 24 hours of harvest. The importance of cooling speed varies greatly with variety. Evidence is increasing for the benefits of faster cooling of apples, especially for long-term storage.

Many apples and some pears are stored under controlled atmospheres (CA), at least for storage beyond 3 months. Most CA storage is done with fruit in field bins, often with unsorted, field-run fruit, but some apples are dumped, sorted, sized, and refilled into their bins before being moved to CA storage.

Relative Humidity

A relative humidity of about 95 percent, considered ideal for pears and apples, is difficult to maintain with large cooling coils alone. Supplemental humidification, especially with fog spray nozzles, is now widely used. Unfortunately the added water increases the problem of coil frosting. Perforated polyethylene bags or liners inside the shipping container, once commonly used, are now seldom seen. Wax or plastic coatings on corrugated containers are used extensively in California to provide a moisture barrier for Bartlett pears.

Handling

Fruits are hand picked into bags, gently transferred into field bins, and transported to packing facilities. Apples are often stored in field bins for future packing, where they may be drenched with a scald inhibitor and fungicide, and sometimes treated with calcium chloride for bitter pit control. Because costs would probably exceed benefits, these treatments are rare in California for short-term storage.

Packing fruit is dumped (water submersion dumps are common), washed, pre-sized to eliminate undersize fruit, and sorted. Often three grades of fruit are distinguished—fresh market, processing, and cull or by-product fruit. Over 50 percent of California Bartlett pears are diverted to processing in a typical harvest.

Apples in the Northwest are mostly tray packed, with fruit in the top trays often individually wrapped. Smaller apples may be bagged. In California many apples are volume-filled into corrugated containers for fresh marketing.

Pears may be wrap packed or tight-fill packed. Most Northwest pears are wrap packed into corrugated containers, although some wood containers are still in use. Most California Bartlett pears are tight-fill packed into corrugated containers.

Packed containers are segregated by fruit size and stacked on unitized pallets. These pallets are designed to move directly to the receiving markets. Final cooling of these unitized pallet loads of fruit follows the completion of packing. Forced-air cooling is now in common use in California.

A flow diagram (fig. 24.1) shows the steps involved in fresh-market handling and processing of California Bartlett pears.

Strawberries

Strawberries are among the most perishable of fresh fruits, yet they are being marketed successfully worldwide in increasing volume. Much of the marketing is at a great distance from the point of production, so effective handling procedures are necessary to prevent excessive deterioration. In California, a large fresh-strawberry shipping industry has been operating since about 1950, based largely on fruit delivery to markets from 3,000 to 5,000 km away.

California strawberries are produced primarily in the coastal valleys and plains of central and southern California. Production starts in late winter in the south, and continues through late fall in the north. The fresh market is the fruit's primary outlet, followed by the freezing and jam manufacturing market. The temperature in production areas is generally cool, although warm summer days with highs of 30° to 35°C (86° to 95°F) sometimes occur.

The primary market for these fresh strawberries extends across the U.S. and into southern Canada. Some fruit is exported by air overseas. At one time, about one-third of the volume was shipped by air, but this has since declined to 10 percent to 12 percent, the balance moving by surface transport (highway trucks). Surface-transported strawberries must have a 5- to 7-day market life when shipped to eastern cities.

Problems in Postharvest Handling

Strawberries have very tender skins and are easily injured. They are subject to invasion by fruit-rotting organisms and have one of the highest respiratory activity rates of all fresh fruits. Each of these factors means the fruit has a severe loss potential.

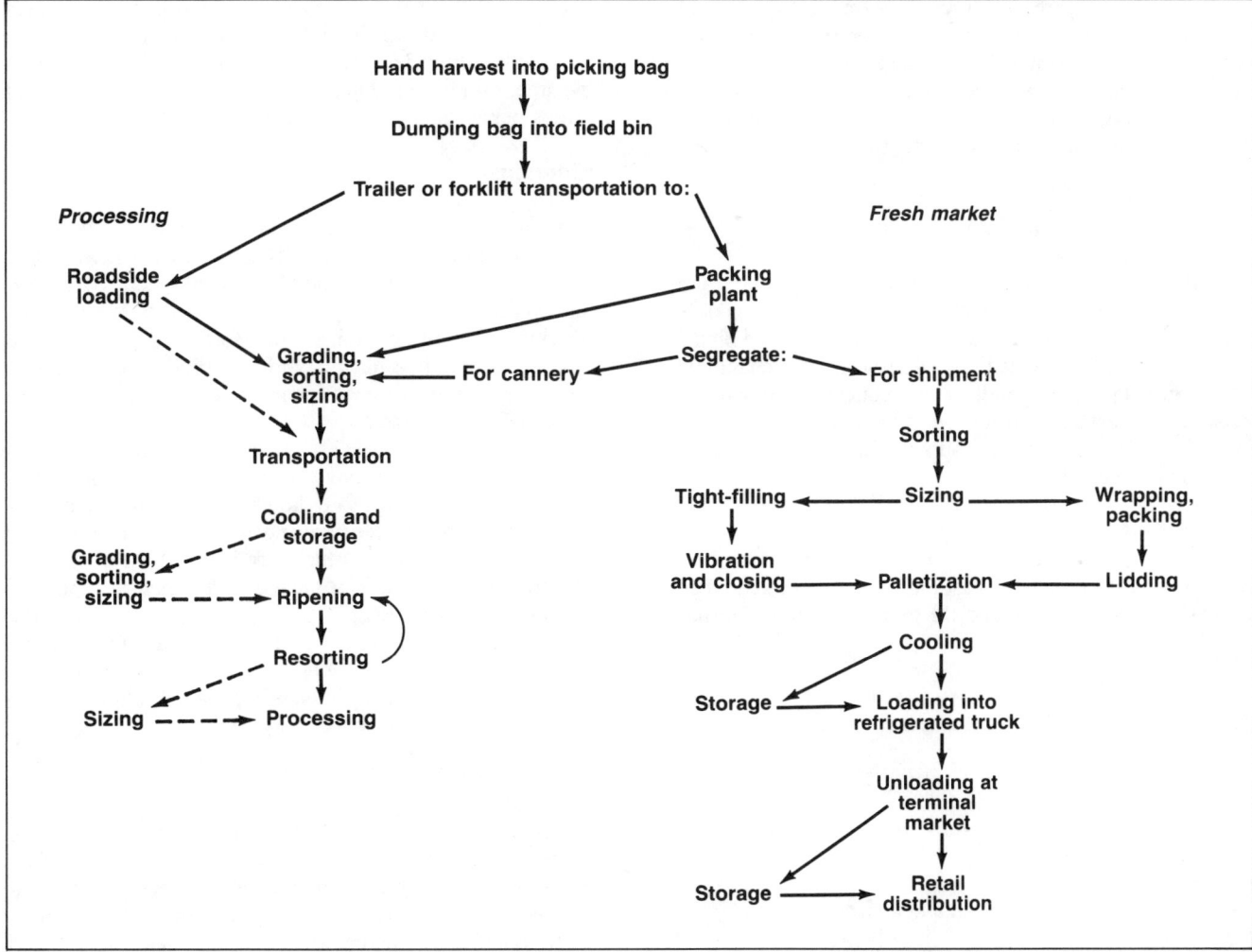

Fig. 24.1. Postharvest handling system for Bartlett pears.

Fruit rots

The greatest single cause of loss to fresh strawberries is "gray mold" or "Botrytis rot," caused by the fungus *Botrytis cinerea*. This organism can invade the berry blossom and remain latent until ripening commences, or it can enter any wound that occurs during ripening or handling. Surface mycelium on infected berries penetrates nearby berries to produce an ever-enlarging "nest" of rotting berries. Harvest wounds are a major source of invasion. The organism continues growth at 0°C but at a very slow rate.

Another fruit-rotting organism, *Rhizopus* sp., can cause severe losses at warm temperatures, but the advent of good temperature management has made it a relatively minor problem.

Fruit shrivel

Strawberries are subject to rapid water loss, which can cause the fruit to shrivel and appear old and deteriorated, and cause the calyx (the green cap or hull) to wilt and dry. These symptoms will affect the fruit's sales appeal before they affect actual eating quality.

Overripeness

Because of their high rate of physiological activity, strawberries quickly pass from ripeness to an overripe or senescent state if held at warm temperatures.

Bruising

Strawberries are subject to serious injury during harvest and postharvest handling. In studies of strawberry losses, differences in the magnitude of mechanical injuries caused by human pickers were so great as to mask any other causes of deterioration.

Curiously, vibration (or roller) bruising has not normally been considered serious for strawberries. Because of its elasticity, the fruit seems to absorb motion rather than transmit it, and thus limit its exposure to damage when in transport.

Often, berries are cut by the tops or sharp ridges of open-mesh plastic baskets that are commonly used for packaging in California. While these baskets allow ventilation for easy cooling and good fruit visibility for the customer, injuries associated with their use cause both a direct loss and subsequent Botrytis rot losses.

Handling Methods

In California, strawberries are harvested, graded, and packed in the field by the picker, who is also responsible for removing any rotted or overripe fruits from the plants (fig. 24.2). This combined operation precludes rehandling of the fruit at packing facilities and reduces the time delay between harvest and cooling. Major attention must, however, be directed to picker supervision if good quality control is to be achieved.

Maturity

Strawberries are usually at least three-fourths colored when harvested for the fresh market. Riper fruit can sometimes be harvested for nearby markets where handling delays can be avoided. Fruit that becomes too ripe for fresh market handling is diverted to the freezer. Depending on weather conditions, harvesting may be necessary several times per week for uniform fruit maturity.

Containers

An open corrugated crate (outer container) that holds 12 1-pint (about 1/2-liter) baskets is used for strawberries. When filled, crates are bundled with special wire ribs into pairs. The ribs then serve as carrying handles and stacking tabs for palletization. Most of the baskets are open-mesh, ribbed plastic. Side ventilation of the crates facilitates air movement for initial cooling and for temperature maintenance.

Because ribbed baskets cause fruit-cutting injury, alternate basket designs have been studied. Solid-sided baskets have been shown to reduce injury, but the industry's concerns with reduced cooling speed, the potential for condensation damage during warming, and the customer's desire to inspect the fruit have delayed their acceptance. A transparent molded plastic basket appears to solve the inspection problem, and recent tests indicate a reasonable cooling rate with no problem resulting from water condensation (fig. 24.3). Thus, a reasonable alternative to the ribbed baskets now seems possible.

Palletization

Most strawberries are shipped on unitized pallets. The "one-way disposable" wooden pallets are placed on flat-bed trucks in the field and stacked with berry crates immediately after harvest, and they remain intact until final distribution to the market area. The tight stacking required for stability and transport-load density must be compensated for in cooling and handling operations.

Cooling delays

Delays of over one hour between harvest and cooling can cause accelerated fruit deterioration. Growers are urged to schedule frequent deliveries of small or partial loads of fruit to the cooler in order to minimize delays.

Cooling method

Forced-air cooling is virtually universal for California strawberries. This is the only method that meets the need for rapid cooling and avoids fruit wetting, which strawberries will not tolerate. Most strawberry coolers are capable of seven-eighths cooling in 2 to 4 hours. Rapid cooling slows the rate of fruit deterioration, and prepares the berries for immediate shipment.

Modified atmospheres

Many loads are shipped under "modified atmosphere transit." The most common procedure involves covering the entire loaded pallet with a plastic bag that is carefully sealed to a plastic pallet cover beneath the berry crates.

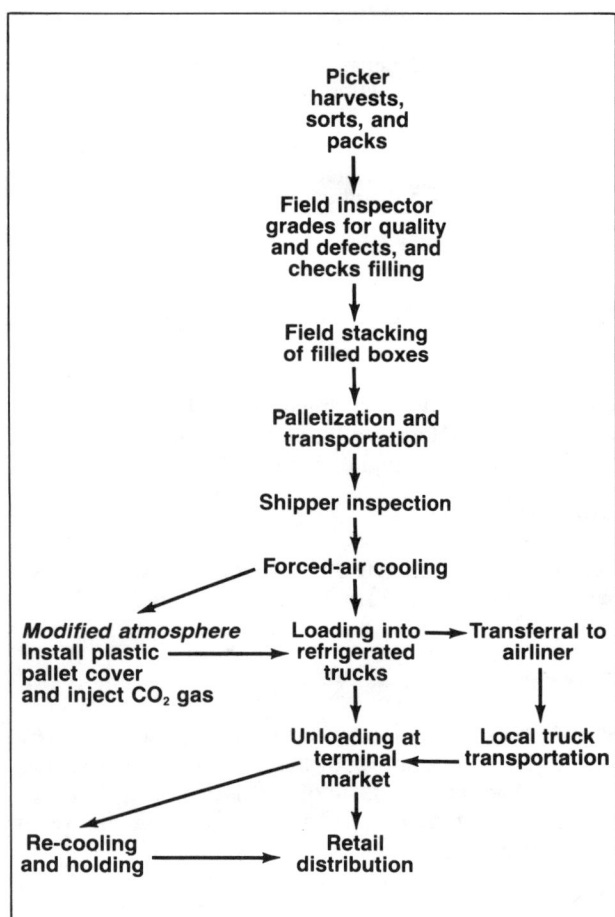

Fig. 24.2. Postharvest handling system for strawberries.

After sealing, the bag is injected with carbon dioxide gas to create an approximate 15 percent CO_2 atmosphere. With satisfactory sealing, this atmosphere can be maintained throughout transit. Pallet bags are installed after cooling, usually just before loading for transport.

Some shippers use the carbon dioxide treatment extensively, others sparingly. One shipper, evaluating a series of arrivals, found significant benefit (primarily fruit-rot reduction) only when the treatment was used after a foggy or rainy period. That shipper has subsequently used the treatment only for shipments judged to have a rot potential. This procedure is apparently successful.

In a recent laboratory study of Botrytis rot on strawberries, elevated-CO_2 atmospheres yielded no measurable benefit at temperatures below 5°C (41°F). Thus, when good transit temperatures (below 2°C [36°F]) can be maintained, the addition of carbon dioxide appears to be unnecessary. Realistically, however, it is common for strawberries to encounter an average 5°C (41°F) temperature during transport.

Loading

All surface transport is now by refrigerated truck vans, and virtually all shipments are of unitized pallet loads. Because of heat leakage through the walls of refrigerated truck vans, the pallets should be "center loaded"; that is, an air space should be left along the two sides of the van, between the walls and the pallets rather than down the center. Special blocks have been designed to prevent loads from shifting during transit. Where side-loading patterns have been used (pallets placed against each sidewall), berry temperatures were sometimes 5° to 10°C (9° to 18°F) warmer near the wall than in the midst of the load. Without pallets under the fruit to facilitate air movement, excessive warming can occur in moving truck vans.

Strawberries are often shipped in mixed loads with other fruits and vegetables. It is vital that other products in the load have storage requirements similar to those of strawberries. Other products' temperatures should always be monitored at time of loading to insure that the strawberries will not be warmed by an improperly cooled product.

Many strawberry loading docks are refrigerated and designed so the refrigerated truck van can be entered directly from the refrigerated dock. The truck van should be thoroughly refrigerated before loading begins. These procedures assure that fruit temperatures will be protected during the loading operation.

Surface transport

Most refrigerated truck vans that transport strawberries are equipped with air-suspension systems that reduce transit vibrations more than 50 percent from spring suspension systems. This greatly reduces the potential for injury to the fruit.

Many truck refrigeration systems are incapable of maintaining fruit temperatures below about 5°C (41°F), too warm for good strawberry protection. Often, temperature-regulating equipment lacks the accuracy to achieve a lower temperature without danger of fruit freezing. While transport conditions need improvement, berries should still be 0°C (32°F) when loaded to achieve the lowest possible average fruit-transit temperature.

Air circulation capacity is also limited in many transport vehicles. If the fruit is not thoroughly cooled before loading and protected from warming during the loading operation, proper cooling may never be achieved.

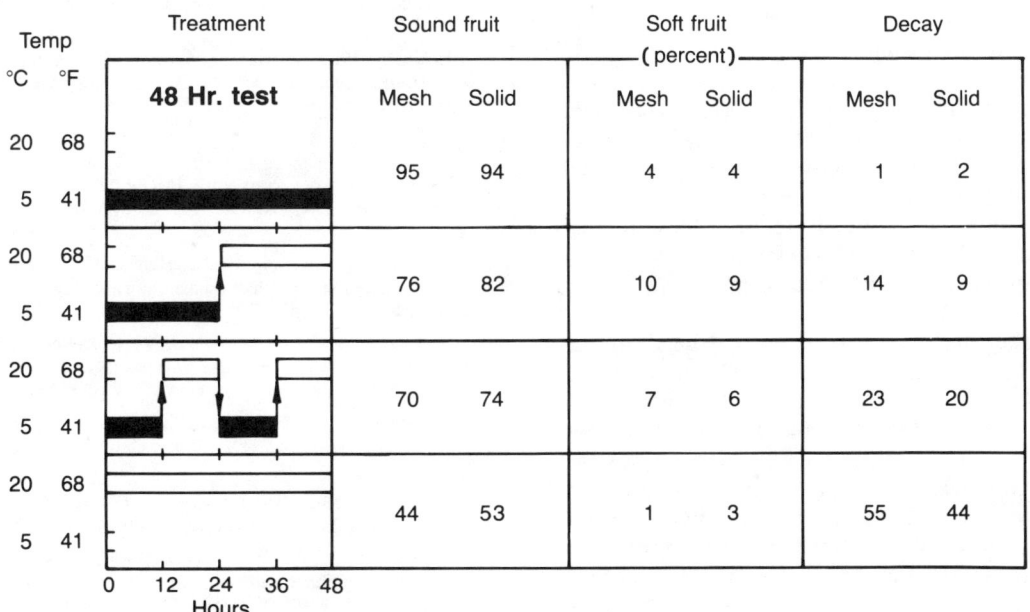

Fig. 24.3. **Effect of temperature-management procedures and basket design on strawberry deterioration.**

Transport time by highway truck can range up to 4 days for the most distant markets. An additional 1 to 3 days are needed for marketing. Generally, carriers servicing those more distant markets are more particular in their transport equipment requirements because they recognize the danger of heavy losses when handling this commodity.

Air transport

The 10 percent to 12 percent of fresh strawberry volume that moves by air is mostly destined for export or distant domestic markets (East Coast cities). The decline in volume of air shipments began before the airfreight rate increases of the 1970s and can be attributed more to the good quality of truck transport.

Strawberries for air transport must be thoroughly cooled before loading. They are often protected with modified atmospheres, using pallet bags and carbon dioxide injection. The berries are normally transported from the production areas to air freight terminals in small refrigerated trucks. After unloading at the terminal they are sometimes placed in cold storage when long delays (several hours) are expected. Normally, however, they remain out of refrigeration until they reach the terminal market receiver.

Studies of the air shipment of strawberries indicate that unrefrigerated transit time is considerably longer than is normally anticipated. The berries are often in flight only about one-third of the total transit time. While in flight, the temperature of the cargo hold is warmed to protect other freight. On the ground at intermediate terminals the temperature is often quite high, as much of the transporting of berries occurs during summer. The fruit is normally warm upon arrival, much warmer than at harvest, and considerable deterioration can occur. Tests have shown that, even though the berries will warm during transport, it is better to provide cooling whenever possible than not to cool at all.

Terminal market handling

Market handlers of strawberries have historically sustained heavy fruit losses (up to 25% in some cases) because of fruit deterioration. Recent improvements in postharvest cooling, combined with the overall improvement of surface transport conditions, have greatly reduced these losses. This has changed the attitude of many receivers toward strawberry handling, and prompted them to seek further improvements.

Some market handlers provide special handling for strawberries in order to minimize losses. This may include establishment of "strawberry rooms" at the distribution warehouse, where refrigeration and air-flow capacities are sufficient to allow reasonably fast recooling of the berries on arrival. Warm berries arriving via air transport benefit greatly from such treatment. Many Berries arriving by highway truck also benefit from this sort of cooling because transport temperatures are often higher than desired.

Terminal market handlers commonly load trucks for store distribution several hours before the start of delivery. Strawberries are often listed among the few items that receive special handling, and are held under refrigeration until just before truck departure. This facilitates continued temperature protection for the strawberries, improves the receiver's chances of making a profit by marketing the berries, and increases the likelihood of consumer satisfaction.

References

Pome Fruits

1. Blanpied, G. D., E. D. Markwordt, and C. D. Londington. 1962. *Harvesting, handling and packing apples*. Cornell Univ. Ext. Bull. 750. 32 pp.

2. Blanpied, G. D., and R. M. Smock. 1982. *Storage of fresh market apples*. Cornell Univ. Inf. Bull. 191. 19 pp.

3. Hulme, A. C., and M. J. C. Rhodes. 1971. Pome fruits. In *The biochemistry of fruits and their products*, ed. A. C. Hulme, Vol. 2, 333-73. New York: Academic Press.

4. Patchen, G. O. 1971. *Storage of apples and pears*. USDA Mark. Res. Rep. 924. 51 pp.

5. Pierson, C. F., M. J. Ceponis, and L. P. McColloch. 1971. *Market diseases of apples, pears, and quinces*. USDA Agric. Handb. 376. 112 pp.

6. Podorny, J. C., R. W. Bohall, and J. Pearrow. 1973. *Harvesting, storing and packing apples for the fresh market: Regional practices and costs*. USDA Mark. Res. Rep. 1009. 49 pp.

7. Porritt, S. W., M. Meheriuk, and P. D. Lidster. 1982. *Postharvest disorders of apples and pears*. Agricultural Canadian Publ. 1737/E. 66 pp.

8. Ryall, A. L., and W. T. Pentzer. 1982. *Handling, transportation, and storage of fruits and vegetables*. Vol. 2, *Fruits and tree nuts*. 2d. ed., Chaps. 7, 13 & 16. Westport, CT: AVI Publ. Co.

9. Smock, R. M. 1979. Controlled atmosphere storage of fruits. *Hortic. Rev.* 1:301-36.

10. USDA. 1965. *A review of literature on harvesting, handling, storage and transportation of apples*. USDA, ARS, Mark. Qual. Res. Div. ARS 51-4. 215 pp.

Strawberries

1. Green, A. 1971. Soft fruits. In *The biochemistry of fruits and their products*, ed. A. C. Hulme, Vol. 2, 375-410. New York: Academic Press.

2. Harvey, J. M., and C. M. Harris. 1973. Strawberries—Market quality in relation to postharvest handling and shipping practices. *ASHRAE Symp*. LO-73-7. pp. 5-9.

3. Harvey, J. M., C. M. Harris, W. J. Tietjen, and T. Serio. 1980. *Quality maintenance in truck shipments of California strawberries*. USDA, SEA Adv. Agric. Tech. AAT-W-12. 13 pp.

4. Mitchell, F. G., E. C. Maxie, and A. S. Greathead. 1964. *Handling strawberries for fresh market*. Univ. Calif. Agric. Exp. Stn. Circ. 527. 16 pp.

5. Mitchell, F. G., and G. Mayer. 1978. Effects of basket design on cooling and holding strawberries. *Calif. Agric.* 32(5):17-18.

6. Ryall, A. L., and W. T. Pentzer. 1982. *Handling, transportation, and storage of fruits and vegetables*. Vol. 2, *Fruits and tree nuts*. 2d. ed., 255-57, 525-29. Westport, CT: AVI Publ. Co.

25

Postharvest Handling Systems: Table Grapes

F. GORDON MITCHELL

Grapes are a nonclimacteric fruit with a relatively low rate of physiological activity. They are subject to serious water loss following harvest, and this can result in stem drying and browning, berry shatter, and even wilting and shriveling of berries. The potential for Botrytis rot infection (gray mold) requires constant attention and treatment during storage and handling. The "bloom" or wax on the grape berry's surface is a very important quality factor. Rough handling and rubbing destroy this bloom, giving the skin a shine rather than the more desirable luster.

California Varieties

While many table grape varieties exist, a few make up most of the commercial shipments. In California the major variety is Thompson Seedless (Sultanina), marketed mostly during the summer months, usually soon after harvest. This variety is incapable of long-term storage (more than two months). Longer storage can be achieved with some fall varieties, including Emperor, Ribier, Almeria, and Calmeria.

Harvesting Maturity

Grapes are harvested when mature based upon the total soluble solids content (or ° Brix or Balling) of the cluster. Titratable acidity and sugar/acid ratio may also be used as maturity indices. The minimum requirements vary with variety and growing area. Colored varieties also have minimum color requirements that vary with variety and quality grade. The grade standards designate the percentage of berries on the cluster that show a set minimum color intensity and coverage.

Sources of Damage

Water loss

Because of the susceptibility of stems and fruit to deterioration from water loss, grapes are normally cooled as soon after harvest as possible. Even a few hours' delay at field temperatures can cause severe drying and browning of cluster stems. The problem is most serious in the hottest growing areas. Rapid cooling is widely used. Because grapes will not tolerate the wetting associated with hydrocooling, most of the fruit is forced-air cooled.

Botrytis rot

The Botrytis rot problem is not entirely eliminated by fast cooling alone. Standard practice in California is to fumigate with sulfur dioxide (SO_2) immediately after packing, and in lower doses each week of storage. Formulas for calculating the initial and subsequent SO_2 fumigation dosages are available (Nelson 1979).

The usual fumigation method in California is periodic introduction of sulfur dioxide into the storage room or fumigation chamber. In some other growing areas, a continuous, very low level of sulfur dioxide is maintained in place of periodic treatments. In recent years SO_2-generating pads have been used, particularly in export marketing when grapes are in ocean transport for extended periods. Sodium metabisulfite is impregnated and/or incapsulated into the pads to allow a slow release of sulfur dioxide throughout the transit and marketing period.

Fumigation effects

One problem associated with sulfur dioxide fumigation of grapes is the constant potential for injury to the fruit and stems. Injured tissue first shows up as a bleaching of color, followed by the sinking of areas where accelerated water loss has occurred. These injuries show first at the site of some other injury such as a harvest wound, transit injury, or breakage at the cap stem attachment. It can also be seen starting around the cap stem, and slowly spreading over the berry. Careful attention to sulfur dioxide treatment procedures is necessary to minimize this damage.

Postharvest Handling

Packing methods

Various methods of handling table grapes in California are summarized in figure 25.1. Most California table grapes are now packed in the field. A few grapes are "vine packed"; that is, the packer does all of the quality selection, trimming, and sorting, and packs the fruit directly into shipping containers. This system minimizes rehandling of the fruit, but makes supervision and consistent quality control difficult.

The most common field-packing system is the so-called "avenue pack." Here the fruit is picked and placed into picking lugs. Usually, the picker is also required to trim the cluster. The picking lug is then transferred a short distance to the packer, who works at a small portable stand in the avenue between vineyard blocks. The packer and several pickers commonly work as a crew. All of the essential packing materials are located at the packing stand, which also provides shade for the packer. Many packing stands around the vineyard make supervision more difficult than it is with a single packing shed, but easier than with vine packing. Lidding is performed in the field in both of these packing systems.

"Shed-packed" fruit is harvested by pickers and placed in field lugs untrimmed. The lugs remain in the shade of the vine to await transport to the shed. At the shed, the field lugs are distributed to packers, who select, trim, and pack the fruit. Often two different grades are packed simultaneously by each packer to facilitate quality selection.

Grapes are nearly always packed on a scale so the packer can precisely determine the net weight. After packing, the lugs are inspected for quality and weighed before lidding.

Some grapes intended for export and some high-quality grapes for domestic markets are "wrap packed." Here, individual clusters are wrapped in tissue before being placed in their shipping containers.

The California grape industry has been slow to accept the use of corrugated containers. Part of the reason for this is undoubtedly the ease of handling wooden containers in the field, where many grapes are packed. Another concern is container strength during long-term storage at high relative humidity (RH). Corrugated containers are increasing in use by the California grape industry, especially for fruit that will be sold without prior storage.

Unitizing

After packing and lidding, grapes are palletized, usually on disposable pallets. Often, pallets coming from the field pass through a "pallet squeeze," a device that tightens

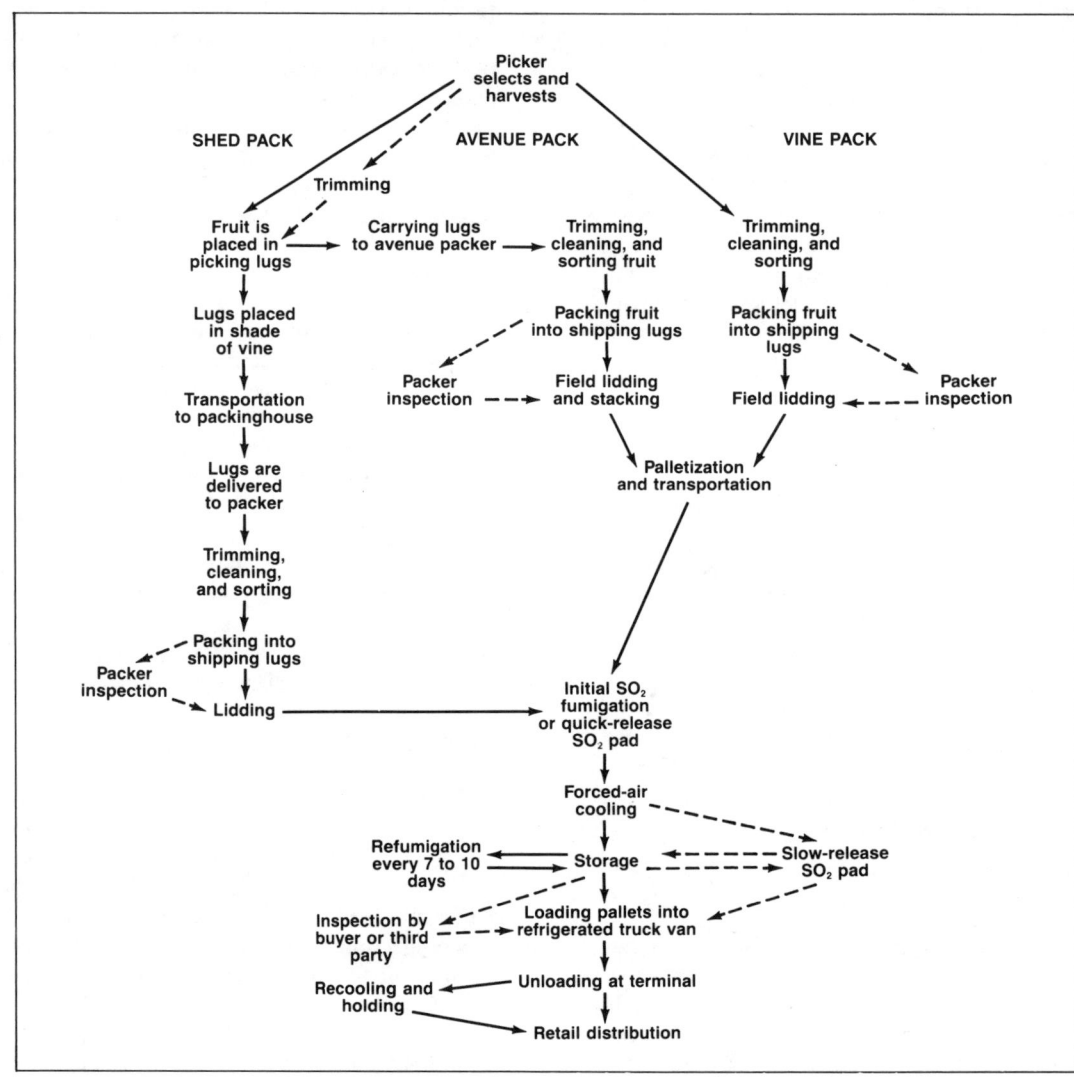

Fig. 25.1. Handling system for table grapes in California.

and straightens the stacks. Pallets are then unitized, usually by strapping. The use of palletizing glue has begun in some shed-packing operations. This glue bonds the containers vertically so that only horizontal strapping is then required.

Cooling

After palletization, the pallets are moved either to a fumigation chamber for immediate sulfur dioxide treatment or to a cooler where fumigation will come with the end of the day's packing. In either case, cooling starts as soon as possible. Many forced-air coolers for California grapes are designed to achieve seven-eighth cooling in 6 to 8 hours.

Once cooling is completed, the fruit moves to a holding area or storage room to await transport. Ideally this room should be kept at $-1°$ to $0°C$ ($30°$ to $32°F$) and 90 percent to 95 percent RH, with a moderate air flow. Fumigation is repeated weekly during the storage period, using a reduced concentration of sulfur dioxide. Grapes should be monitored regularly during storage for physiological deterioration, fruit rot, and sulfur dioxide injury.

Shipment

When grapes are loaded for transport they often receive one more sulfur dioxide fumigation. Their market life is limited, however, because further fumigation is seldom available in receiving markets. Unless sulfur dioxide fumigation is available, the receiver must order grapes only for immediate needs, and must distribute and market them within a reasonable time after receipt. Containers with slow-release SO_2-generating pads are exceptions to this rule.

References

1. Harvey, J. M., and W. T. Pentzer. 1960. *Market diseases of grapes and other small fruits*. USDA Agric. Handb. 189. 37 pp.
2. Harvey, J. M., and M. Uota. 1978. Table grapes and refrigeration: Fumigation with sulfur dioxide. *Int. J. Refrig.* 1(3):167-71.
3. Nelson, K. E. 1978. Pre-cooling—Its significance to the market quality of table grapes. *Int. J. Refrig.* 1(4):207-15.
4. _____. 1979. *Harvesting and handling California table grapes for market*. Univ. Calif., Div. Agric. Nat. Resour. Publ. 4095. 67 pp.
5. _____. 1980. Improved harvesting and handling benefit table grape markets. Calif. Agric. 34(7):34-36.
6. Reynaud, E., and P. Ribereau-Gayon. 1971. The grape. In *The biochemistry of fruits and their products*, ed. A. C. Hulme, Vol. 2, 172-206. New York: Academic Press.
7. Ryall, A. L., and W. T. Pentzer. 1982. *Handling, transportation and storage of fruits and vegetables*. Vol. 2, *Fruits and tree nuts*. 2d. ed., 257-62, 529-42. Westport, CT: AVI Publ. Co.
8. Winkler, A. J., J. A. Cook, W. M. Kliewer, and L. A. Lider. 1974. *General viticulture*, Chaps. 8, 20, 21, 22. Berkeley and Los Angeles: Univ. of Calif. Press.

26
Postharvest Handling Systems: Subtropical Fruits

ADEL A. KADER

Introduction

Subtropical fruits include avocado, carob, cherimoya, citrus fruit, date, fig, jujube, kiwifruit, loquat, lychee, olive, persimmon, and pomegranate. Some of these fruits are also grown in tropical- and temperate-zone areas. Subtropical fruits are diverse in their morphological and compositional characteristics and their postharvest requirements. The fruits can be grouped according to their relative perishability, as follows:

- Highly perishable—fresh fig, loquat, lychee
- Moderately perishable—avocado, cherimoya, olive, persimmon
- Less perishable—citrus fruit, carob (dry), dried fig, date, jujube (Chinese date), kiwifruit, pomegranate

In this chapter we will review the general characteristics of subtropical fruits in relation to their postharvest biology and handling requirements, with emphasis on avocado and citrus fruits, the most important fruits commercially.

Commercial production of citrus fruits in the U.S. is limited to Arizona, California, Florida, and Texas. Almost all U.S. production of dates, figs, kiwifruits, olives, persimmons, and pomegranates is in California. There are three strains of cultivated avocados: Mexican (e.g., Bacon), Guatemalan (e.g., Hass, Reed), West Indian (e.g., Pallock, Waldin), and their hybrids (e.g., Fuerte). Fresh avocados from Florida (West Indian cultivars) are available July through February, while California avocados (Mexican, Guatemalan, and hybrid cultivars) are available year round.

U.S. production of citrus fruits was 15 million tons in 1980-81. The U.S. produces about 30 percent and 40 percent of the world's production of lemons and oranges, respectively. Florida is the leading U.S. producer of citrus fruits, most of which (>80%) are processed. Most California citrus fruits are marketed fresh. Citrus consumption per capita in the U.S. increased by about 33 percent between 1960 and 1977 (primarily because of the increase in frozen, concentrated orange juice consumption), and it has continued to increase slightly every year since then.

Morphological and Compositional Characteristics

Avocados are one-seeded berries. Cultivars vary in size. Avocados are usually pear shaped, but they can be round or oval. An avocado has more energy value than meat of equal weight, and is a good source of niacin and thiamin. Avocados and olives are the highest in protein and fat content of all tree fruits. In California, minimum maturity of the avocado is defined in terms of oil content—not less than 8 percent.

Citrus fruits are very good sources of vitamin C, ranking first in their contribution of vitamin C to human nutrition in the U.S. All citrus fruits are berry-like fruits classified as hesperidium, having separate rinds. The pigmented part of the rind is called the flavedo (epidermis and several subepidermal layers), while the whitish part of the rind is called the albedo. The juicy part of the fruit consists of segments filled with juice sacs. Minimum maturity requirements of citrus fruits are based on juice content (lemons and limes) or soluble solids content, titratable acidity, and the ratio of the two (oranges, grapefruits, and tangerines).

Postharvest Physiology

Avocado fruits have a relatively high respiration and ethylene production rate, and exhibit a climacteric pattern for both. Citrus fruits are nonclimacteric and their respiration and ethylene production rates are low. Postharvest compositional changes in citrus fruits are minimal, whereas avocados undergo many changes in composition, texture, and flavor associated with their ripening.

Avocados do not ripen on the tree. The exact nature of the ripening inhibitor is not known. Factors that prevent avocado ripening on the tree continue to exert their effects for about 24 hours after harvest. Avocado ripening can be hastened by exposure to 10 ppm ethylene at 15° to 17°C (59° to 62.6°F). On the other hand, removal or exclusion of ethylene from the storage environment helps extend the storage life of avocados by delaying decay incidence and softening.

Cold nights followed by warm days are necessary for loss of green color and development of yellow or orange color in citrus fruits. This is why citrus fruits remain green after attaining full maturity and good eating quality in tropical areas. Some regreening of Valencia oranges may occur in certain production areas after the fruit has reached full orange color.

The degreening of citrus fruits constitutes the removal of chlorophyll from the flavedo. Degreening does not influence the fruits' composition. The need for and duration of degreening treatments depend on the cultivar and the fruit's condition at harvest—the amount of chlorophyll to be removed. Lemons are usually degreened at 16°C (60.8°F) with or without added ethylene; higher temperatures may be used for faster degreening. Recommended conditions for degreening California oranges and grapefruits are:

Temperature: 20° to 25°C (68° to 77°F)

Relative humidity: 90 percent

Ethylene concentration: 5 to 10 ppm

Air circulation: One room volume per minute

Ventilation: One to two air changes per hour, or sufficient changes to keep CO_2 below 0.1 percent

In Florida, a temperature of 27° to 29°C (80.6° to 84.2°F) and an ethylene concentration of 1 to 5 ppm are recommended.

Physiological disorders

Chilling injury. Subtropical fruits vary with species and with cultivar within a species in relative susceptibility to chilling injury. For example, grapefruits, lemons, and limes are much more susceptible to chilling injury than are oranges and mandarines. Dates, figs, kiwifruits, and Hachiya persimmons are not sensitive to chilling injury. Fuyu persimmons, pomegranates, olives, and other subtropical fruits are chilling sensitive. Ripe avocado fruits tolerate lower temperatures than unripe fruits, without danger of chilling injury. Orange cultivars grown in Florida are reported to be less sensitive to chilling injury than those grown in California and Arizona. Symptoms of chilling injury on selected subtropical fruits are summarized in table 26.1.

Other disorders. Freezing injury can be a problem for fruits exposed to temperatures below their freezing point before or after harvest. Citrus fruits injured by freezing become dry and useless, and are separated in the packinghouse using flotation or X-ray techniques. High-temperature disorders resulting from preharvest exposure to the sun can result in sunburn of avocados and citrus fruits. Exposure to temperatures above 25°C (77°F) causes uneven softening, skin discoloration, flesh darkening, and development of off-flavors in avocados.

Citrus fruit peel disorders other than chilling injury include oil spotting or oleocellosis (breaking of oil cells, causing the oil to extrude and damage surrounding tissue), rind staining of navel oranges (peel overmaturity; can be controlled by preharvest application of gibberellin), stem-end rind breakdown of oranges, stylar-end breakdown of limes, and aging (shriveling and peel injury around the stem end).

Pathological breakdown

Avocados. Avocado fruits can be affected by one or more of the following pathogens. *Dothiorella gregaria* (probably the asexual state of *Botryosphaeria ribis*) is an important postharvest rot of California avocados. Anthracnose is another problem, especially in humid areas such as Florida. It is usually not serious in California unless the weather has been unusually wet at or near harvest time. Stem-end rots (*Diplodia natalesis*, *Phomopsis citri*) can also be serious in Florida and other humid growing areas.

Citrus fruits. Postharvest diseases also play a major role in limiting the postharvest life of citrus fruits. Blue mold (*Penicillium italicum*) and green mold (*Penicillium digitatum*) are the most important postharvest diseases of citrus fruits in all production areas. In humid areas, stem-end rots (*Diplodia* spp. and *Phomopsis* spp.) and anthracnose (*Colletotrichum gloeosporioides*) are important problems. Sour rot (*Geotrichum candidum*) can be a problem during wet seasons, and phytophthora brown rot is seldom seen in California except following wet weather. Phytophthora brown rot can be controlled by heat treatment. Alternaria stem-end rot (*Alternaria citri*) usually follows senescence of buttons.

Control of citrus diseases can be accomplished by the following procedures:

- Reduce the pathogen population in the environment—use an effective preharvest disease control program (can reduce postharvest incidence of stem-end rots and anthracnose); use chlorine (e.g., sodium hypochlorite) in wash water; regularly disinfest (e.g., using a fog of 1% formaldehyde solution) field containers, packinghouse equipment, and storage facilities; circulate *Penicillium* spore-laiden air through filters in a special box-dumping room for stored lemons.

Table 26.1. Chilling injury symptoms on avocados and citrus fruits

Fruit	Minimum* safe temperature		Symptoms
	Celsius	Fahrenheit	
	(degrees)		
Avocado	5-10	41-50	Grayish brown discoloration of flesh, softening, pitting, development of off-flavors
Grapefruit	10-13	50-55	Pitting, scald, watery breakdown
Lemon	10-13	50-55	Pitting, membranous stain, red blotch
Lime	10-13	50-55	Pitting, accelerated decay
Orange	3-5	37-41	Pitting, brown stain

* Varies with cultivar, maturity stage, and duration of storage.

- Maintain fruit resistance to infection—minimize all kinds of mechanical injuries during harvesting and postharvest handling; use proper temperature and relative humidity management throughout the postharvest handling system; use 2,4-D treatment (200 ppm) on lemons to maintain vitality of button tissue and reduce development of stem-end rots; use postharvest fungicides, including biphenyl (diphenyl), sodium orthophenylphenol (SOPP), thiabendazole (TBZ), benomyl, sec. butylamine, and imazalil.

New fungicides are continuously being evaluated. Fungicide choice depends upon approval for use, and acceptance of the chemical by importing countries.

Postharvest Handling Procedures

Harvesting

Research into the mechanical harvest of citrus fruits (especially for processing fruits) has been extensive for many years, but no satisfactory system is available yet. Chemicals that promote abscission will probably be part of any mechanical harvesting system. Less than 2 percent of processing oranges were harvested mechanically in Florida during 1980. Several harvest aids such as mobile ladders and picker platforms have been tested, but few are in commercial use. California avocados and citrus fruits are cut with hand clippers. Some Florida citrus fruits are snap-picked (twist and pull method), but this may increase their susceptibility to decay. Some Florida processing oranges and grapefruits are picked and dropped on the ground. This practice is detrimental to the fruits even though they are processed within a day or two after harvest.

Packinghouse operations

The packinghouse operations for avocados and citrus fruits are summarized in figures 26.1 and 26.2, respectively. Lemons are usually sorted into four color classes (dark green, light green, silver, and yellow) by electronic sorters discerning variations of light reflectance. Some orange and tangelo cultivars receive color-addition treatment, with a certified food dye, in Florida, but this treatment is not allowed in California. Cooling methods include hydrocooling, forced-air cooling, and room cooling. Attention to proper and fast cooling for citrus fruits is badly needed in most citrus handling facilities.

During the past few years, *seal-packaging* (wrapping with various types of plastic film) of individual citrus fruits has been extensively tested, and this technique is currently used by a few shippers. This treatment reduces water loss and maintains the vitality of the peel because of the high relative humidity maintained around the fruit. It also prevents the spread of decay from fruit to fruit. For decay control, fruits must be treated with fungicides before wrapping. While seal-packaging of individual fruits may allow short-term holding of citrus fruits without refrigeration, it must be combined with refrigeration for long-term storage in order to maintain good quality and reduce losses.

Citrus fruits produced in certain areas must be treated for insect control before shipment to some markets. The main disinfestation method in use was, until recently, fumigation with ethylene dibromide (EDB) against fruit flies. Because EDB was withdrawn from the Environmental Protection Agency's list of approved chemicals in September, 1984, alternatives to EDB fumigation are being evaluated. These include cold treatments, irradiation, and controlled atmospheres.

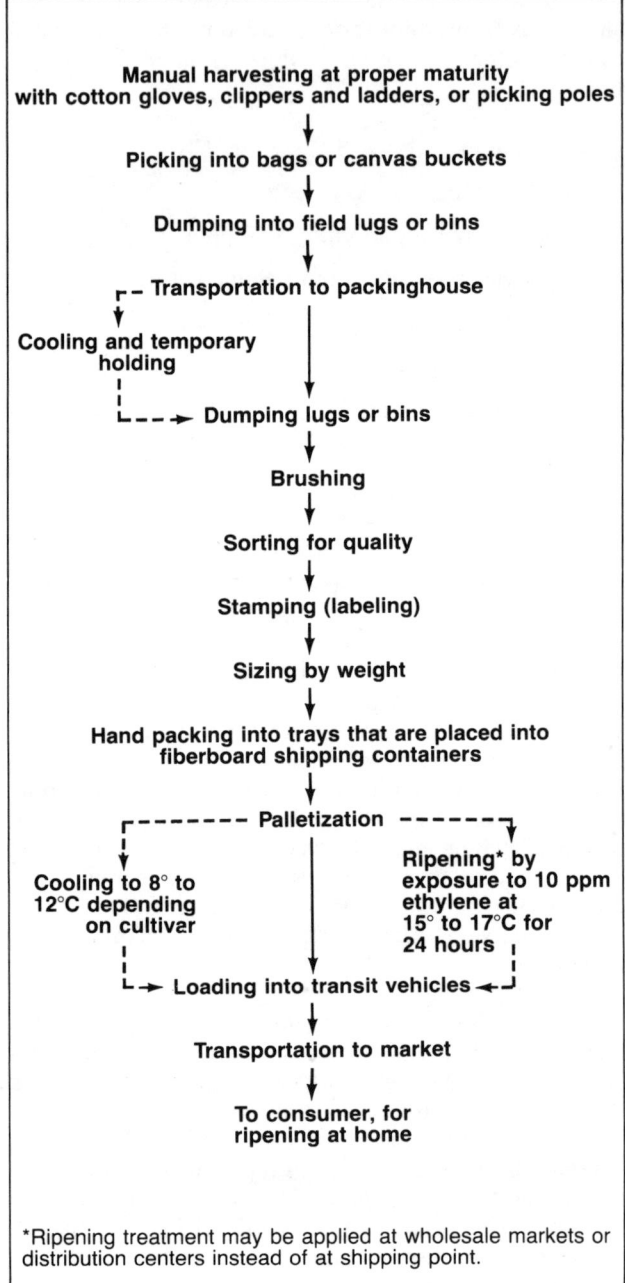

Fig. 26.1. Postharvest handling system for avocados.

Quality and Storage Life of Citrus Fruits

The composition and quality of citrus fruits at harvest and the fruits' potential for storage are influenced by many pre- and postharvest factors. Preharvest factors include rootstock and cultivar, fruit maturity at harvest, harvesting season, tree condition (vigor), weather conditions (temperature, relative humidity, rain), and cultural practices (fertilization, irrigation, pest control, growth regulators). Harvesting methods influence the uniformity of maturity among fruits and the extent of mechanical injuries due to rough handling.

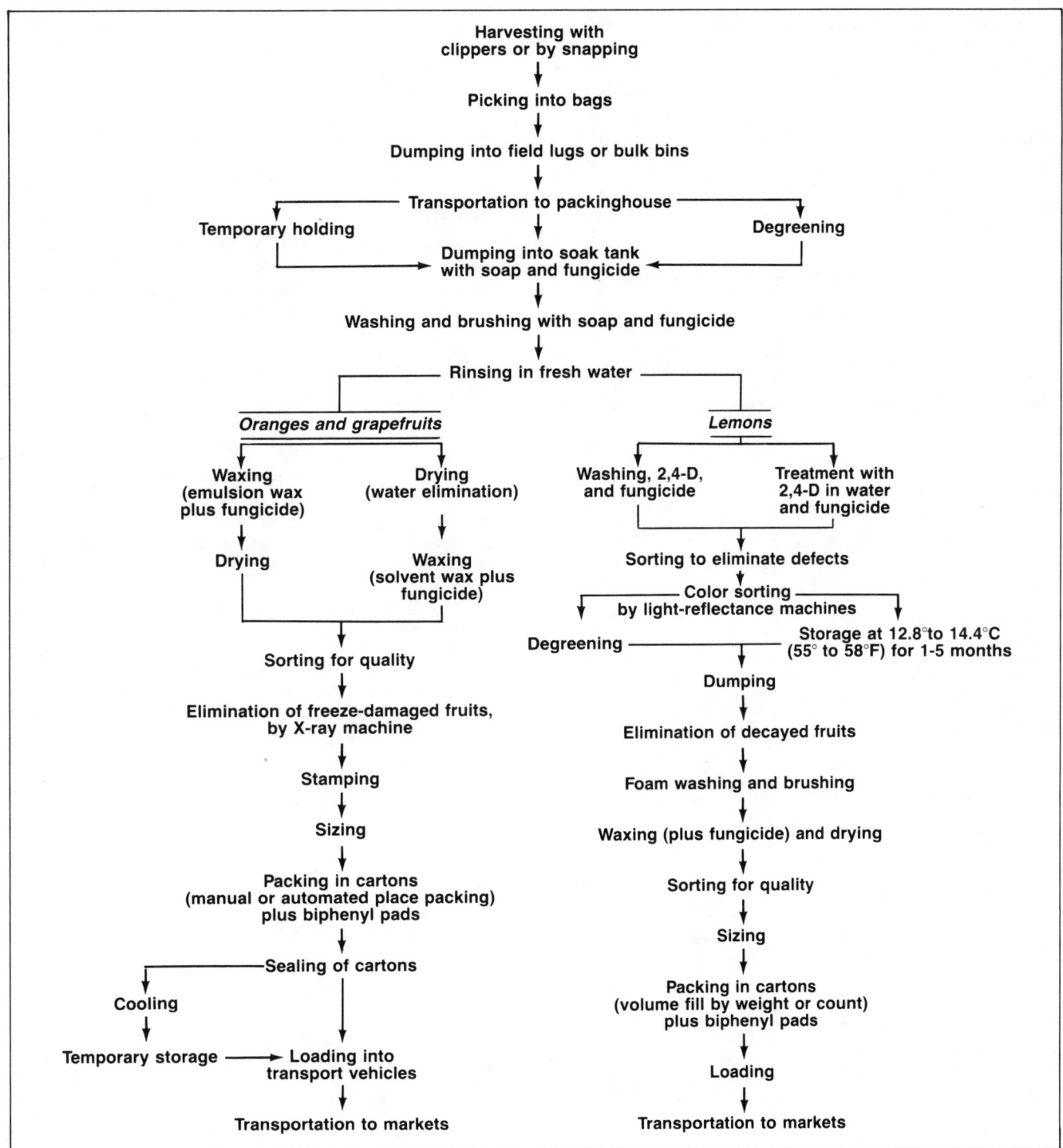

Fig. 26.2. Postharvest handling systems for citrus fruits.

Postharvest factors that influence the postharvest life-span of citrus fruits include delays between harvest, packing, and cooling; degreening conditions; fungicidal treatments; waxing; seal-packaging; growth-regulator treatments; temperature and relative humidity management; and presence of ethylene and other volatiles in storage.

Storage

Some citrus cultivars may be left on the tree for up to 5 months after attaining legal maturity. Depending on the cultivar, avocados will remain attached to the tree for 3 to 12 weeks after maturity before excessive abscission begins. Citrus fruit and avocado quality may, however, deteriorate during on-tree storage. For successful postharvest storage, the conditions summarized in table 26.2 should be maintained. These recommendations also apply for optimum transport and temporary storage conditions.

Table 26.2. Optimum storage conditions for avocados and citrus fruits

Commodity	Temperature Celsius (degrees)	Temperature Fahrenheit (degrees)	Approximate storage-life* (weeks)	Modified atmospheres if used† O_2 (%)	Modified atmospheres if used† CO_2 (%)
Avocado, unripe	8-12	46-54	2-4	2-5	3-10
Avocado, ripe	5-8	41-46	1-2	§	§
Grapefruit	12-14	54-57	4-8	5-10	5-10
Lemon	12-14	54-57	16-24‡	5	0-5
Lime	10-12	50-54	6-8	5	0-10
Orange	4-8	39-46	4-8	5-10	5
Tangerine	5-8	41-46	2-4	5-10	5

* Under optimum temperature and 85% to 90% relative humidity.
†MA use on citrus is very limited; 5% to 10% CO may be added to MA for decay control during transport to export markets.
‡Storage life for dark green lemons; for other stages: light green (8 to 16 weeks), silver (4 to 8 weeks), yellow (3 to 4 weeks).
§Not used.

References

1. Biale, J. B., and R. E. Young. 1971. The avocado pear. In *The biochemistry of fruits and their products*, ed. A.C. Hulme, Vol. 2, 2-64. New York: Academic Press.

2. Bowman, E. K., A. H. Spurlock, S. Hedden, and W. Grierson. 1971. *Modernizing handling systems for Florida citrus from picking to packing line*. USDA Mark. Res. Rep. 914. 54 pp.

3. Eaks, I. L. 1977. Physiology of degreening—Summary and discussion of related topics. *Proc. Int. Soc. Citriculture* 1:223-26.

4. Eckert, J. W. 1978. Postharvest diseases of citrus fruits. *Outlook Agric.* 9(5):225-32.

5. Grierson, W., and T. T. Hatton. 1977. Factors involved in storage of citrus fruits: A new evaluation. *Proc. Int. Soc. Citriculture* 1: 227-31.

6. Grierson, W., W. M. Miller, and W. F. Wardowski. 1978. *Packingline machinery for Florida citrus packinghouses*. Inst. Food Agric. Sci., Univ. Fla. Coop. Ext. Serv. Bull. 803. 30 pp.

7. Hatton, Jr., T. T., and D. H. Spalding. 1974. Maintenance of market quality in Florida avocados. *ASHRAE Trans.* 80:335-40.

8. Lutz, J. M., and R. E. Hardenburg. 1968. *The commercial storage of fruits, vegetables, and florist and nursery stocks*. USDA Agric. Handb. 66. 94 pp.

9. Nagy, S., and J. A. Attaway, eds. 1980. *Citrus nutrition and quality*. Symposium Series 143. Washington: American Chemical Society. 456 pp.

10. Nagy, S., and P. E. Shaw. 1980. *Tropical and subtropical fruits: composition, properties, and uses*. Westport, CT: AVI Publ. Co. 570 pp.

11. Ryall, A. L., and W. T. Pentzer. 1974. *Handling, transportation and storage of fruits and vegetables*. Vol. 2, *Fruits and tree nuts*. Westport, CT: AVI Publ. Co. pp. 242-64, 490-512.

12. Smoot, J. J., L. G. Houck, and H. B. Johnson. 1971. *Market diseases of citrus and other subtropical fruits*. USDA Agric. Handb. 398. 115 pp.

13. Ting, S. V., and J. A. Attaway. 1971. Citrus fruits. In *The biochemistry of fruits and their products*, ed. A.C. Hulme, Vol. 2, 107-71. New York: Academic Press.

14. United Fresh Fruit and Vegetable Association, various dates. *Fruit and vegetable facts and pointers*. Sections on avocado, grapefruit, lemon, lime, olive, and orange. Alexandria, VA.: United Fresh Fruit and Vegetable Association.

27

Postharvest Handling Systems: Tropical Fruits

NOEL F. SOMMER

Introduction

The most important tropical fruit by far in temperate markets is the banana, having a per capita consumption in the U.S. higher than any other fruit. Popularity among consumers is enhanced by a reasonable price, high fruit quality, and availability throughout the year. Furthermore, bananas have been marketed regularly in the U.S. for several generations, and the result is widespread familiarity with the commodity among consumers. Bananas were occasionally shipped to eastern U.S. markets from Central America in the days of sailing ships, but systems capable of maintaining high fruit quality in markets throughout the year were developed only toward the end of the last century and the beginning of this one.

The other fresh tropical fruits most commonly found in the temperate zone markets—mango, papaya, and pineapple—are more recent arrivals. They are categorized as gourmet fruits, and purchased only occasionally by most consumers.

The regular appearance of papayas in mainland U.S. markets was preceded by the development of small-fruited Solo cultivars weighing only about 1/2 kg, and the establishment of regular air cargo service. The small size permitted the fruit to be marketed profitably at prices comparing favorably to those of other gourmet foods. To develop markets for this unfamiliar and expensive fruit, advertisement and other market-development activities were concentrated in certain chosen cities. Market development was further aided by increased tourism to Hawaii. Early mainland consumers of papaya were commonly found to have been introduced to the fruit on vacations to the Islands. High air transport costs and expanding markets have stimulated increased shipments via refrigerated marine containers.

The introduction of fresh pineapples to the mainland market was aided by consumers' familiarity with the canned product. The mango, however, remains a fruit that is inadequately appreciated by most consumers. Tourism in producing areas has been an important factor in increasing consumers' awareness of the fruit. The influx of many people from tropical countries into the U.S. has also increased the potential demand for mangoes. Although it seems necessary, market development for mangoes is complicated by the fruit's seasonal nature.

Banana Handling

The banana is a large herbaceous plant. The underground tuberous stem or "bulb" gives rise to leaves and the fruit bunch. The aboveground "trunk" of the banana tree is in reality a pseudostem consisting of tightly appressed leaf bases. The pseudostem dies after a fruit bunch has been produced and is replaced by one of the young pseudostems that have emerged. The banana inflorescence contains three kinds of flower. The fruits originate from female flowers that are produced first, followed by hermaphrodites, and then male blossoms.

The postharvest operations required for bananas include transportation to packinghouse, dehanding, washing to remove dirt and latex, disease control, packaging, transportation to market, ripening, and retail sale (fig. 27.1).

Harvesting

Bunches are examined about 3 months before harvest. Those that have completed their female (fruit-producing) stage have their buds removed to prevent further floral development. One or two apical hands are also removed at this time to promote development of the remainder. Removed buds may be consumed or discarded. A single finger is retained, apical to the position of the hands that were removed. This terminal finger continues to grow, preventing the stalk from rotting.

Each bunch is covered with a perforated polyethylene bag. The top of the bag may be secured to the stalk with a colored ribbon. With different colors used each week a ready record of age is maintained. The plastic film protects fruits from leaf scarring and keeps dust off, and an insecticide may be placed within the bag to reduce insect damage.

During fruit development, the bunches may require props to support their weight (fig. 27.2). Sometimes a pseudostem is provided with guys and twine from its crown to the bases of nearby pseudostems.

Bananas are harvested green and are ripened in market areas. Fruits that are allowed to ripen on the tree often split, and tend to be mealy. The maturity at which bananas are harvested depends on the time required to get them to market. Fruits shipped from Central America to Europe are usually harvested less mature than those shipped to

North America. A penalty in lost yield is, of course, paid when bunches are harvested before fingers are fully developed. Sizes and shapes of finger sections at various stages of maturity are illustrated in figure 27.3.

At harvest, crews pass through the plantation, usually at 3 or 4 day intervals, selecting bunches for harvest. Colored ribbons provide information regarding age. The diameter (caliper) of fruits is monitored.

Harvesting and field handling vary with location. However, harvesting is usually a two-person operation. The *cutter* makes a cut with a machete, partially severing the pseudostem at about its midpoint. A *backer,* positioned under the bunch, catches and braces the bunch firmly. The cutter then severs the bunch from the pseudostem, just below the basal hands.

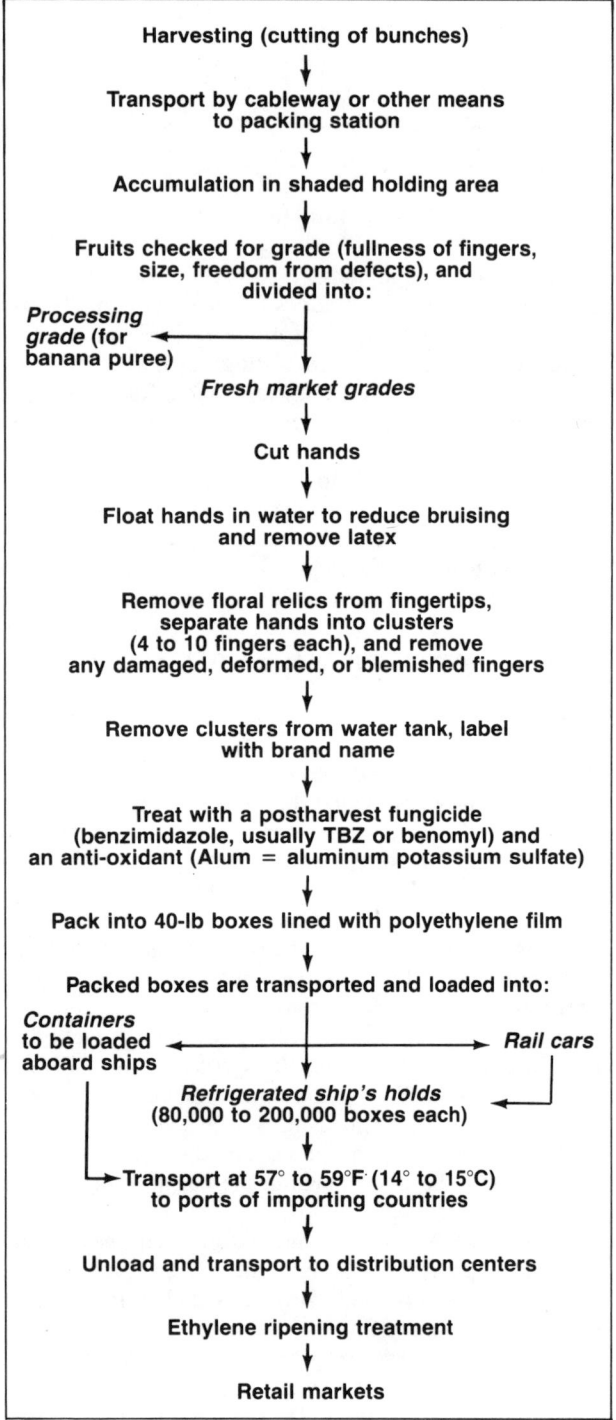

Fig. 27.1. A flow diagram of a postharvest handling system for bananas.

Fig. 27.2. Banana fruit bunches are enclosed in plastic bags, and guys are run to prevent pseudostems from falling (near Puerto Limon, Costa Rica).

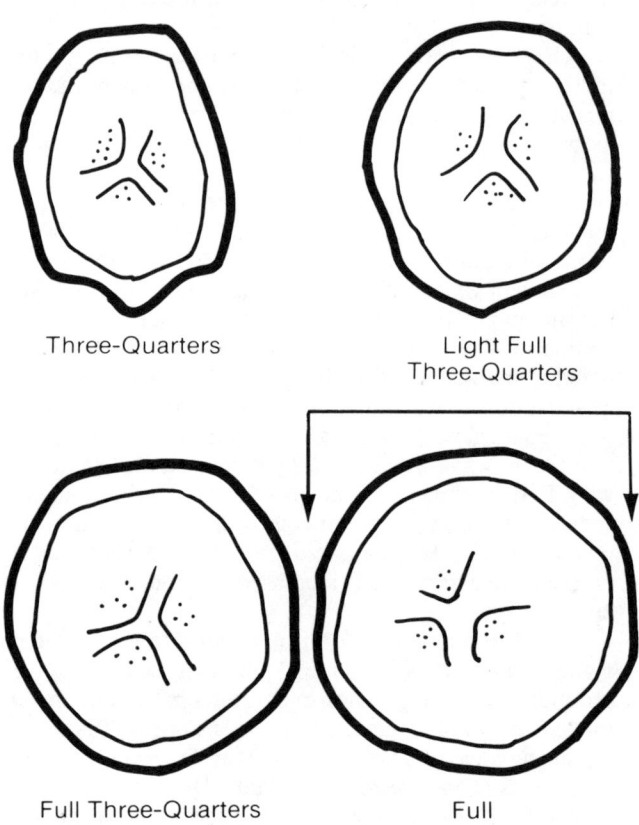

Fig. 27.3. Changes in size and shape of banana fruit sections at various stages of maturity.

Fig. 27.4. A train of banana stems moves to the packinghouse by aerial tramway (LaLima, Honduras).

Fig. 27.5. Inspecting banana stems (LaLima, Honduras).

Transportation to packing station

Many banana plantations have been equipped with a system of cableways. A backer carries a cut bunch to the nearest cableway, where the bunch is attached by its base to a roller on the cable. The bunches are separated by spacer bars to prevent contact. A train of up to 75 or 150 bunches of 30 to 60 kg each forms and is pulled along the cableway to the packing area by a small, garden-type wheeled tractor. Shaded areas are provided for bananas waiting to be packed (fig. 27.4).

A tractor that hangs from the cable has the advantage of not requiring a roadway or bridges to cross drainage ditches. Such tractors have been developed and are used in some areas.

In Queensland and New South Wales, Australia, winter temperatures limit banana production to sunny, northern exposures, often on very steep hills. Some cableway systems have been installed that use gravity as the means of locomotion. Otherwise, bunches accumulate at roadways that have been bulldozed across the slopes. Bunches are placed on small trucks or on trailers, usually two or three bunches deep. Padding is provided on the vehicle bed and between bunches to limit damage. Low tire pressure and low speed during transit are important injury avoidance measures.

Packinghouse operations

Upon entering the packinghouse, bananas are checked for finger fullness and length, and for blemishes from leaf rub, insect activity, pathogens, and handling bruises (fig. 27.5). Those not meeting a fresh fruit grade are processed as puree or discarded.

Most bananas from the American tropics are shipped as hands in fiberboard cartons. Hands are removed from the stalk by the *dehander,* using a sharp curved knife (fig. 27.6). When the hand is cut away, latex flows from the wound for a time. If allowed to coat the surface of the fingers, latex deposits and the resulting stains would seriously detract from appearance. Consequently, the hands are immediately placed in water to coagulate the exuded latex, and reduce staining of the fruit surface (fig. 27.7). Dust and dirt are removed at the same time.

Selectors at the dehanding tank remove dead floral parts still adhering to the fruit and sort out undersized, damaged, deformed, and blemished fingers. Large hands are divided into smaller clusters to facilitate packing and provide a convenient unit for the consumer. A second water tank may be provided, where the bananas are floated for an additional 10 or 15 minutes to permit further exudation of latex.

The water tank is a potential source of serious disease problems. Fungus spores on dead floral parts may accumulate in the water and contaminate the cut surfaces of freshly cut hands. In particular, crown rot results from inoculation by mixed spores of *Colletotrichum musae* (Berk. and Curt.) Arx, *Fusarium roseum* (LK) Snyd. and Hans., *Nigrospora sphaerica* (Sacc.) Mason., *Thielaviopsis paradoxa* (de Seynes) Dade, *Botryodiplodia theobromae* Pat., and other fungi.

Sodium hypochlorite (75 to 125 ppm) in the water will kill spores, reducing the likelihood of inoculation. Concentrations of the chemical must be carefully monitored and maintained. Indicator papers provide a simple and inexpensive means of checking chlorine concentrations.

As an alternative, a fungicide such as TBZ, benomyl, or another benzimidazole is added to the wash water instead of as a separate treatment. Relatively large amounts of the fungicide are usually required for this practice, because the wash water becomes dirty and must be changed from time to time. The fungicide must be included with fresh water.

Fig. 27.7. Bananas are floated in water to coagulate latex exuding from cut surfaces. Withered floral parts are removed from the fingers (LaLima, Honduras).

Fig. 27.6. A worker cuts the hands from banana stems. Hands are placed in water to clean the fruits and avoid latex stains (LaLima, Honduras).

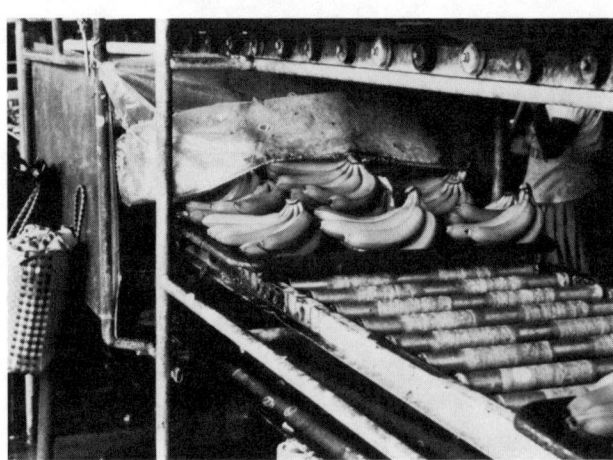

Fig. 27.8. Banana hands emerge from a plastic enclosure where the fruits had been sprayed with TBZ for disease control and alum as an added precaution against latex staining (LaLima, Honduras).

A benzimidazole fungicide may be applied as a separate dip or spray treatment after washing (fig. 27.8). Alum, which may be applied at the same time, serves as an antioxidant to prevent subsequent latex exudations from darkening and staining the surface of the fruit. This is the last step before packing.

Packing and shipping

In the American tropics, clusters of fruit are packed in corrugated fiberboard cartons lined with polyethylene (fig. 27.9). Individual packages are moved to conveyors for loading onto trucks or trains for transportation to the wharf. Cartons move by conveyer to an elevated conveyer, or gantry, that lowers them into the hold of a ship. The ship's refrigeration system cools the fruit and holds the temperature at 13° to 14°C (55° to 57°F).

Marine shipment of bananas in refrigerated containers on container ships will probably continue to increase. Less time required for loading and unloading is an important economic advantage for container ships. Containers allow the use of pallets and forklift trucks, and are adaptable to intermodal transportation. Fruits can be loaded into containers at the packing stations and remain untouched until arrival in the market area. Presumably, that reduced handling minimizes fruit bruising.

Handling in market areas

Banana fingers should still be green on arrival in market areas in order to provide consumers with a constant supply of fruit at the desired yellow color. Facilities for controlled ripening of the fruit are often provided at produce distribution centers (figs. 27.10 and 27.11). Ripening is triggered by exposure to the ripening hormone, ethylene (C_2H_4) gas. By scheduling the time at which fruit lots are exposed to ethylene and adjusting subsequent temperatures, distributors have a measure of control over the stage of ripeness of fruits delivered to retail stores.

Fig. 27.9. Bananas are hand packed into polyethylene-lined corrugated paper containers (LaLima, Honduras).

Ripening facilities

Ripening rooms require close temperature control. The rooms must be well-insulated and provided with heating and refrigeration. Vigorous air circulation is required to

Fig. 27.10. Ripening rooms for banana fruits (San Francisco, California).

Fig. 27.11. Cartons of bananas are spaced to ensure air circulation and uniform exposure to ethylene gas (San Francisco, California).

thoroughly disperse the ethylene and facilitate removal of respiration heat from the fruit. Rooms are as nearly airtight as possible to contain the ethylene. High humidity is essential (ca. 90% RH) in ripening rooms to avoid fruit dehydration. The introduction of water is usually required. Moisture may be introduced automatically in the form of steam or a mist or spray of water at ambient temperature. In the absence of a special system, walls and floors may be wetted before closing the room to initiate ripening.

The requirements for ripening rooms closely resemble those for refrigeration rooms. Commonly, rooms are prefabricated, have polystyrene foam insulation between sheets of steel or aluminum and are installed on a concrete slab. Insulated wood or steel frames and concrete block construction are common. Any wood must be treated to prevent rot in the high-humidity of the room.

The size of ripening rooms varies with the volume of bananas handled. In modern facilities, bananas are normally handled in palletized boxes. Often, all pallets are placed on the floor. Valuable floor space is better used if rooms are designed for pallets stacked two or three high. Supporting framework is required to support the upper pallets.

In order to provide ripening bananas continuously to retail stores, a minimum of three ripening rooms is required; bananas can be delivered from each room on two successive days. Fruits leaving on the second day would be somewhat riper than the first day's output. With six or more ripening rooms, all fruits from a single room can be delivered the same day.

Controlled ripening procedures

The objective of controlled ripening is to provide retail stores with bananas at a stage of ripeness desired by consumers. The state of ripeness is judged primarily by color, using a 1 to 7 scale common in the industry. At color No. 1, the finger is hard and completely green; No. 2, green but with some traces of yellow color; No. 3, about half green and half yellow; No. 4, more yellow than green; No. 5, yellow but with green tips; No. 6, fully yellow; and No. 7, yellow with sugar spot development.

Fruits should be ripened at least to color No. 3 before delivery to retail stores, or ripening may not continue normally. Generally, the fruits are not riper than color No. 4 when shipped from the distribution center to the retail store, because fruits may suffer handling injury if they are too ripe.

Ripening is initiated by release of ethylene gas into the ripening room for 24 hours with fruit pulp temperatures at 15.5° to 16.5°C (60° to 62°F). Sufficient gas is introduced into the room to provide a concentration, by volume, of 100 to 1,000 ppm in air with vigorous air circulation.

The ripening rate varies to some extent between lots. Cloudy conditions or low temperature during growth may slow the rate of ripening. Temperature conditions during handling and transit may also affect the ripening rate. In particular, if elevated temperatures have caused some bananas to ripen enroute, as shown by slight color changes or slight softening of pulp, the rate of ripening will be much faster than average. The maturity of fruit affects ripening time—hard-green fruits at the three-quarter stage require a noticeably longer time to ripen than fruits at full three-quarters.

Steady shipments of ripening bananas over a period of 2 weeks or more require considerable judgment based on experience. Bananas can be held at 13° to 14°C (56° to 58°F) for a number of days, depending on condition, before being placed in ripening rooms. After ripening has begun, the speed of ripening can be slowed by lowering the temperature to 13°C (56°F) or speeded by raising the temperature to 18.5°C (65°F).

Special considerations

Ethylene gas in concentrations greater than 30,000 ppm in air (3.02% to 34% by volume) is explosive. Care must be exercised to insure that dangerous concentrations are not reached.

Fruits should not be exposed for extended periods to temperatures below about 13°C (56°F), for fear of chilling injury. Adverse affects of chilling injury are shown as a muddy yellow instead of bright yellow color of ripened fruit. The color difference is the consequence of the necrosis of vascular tissues in the peel. Symptoms are most severe if exposure occurs before ripening has started. Length of exposure affects the severity of symptoms.

Papaya Handling

Intensive culture of papayas for long-distance marketing has been developed mainly in Hawaii. Production is mostly of small-fruited Solo cultivars. Fruits generally average about 1/2 kg, a size that facilitates harvesting, packing, transportation, and marketing operations. Consumer acceptance is also easier as the size is appropriate for small North American families.

Handling and market preparation requirements are influenced greatly by the susceptibility of papayas to certain diseases. The most important of these is anthracnose, caused by the fungus *Colletotrichum gloeosporioides* (see chap. 15). Infections may occur by direct penetration during fruit development on the tree, but disease development does not proceed at this time because of the almost-complete immunity of the fruit flesh. Infections remain latent until the start of the climacteric rise in respiration rate. The disease becomes evident in ripe or ripening fruits. Satisfactory control requires the use of heat treatments. Fungicidal sprays in the orchard during fruit development are not sufficiently successful to permit elimination of the postharvest heat treatment, but they may reduce the disease pressure considerably.

Other important diseases include stem-end rots caused by *Ascochyta caricae-papayae* Pat. (see chap. 15), *Phomopsis caricae-papayae* Petri and Cif., and *Phytophthora nicotiana* var. *parasitica* (Dast.) Waterh. Although characteristically colonizing the fruit stem, these fungi often invade wounds caused by handling. Rhizopus rot (*Rhizopus stolonifer* and, possibly, related species) is an important cause of rot among fruit marketed in Hawaii. It has not been observed commonly among Hawaiian fruit marketed in North America.

The presence of the Mediterranean, oriental, and melon fruit flies in Hawaii necessitates fumigation or other treatments to eliminate flies before shipment to noninfested areas. Post fumigation handling must be in screened areas. Packages must be sealed to avoid reinfestation, unless they can be loaded into marine containers with screen protection.

Harvesting

Maturity. Papaya trees are normally trained to a single trunk. Buds form progressively higher as the tree grows. Consequently, the lowermost fruits are the oldest. The fruit's position on the tree, therefore, provides an indication of relative maturity (fig. 27.12).

Fruits are generally picked according to the change in color from mature green (deep green) to color break (light green), exhibiting a slight overall loss of green color, with some hint of yellow at the blossom end. More mature fruits are separated into three groups, i.e., one-quarter, one-half, or three-quarters yellow color.

For long-distance shipment, fruits are generally harvested at color break or between color break and one-quarter color. To obtain maximum fruit life for long-distance surface shipments to mainland North America, the fruits are better harvested at the mature-green stage, but the difficulty of distinguishing mature-green from immature-green fruits makes this impractical. Immature-green fruits are incapable of ripening after long-distance refrigerated transit.

Fruit removal. Fruits are harvested by pickers standing on the ground while trees are small and fruits can be reached. As the fruit-bearing area progresses higher, a long-handled suction cup (plumber's friend) is positioned by the picker over the stylar end of the fruit. A twist of the handle breaks the fruit's pedicel, and the picker catches the fruit if it falls from the suction cup. Fruits are placed in a pail and transferred to field boxes or bins.

In recent years various picking aid machines have been used to elevate the pickers to fruit level. Bins or other containers on the picking aid machines permit accumulation of the fruit.

Transportation—orchard to packinghouse

Bulk bins have largely replaced 40-pound-capacity field boxes. Extreme care is essential when filling bins or boxes and transporting them to the packinghouse, because of the fragile nature of the fruit. Compression or impact bruises result from careless handling or transportation over rough roads. Sometimes equally serious is abrasion damage to the fruit's tender skin. Damaged skin generally does not degreen when the fruit ripens.

Packinghouse operations

Some operations include preliminary washing and sorting to separate cull fruits and ripe fruits (which are diverted to processing) from those suitable for fresh marketing. This reduces the volume of fruit that must be heat-treated and handled within the insect-proof packinghouse.

Heat treatment, vapor heat. The application of heat was originally intended to eliminate fruit fly infestations and to permit the marketing of fruit within the U.S. mainland. Treatment was administered in a room with accurate temperature control and adequate air circulation. Fruits in the room were subjected to air temperatures of 44°C (111.2°F) for 6 to 8 hours without a high relative humidity. After warm-up, fruit was exposed to saturated air for 4 hours at 48°C (118.4°F). Handling and fruit-injury problems caused this treatment to be abandoned in Hawaii, although it is still a legal treatment for fruit fly disinfestation.

Hot water treatment. When ethylene dibromide fumigation was first substituted for the vapor heat treatment, a serious increase in disease resulted. Once handlers realized that the vapor-heat treatment was fungicidal as well as insecticidal, substitute heat treatments were explored.

Bins or pallet loads of fruit are commonly submerged for 20 minutes in 46° to 50°C (114.8° to 122°F) water

Figure 27.12. Developing papayas are produced successively on the stem. The lowest fruits are the oldest. Size is variable despite position (Kahului, Hawaii).

with vigorous circulation. As the fruit warms, the water may cool several degrees.

Integration of the heat treatment operation into the packinghouse line has been attempted in recent years. At the speed at which fruit moves through such lines, the dwell period in the hot water (without excessively long tanks) would necessarily be very brief, perhaps only 20 to 30 seconds. To obtain sufficient heat penetration for fungicidal effectiveness, the temperature of the water must be very high—60.6°C (141°F) for 20 seconds. Positive movement of fruits through such a bath to insure that each fruit is exposed to an exact dwell time is essential. A fruit with a short dwell period would not receive an effective treatment, while those with long periods would sustain heat injury. In papayas, heat injury results in failure to degreen. A rot caused by *Dothiorella* sp. often develops, usually from the blossom end, in heat-injured fruit.

Fumigation requirements. A hot water dip at 46.1° to 48.9°C (115° to 120°F) for 20 minutes may be followed by fumigation with ethylene dibromide (EDB) at normal atmospheric pressure and a fruit temperature of 21°C (70°F), or above. With the fumigation chamber volume filled three-quarters or less with fruit, dosage is 8 g EDB per m³ of chamber space (8 oz/1,000 ft³). The maximum time between hot-water dipping and fumigation is 6 hours, or until the fruit temperature reaches 26.7°C (80°F). Sufficient time should be allowed for the papayas skin surface to dry. Packing boxes, if used, must be dripped dry.

With fumigation preceded or followed by hot water dip, the following water temperatures and treatment times are suggested:

Minimum water temperature	Minimum treatment time
60.6°C (141°F)	20 seconds (can use spray)
56.7°C (134°F)	80 seconds
54.4°C (130°F)	3 minutes
52.5°C (126.5°F)	6 minutes
49.5°C (121.1°F)	11 minutes
46.1°C (115°F)	20 minutes

If fumigation precedes hot water treatment, an aeration period of 1 to 6 hours is required. If fumigation is not accompanied by heat treatment, the EDB dosage at normal atmospheric pressure is 16 g/m³ (16 oz/1,000 ft³.) for two hours, at 21°C (70°F) or above. The chamber space occupied by the fruit must not exceed three-quarters.

Use of ethylene dibromide (EDB) within the United States as a fumigant on fruits and vegetables for consumption in the United States was prohibited effective September 1, 1984. Papayas can no longer be treated with EDB if destined for U.S. markets, but the fumigant can be used for export fruit.

An alternative quarantine treatment under commercial testing is a heat treatment. Commonly called the *double-dip* treatment, fruits are submerged in water at 42°C for 40 minutes, followed by 20 minutes at 49°C.

Fruit segregation. Fruits are sorted for undersize or defective fruits as they move on belts. Fruits can be sized by eye, and uniformly sized fruits are packed into corrugated containers by count. Automatic weight sizers are increasingly replacing eye sizing (fig. 27.13). Fruits are hand packed into shipping cartons (fig. 27.14).

Fig. 27.13. A weight sizer is often used with papayas (Hilo, Hawaii).

Fig. 27.14. Workers hand pack papaya fruits (Hilo, Hawaii).

Transportation and storage temperatures. Papayas are sensitive to injuries when chilled. Symptoms include sensitivity to *Alternaria* sp., failure to ripen normally, and, sometimes, water-soaked tissues. The greenest fruit is most sensitive to chilling temperatures. Lowering the temperature below a critical point and lengthening the exposure period increases the injury. The critical temperature above which injury will not occur can be as low as 6°C (43°F), but Alternaria rot has occurred at that temperature in fruits harvested at the color-break stage and shipped distances requiring up to 2 weeks in transit. Under these conditions, temperatures above about 13°C (55.4°F) appear to be required.

Mango Handling

The essential operations that prepare mangoes for market include harvesting, transport to packing area, washing, sorting, sizing, packaging, and transport to market. Frequently mangoes must be fumigated for quarantine purposes before fruits can enter certain markets.

Harvesting

Mangoes picked for nearby markets may exhibit color changes from dark green to light green or even to light yellow. Fruits are harvested while still dark green if destined for distant markets requiring several days' transit. Selection of fruit before color changes have occurred makes it difficult to discriminate between fruits capable of ripening to an acceptable maturity and those without that capability. In general, dark-green mature fruits are harvested when the cheeks have filled out. In some varieties, maturity is indicated by the degree to which the shoulder extends out at the stem end.

Maturity indices commonly used with other fruits have not proved adaptable to dark-green mangoes. Neither changes in flesh texture (softening), total soluble solids, nor acidity have proved highly useful. Softening increased soluble solids, and decreased acidity occur mostly after the proper harvest time for distant markets. Studies suggest that starch content and specific gravity might be useful indices for some varieties.

Mangoes are harvested by hand if the pickers can reach them. The fruit is twisted sharply sideways or upward to break the pedicel. To avoid stem punctures, any long pedicels are trimmed flush with the stem end of the fruit. Fruits on high branches are harvested with a picking pole having a cloth bag and cutting knife at the top (fig. 27.15). When the pedicel is severed, the fruit drops into the bag, the picking end of the pole is gently lowered to the ground, and fruits are placed into 18-to 23-kg field boxes or bins for transport to the packing shed.

Packinghouse operations

Washing and fungicidal treatment. Mature-green mangoes exude a large amount of latex from the cut stem. This is washed off with water in a tank. The water may contain benomyl or thiobendazole (TBZ) fungicide, mainly

Fig. 27.15. A picking pole of the type commonly used for mangoes or fruits of other large trees (Colima, Mexico).

to control anthracnose caused by the fungus *Colletotrichum gloeosporioides* (Penz.) Arx. Disease control is more effective if the solution is heated (fig. 27.16), so postharvest fungicidal treatments commonly consist of water at 52°C (125.6°F) containing 0.1 percent TBZ. Fruits are commonly in the water for 1 to 3 minutes. Alternatively, benomyl may be applied as a spray after fruits have been elevated from the water dump-wash tank.

Sorting. Sorters generally examine fruit as it passes on moving belts. Fruits that are judged immature, overmature, or undersized, and those exhibiting limb-rub or other scarring or defects, are diverted.

Sizing. Sizing is usually by eye. Individuals, often the packers, select fruits of the desired size from a moving return-flow belt. Use of weight sizers and other mechanical sizers is increasing.

Packing. Although various containers are used, mangoes are commonly packed in one or more layers in fiberboard

Fig. 27.16. Mango packinghouse. A hot- water-thiabendazole bath is in the background, and a sorting table in the foreground (Colima, Mexico).

boxes that may have individual compartments for separating fruits.

Fumigation. Fumigation is required if the fruit has been grown in areas of infestation by one or more fruit fly species. Dosage of ethylene dibromide (EDB) for entrance into the U.S. varies with producing area and the nature of the fly problem. A dosage of 16 g/m^3 (16 oz/1,000 ft^3) of the total chamber space is common, if the fruit does not occupy more than 50 percent of the chamber volume. The fumigation period is 2 hours and the fruit must be at 21°C (70°F) or above. With some insect problems a lower dose will suffice.

Use of ethylene dibromide (EDB) within the United States as a fumigant on fruits and vegetables for consumption in the U.S. was prohibited effective September 1, 1984. Mangoes may be fumigated with EDB under USDA, Plant Protection and Quarantine supervision outside the United States for shipment into the U.S. Studies of alternative treatments, such as heat, are underway for use with Florida-grown mangoes.

Ripening. Mature-green mangoes may be subjected to an ethylene treatment—100 ppm for 24 to 48 hours at 20°C (68°F)—to promote faster and more uniform ripening. The treatment may be applied at the shipping point if transit time to market is less than 5 days, or at the destination if transit times are longer.

Pineapple Handling

The pineapple (*Ananas comosus* L., Merr.) fruit is a multiple structure that evolved, presumably, from a racemose inflorescence. Berry-like fruitlets, generally 100 to 200 in number, and their subtending leafy bracts are together fused to the core, a continuation of the fibrous peduncle. The fruitlets are in a regular spinal pattern on the fruit axis, the pattern usually consisting of two distinct spirals, one turning to the left and the other to the right.

The short, shoot-like, leaf-bearing growth at the top of the fruit is called the crown. The crown is a continuation of the original meristem of the plant's main axis extending through the fruit. Crowns are frequently used as planting materials after they are removed from fruits that are to be processed. Slips and suckers developing from axillary buds at the base of the leaves below the fruit are other choices for planting materials.

The most widely grown pineapple variety is the Smooth Cayenne, although other varieties are locally important, especially for the fresh market. These include Red Spanish, Queen, Singapore, Spanish, Selangor Green, Sarawak, and Mauritius. Postharvest handling operations are shown in figure 27.17.

Maturity

During maturation pineapples increase in weight, flesh soluble solids, and acidity. During ripening, carotenoid pigments and soluble solids of the flesh increase dramatically and the fruit attains its maximum aesthetic and eating quality. During ripening, the shell of the pineapple loses chlorophyll rapidly, starting at the fruit base, in a process similar to that of degreening citrus fruit. Approximately 110 days elapse between the end of flowering and ripeness.

Harvest time is often when the base of the fruit has changed from green to yellow or light brown. Fruits may be harvested for fresh market before striking color changes have occurred. Acceptable quality may develop before color changes occur in the shell. Since the pineapple fruit has no accumulation of starch, there is no reserve for major postharvest quality improvements. As a nonclimacteric fruit, obvious compositional changes after harvest are mostly limited to degreening and a decrease in acidity.

Harvesting

Pickers select fruit by size, color, or both, and twist it from the stalk. In small operations the harvested fruits may be placed in sacks or baskets that are carried to the end of the row to be picked up.

Modern, large-scale production avoids hand carrying by using a harvesting aid consisting of an endless belt extending across a number of rows (fig. 27.18). Fruits

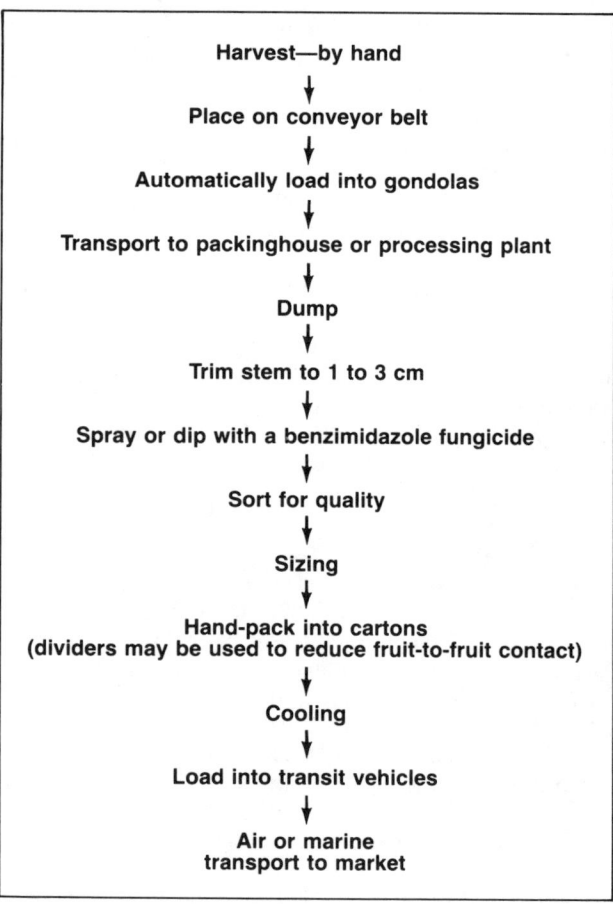

Fig. 27.17. A flow diagram of a postharvest handling system for pineapples.

picked by hand are placed on the belt, and it carries them to the machine where they accumulate in a bin or truck gondola. The machine and belts, usually mounted on a truck chassis, move slowly across the field at a speed determined by pickers.

Packinghouse operations

Fruits are sorted at the packinghouse to eliminate those that are defective. Sizing may be by eye, but weight sizers are increasingly used (fig. 27.19).

A fungicidal treatment consisting of a benomyl or TBZ dip or spray is commonly applied before packing to control water blister disease, caused by the fungus *Thielaviopsis paradoxa* (de Seynes) Hohnel. If uncontrolled, the disease is serious in fruits after harvest. The fungus enters the fruit via wounds or at the stem and grows rapidly throughout the fruit flesh. A very watery soft rot is produced. In the past, the stem surface was smeared with a paste containing a fungicide such as benzoic acid to prevent water blister.

Pineapples are sometimes waxed to reduce physiological disorders (fig. 27.20) and improve appearance.

Fig. 27.18. A pineapple harvest aid in use (Hawaii).

Fig. 27.19. Weight-sizing pineapples (Hawaii).

Fig. 27.20. Wax is applied to pineapple fruits.

Fruits are packed into full telescoping cartons (figs. 27.21 and 27.22) with inside dimensions of approximately 30.5 cm wide by 45 cm long by 31 cm deep. From 8 to 14 or 16 fruits of uniform size are placed in each container.

Holding and transit

The pineapple is a chilling-sensitive tropical fruit. The consequences of chilling include darkening of flesh tissues, particularly around the central cylinder. Temperatures below about 6°C (42.8°F) may result in chilling injury. Pineapple fruits are shipped in marine containers (fig. 27.23) or by air.

A physiological condition called endogenous brown spot (EDS) is frequently seen in pineapples (fig. 27.24). This disorder is more common in the winter, and may be associated with chilling in the field. EDS may however occur without exposure to chilling.

Fig. 27.22. Pineapples are packed in shipping cartons.

Fig. 27.23. A refrigerated marine container is loaded with pineapples.

Fig. 27.21. Workers hand pack pineapple fruits.

Fig. 27.24. Endogenous brown spot discolors pineapple fruits.

References

1. Akamine, E. K. 1967. History of the hot water treatment of papayas. *Hawaii Farm Sci.* 16(3):4.

2. _____. 1975. Hawaii: Papaya and pineapple handling for local and export markets. In *Postharvest physiology, handling and utilization of tropical and subtropical fruits and vegetables*, ed. E. B. Pantastico, 538-41, Westport, CT: AVI Publ. Co.

3. Barmore, C. R. 1974. Ripening mangoes with ethylene and ethephon. *Proc. Fla. State Hortic. Soc.* 87:331-34.

4. Bhullar, J. S., and H. L. Farnhan. 1980. Studies on the ripening and storage behavior of Safeda guava (*Psidium guajava* L.). *Indian Food Packer* 34(4):5-7.

5. Caygill, J. C., R. D. Cooke, D. J. Moore, S. J. Read, and H. C. Passam. 1976. *The mango (Magnifera indica L.). Harvesting and subsequent handling and processing: An annotated bibliography*. London: Trop. Prod. Inst. 124 pp.

6. Champion, Jean. 1963. *Le bananier*. Paris: Maisonneuve & Larose. 263 pp.

7. Collins, J. L. 1960. *The pineapple: Botany, cultivation and utilization*. New York: Interscience Publications.

8. French, C. D. 1972. *Papaya—The melon of health*. New York: Exposition Press. 96 pp.

9. Gortner, W. A., and V. L. Singleton. 1965. Chemical and physical development of the pineapple fruit. III. Nitrogenous and enzyme constituents. *J. Food Sci.* 30:24-29.

10. Krishnamurthy, S., and H. Subramanyam. 1970. Respiratory climacteric and chemical changes in the mango fruit (*Mangifera indica* L.). *J. Am. Soc. Hortic. Sci.* 95(3):333-37.

11. _____. 1973. Pre- and post-harvest physiology of the mango fruit: A review. *Trop. Sci.* 15:167-93.

12. Marriott, J. 1980. Bananas—Physiology and biochemistry of storage and ripening for optimum quality. *CRC Crit. Rev. Food Sci. Nutr.* 13(1):41-88.

13. Mukerjee, P. K. 1972. Harvesting, storage and transport of mango. *Acta Hortic.* 24:251-58.

14. Nagy, S., and P. E. Shaw. 1980. *Tropical and subtropical fruits—Composition, properties and uses*. Westport, CT: AVI Publ. Co. 570 pp.

15. Pantastico, E. B. 1975. *Postharvest physiology, handling and utilization of tropical and subtropical fruits and vegetables*. Westport, CT: AVI Publ. Co.

16. Py, Claude. 1965. *L'Ananas*. Paris: Maisonneuve & Larose. 298 pp.

17. Popenloe, J. 1960. Determinacion de madurez en los mangos. *Proc. Caribb. Reg., Ann. Soc. Hortic. Sci.* 4:31-32.

18. Rosenbaum, Helen. 1975. *How to grapple with the pineapple: From planting pineapple tops to baking up-side-down cakes*. New York: Hawthorn Books. 129 pp.

19. Shillingford, C. A. 1978. Postharvest banana fruit rot control with systemic fungicides. *Turrialba* 28(4):275-78.

20. Simmonds, N. W. 1962. *The evolution of the bananas*. London: Longman. 170 pp.

21. _____. 1966. *Bananas*. 2d ed. London: Longman. 512 pp.

22. Singh, Lal Behari. 1960. *The mango: Botany, cultivation and utilization*. New York: Interscience Publications.

23. Singleton, V. L. 1965. Chemical and physical development of the pineapple fruit. I. Weight per fruitlet and other physical attributes. *J. Food Sci.* 30:98-104.

24. Singleton, V. L., and W. A. Gortner. 1965. Chemical and physical development of the pineapple fruit. II. Carbohydrate and acid constituents. *J. Food Sci.* 30:19-23.

25. Slabaugh, W. R., and M. D. Grove. 1982. Postharvest diseases of bananas and their control. Plant Dis. 66(8):746-50.

26. Spalding, D. H., and W. F. Reeder. 1974. Current state of controlled atmosphere storage of four tropical fruits. *Proc. Fla. State Hortic. Soc.* 87:334-37.

27. Stover, R. H. 1972. *Banana, plantain, and abaca diseases*. Kew: Comm. Mycol. Inst. 316 pp.

28. Subremanyam, H., N. V. N. Moorthy, S. Lakshminarayana, and S. Krishnamurthy. 1972. Studies on harvesting, transport and storage of mango. *Acta Hortic.* 24:260-64.

29. Thompson, A. K. 1971. Transport of West Indian mango fruits. *Trop. Agric. (Trinidad)* 48(1):71-77.

30. Wardlaw, C. W. 1972. *Banana diseases; including plantains and abaca*. 2d. ed. London: Longman. 878 pp.

31. Wardlaw, C. W., E. R. Leonard, and R. E. Baker. 1934. Observations on the storage of various fruits and vegetables. II. Papayas, pineapples, grandadillas, grapefruit, and oranges. *Trop. Agric. (Trinidad)* 11:230.

28

Postharvest Handling Systems: Tree Nuts

ADEL A. KADER

Proper management of harvesting and postharvest handling procedures is important in attaining a maximum yield of good quality tree nuts, and that determines marketability and profit. This chapter includes a brief discussion of the steps involved in harvesting and postharvest handling of the three major tree nuts in California—almond, pistachio, and walnut—and their impact on quality and safety attributes.

Harvesting

When to harvest

Tree nuts should be harvested as soon as possible after maturation to avoid quality loss and to minimize problems involving fungal attack and infestation with insects, especially the navel orange worm. The following indices determine optimum harvesting dates.

- Almonds: dehiscence (splitting) of the hull; separation of hull from shell; decrease in fruit removal force (development of abscission zone); drying of hull and kernel
- Pistachios: ease of hull separation from the shell; shell dehiscence (splitting) and color; decrease in fruit removal force; kernel dry weight and crude fat content
- Walnuts: ease of hull removal (hullability); packing tissue browning (when the packing tissue between and around the kernel halves has just turned brown)

Uneven maturation presents a problem in once-over harvesting. Nuts on the tree's periphery usually mature earlier than those at the center. The use of ethephon to overcome this problem or accelerate maturation has produced mixed results that vary with the species. Although ethephon application accelerates almond maturation, it does not improve uniformity of maturity, and it may reduce yield and induce tree gummosis. Consequently, ethephon is not used on almonds. Studies with pistachio nuts indicate that ethephon application does not influence the nuts' maturation processes. Ethephon applied at packing-tissue browning will allow walnut harvest within 7 to 10 days (instead of 15 to 20 days without ethephon). This treatment is used commercially in walnut harvesting.

Harvesting procedures

Harvesting season in California depends on the cultivar, production area, and cultural practices in use. Almonds are harvested between July and October. Pistachios (primarily the Kerman cultivar) are harvested during the first 2 or 3 weeks of September. Walnuts are harvested during September and October. Orchard floor management is important in preparation for the harvest of almonds and walnuts because they are knocked to the ground during harvesting.

Most harvest operations are mechanized. Hand harvesting (knocking) is used on young trees and where steep terrain makes mechanical harvesting difficult. Pickers knock the nuts loose by hitting the branches with poles (mauls). Almonds and walnuts are usually knocked or shaken to the ground by mechanical tree shakers. Pistachios are harvested with a shake-catch mechanical harvester and the nuts are placed in bins (1.2 by 1.2 by 0.6 meter [4 by 4 by 2 feet]). Pistachios should not be shaken to the ground because of their open shells and high moisture content, relative to almonds and walnuts, at harvest.

Postharvest Handling

A comparison of the handling systems for almond, walnut, and pistachio nuts is illustrated in figure 28.1. Some important considerations:

1. Almonds should be picked up from the orchard floor as soon as they are dry to avoid exposure to adverse weather conditions, especially rain, and to minimize fungal infection and insect infestation.

2. Exposure of almonds to wet and hot conditions results in an internal disorder called *concealed damage*, characterized by a rust-brown to black discoloration of the nut meat and, in extreme cases, an unpalatable off-flavor.

3. It is very important to pick up, hull, and dry walnuts as soon as possible after harvest. Walnuts left on the ground can deteriorate (the kernels darken) rapidly especially at high ambient temperatures (e.g., 3 hours at 32°C [90°F]).

4. Pistachios should be hulled and dried soon after harvest to minimize shell staining and decay. If temporary storage of fresh pistachios at the dehydration plant is necessary, they should be cooled and held before hulling at 0°C (32°F) and a relative humidity lower than 70 percent. Sorting the nuts before cold storage to remove defective nuts, leaves, twigs, and other foreign materials (which are usually much more susceptible to decay) is important to minimize losses during cold storage.

5. Fumigation with methyl bromide, phosphine, or other chemicals to control insects is an essential step in postharvest handling, and it may have to be repeated periodically depending on the conditions and duration of storage.

Drying

Water removal by dehydration

Water is present in plant tissues in three forms: *bound water,* which is bound with other constituents by strong chemical forces; *adsorbed water,* which is held by molecular attraction to adsorbing substances; and *absorbed water,* which is held loosely in the extracellular spaces by the weak forces of capillary action. The absorbed and adsorbed waters constitute the "free water" that is removed by drying. Bound water is not removed except at very high temperatures that would also decompose some organic matter.

The moisture contents of almonds, pistachios, and walnuts at harvest range between 5 percent and 15 percent, 40 percent and 50 percent, and 10 percent and 20 percent, fresh weight basis, respectively. To improve stability and insure the safety of the nuts, they should be dried to 5 percent or 8 percent moisture as soon after harvest as possible.

Drying methods

Most of the final drying of almonds occurs naturally while they are on the orchard floor. However, when rainy conditions prevail during harvesting, heated-air drying may be used to complete their dehydration.

Walnuts are dried with heated air. Drying temperatures should not exceed 43°C (110°F), because higher temperatures induce rancidity in walnut oil.

Most commercial pistachio dehydration uses 71°C (160°F) air for 10 to 14 hours, depending on initial moisture content and efficiency of the drying system used. Air heated to 93°C (200°F) can be used without detriment to the quality or stability of pistachio nuts. This reduces drying time by about 50 percent and increases utilization of the dehydration facilities. The only disadvantage of using temperatures between 71° and 93°C (160° and 200°F) is an increase in the number of nuts with excessive shell opening from about 2 percent to 6 percent.

The use of ambient, unheated air is one alternative to heated-air drying. It is cheaper in terms of initial facility cost and energy cost. The main disadvantage of this method is the longer duration required, especially when ambient air temperatures are low and relative humidities are high. This may increase the potential for fungal growth and aflatoxin development.

Another alternative for pistachio nuts is to dry them in stages: to about 20 percent moisture content using heated air, and then to completion using ambient air at a flow of 70 feet per minute past the nuts. The rapid removal of water to attain less than 20 percent moisture greatly decreases the potential for fungal growth and aflatoxin development.

Post-drying storage

Dried nuts are usually stored in bins, silos, or other bulk-storage containers for a few weeks or several months before final processing and preparation for market. Optimum storage conditions of 10°C (50°F) or lower and 65 percent to 70 percent relative humidity must be maintained to minimize deterioration during storage. Protection against insects is also essential during the storage period.

Preparation for Market

Quality and safety factors for nuts

Appearance. Important factors in a nut's appearance include freedom from defect (shell staining, insect damage,

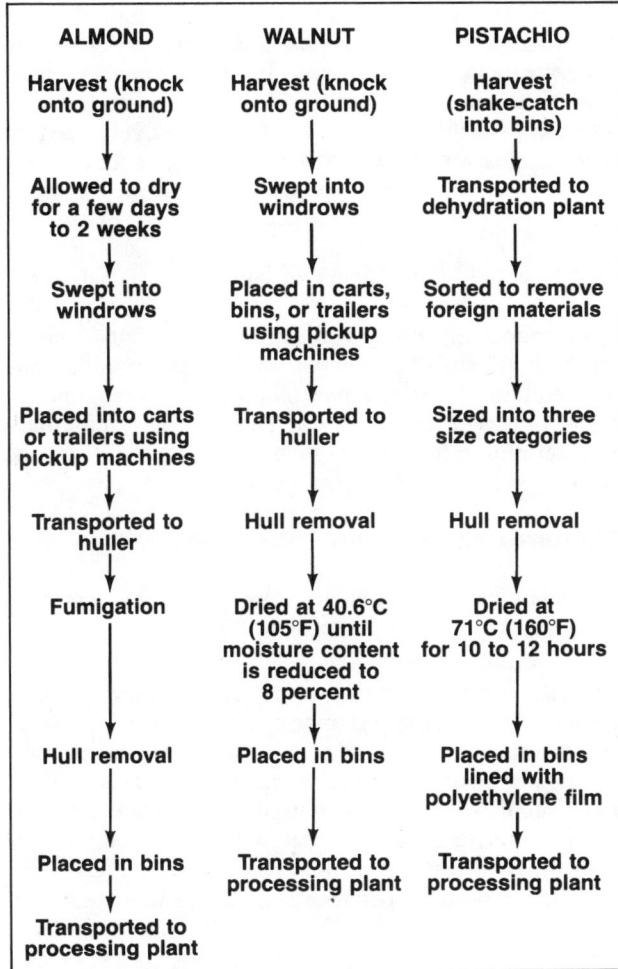

Fig. 28.1. Handling systems for almonds, walnuts, and pistachio nuts.

mold, adhering hull, kernel discoloration, shriveling), kernel size relative to nut size (percent edible portion), and cleanliness.

Texture. Crispness, chewiness, and other textural quality attributes are influenced by the degree and uniformity of dryness within the nut and among nuts.

Flavor and nutritive value. Sweetness, oiliness, and roasted flavor are usually related to good flavor, while rancidity is a major factor in poor quality. Nuts are excellent sources of protein, but they are also very high in fat content.

Safety. Aflatoxin contamination must be avoided by protecting the nuts against growth of *Aspergillus flavus* before harvest, and after harvest prior to drying. Aflatoxin is a highly carcinogenic secondary metabolite produced by the fungi *Aspergillus flavus* Link and *A. parasiticus* Speare. Often *A. flavus* is used in a collective sense to also include *A. parasiticus*.

Most prone to contamination in the orchard are the nuts poorly protected by hulls. In pistachios, *early-split* nuts vary between 1 percent and 5 percent from tree to tree. Early splits are nuts in which the shell splits before the hull has dehisced. Consequently, the entire pericarp splits to expose the kernel to the elements. In a recent study, about 20 percent of each 50-nut sample of early-split but otherwise good nuts were found to be contaminated with aflatoxin, as opposed to no contamination in regular-split nuts, where the shell had split within the loose, intact hull. Thin-shelled almonds are much more likely to be contaminated than thick-shelled nuts, probably because of poorer sealing of the sutures. Walnuts are less likely to develop aflatoxin than other nuts, but all nut species exposed to moist soil on the orchard floor risk infection.

All nuts infested with navel orange worm or certain other insects risk aflatoxin contamination, but these nuts pose less of a danger because they are easier to eliminate by sorting, and consumers are likely to reject them. In contrast, nuts not infested with insects cannot usually be eliminated by normal sorting methods. Furthermore, consumers will not necessarily reject them.

Early-split nuts not infected in the orchard may become infected during transport and handling. High humidities and temperatures within bulk bins provide ideal conditions for the infection of early-split nuts. The incidence of nuts with aflatoxin increases and aflatoxin levels rise dramatically, until nuts are mycologically stabilized by drying or refrigeration after removal of field heat. Fungicidal sprays in the orchard at the time of early splitting might materially reduce the potential for aflatoxin development, but no tests have demonstrated their effectiveness.

Processing operations

- sorting for elimination of defects—nuts with adhering hulls, stained nuts, moldy nuts—using visual or light-reflectance electronic sorting techniques
- separation of blank or empty pistachios by an air stream, and unsplit pistachios by flotation in water
- cracking shells to extract kernels when desired
- sorting for elimination of nuts with possible aflatoxin contamination, a labor-intensive and costly, but necessary, operation (Research efforts to develop a reliable automated sorting method are continuing.)
- sizing of in-shell or shelled nuts into size categories using mesh screens
- sorting by color of shell or kernel; chemical bleaching of in-shell almonds and walnuts to improve shell color
- treatment with antioxidants to slow rancidity resulting from the oxidation of fatty acids
- salting, flavoring, and roasting
- packaging of in-shell, shelled, nut meats (broken kernels) in various types and sizes of package—important considerations in packaging include protection against insects, providing an effective moisture barrier, exclusion of oxygen to slow down rancidity, and exclusion of light to minimize color deterioration of some nuts

Marketing and utilization

Tree nuts are marketed in shell or as shelled intact kernels or kernel pieces for use as snacks or in confectionary and bakery products. U.S. per capita consumption in 1980 was 0.42 pounds, 0.10 pounds, and 0.52 pounds for almonds, pistachios, and walnuts, respectively. Export marketing is very important since it accounts for as much as 50 percent of the U.S. production for some tree nuts.

Storage of Nuts

Maintenance of quality and storage life of nuts depends on the nuts' moisture content, the relative humidity and temperature in storage, the exclusion of oxygen, and the effectiveness of insect control. The role each factor plays in determining the storage potential of nuts is discussed below.

Moisture content

The relationship between moisture content and water activity (a_w) of nuts can be expressed as

$$a_w = 0.01 \times RH,$$

when RH (relative humidity) is in equilibrium with moisture content. Water activity is important in relation to the nuts' susceptibility to fungal attack, including mycotoxin-producing organisms (*Aspergillus flavus* and *A. parasiticus*). FDA regulations for tree nuts define a "safe moisture level" (moisture content that will not support fungal growth) as an a_w that does not exceed 0.70 at 25°C (77°F).

The relationship between a_w and moisture content of selected tree nut meats at 21°C (70°F) is shown below.

Nut	Moisture content at a_w 0.2 to a_w 0.8
Almond	3.0% - 8.7%
Pistachio	2.2% - 8.2%
Walnut	2.8% - 7.0%

Effect of relative humidity

The amount of water sorbed by a product as a function of relative humidity (RH) at a given temperature can be illustrated by the sorption curve (fig. 28.2) which is usually sigmoid and has three distinct sections:

1. The lowest section (labeled "a") of the curve, representing bound water, is concave to the RH axis. As the moisture content rises, the water molecules form a monolayer that coincides with about 20 percent RH.

2. As the RH increases, water molecules form successive layers of diminishing bond to the adsorbing substance of the commodity. This section of the curve (labeled "b") is almost linear, curving gently and consistently.

3. Above 75 percent RH (section "c" of the curve), the product absorbs water to saturation (large increases in moisture content result from small increases in RH). This encourages deterioration and attack by microorganisms.

A relative humidity of 70 percent at 25°C (77°F) is the lower limit for significant mold growth.

Effect of temperature

The relationship between moisture content and equilibrium relative humidity (ERH) is temperature-dependent. Between 20 percent and 80 percent ERH for any given moisture content, ERH rises about 3 percent for every 10°C (18°F) rise in temperature. For any given RH, air contains more water vapor at a high temperature than at a low temperature.

Lower temperatures reduce insect activity and retard mold growth and deterioration, including lipid oxidation (which results in rancidity). Temperatures between 0° and 10°C (32° and 50°F) are recommended for tree nuts, depending on expected storage duration. Duration can exceed 1 year; the lower the temperature, the longer storage life.

Effect of oxygen level

Low oxygen (less than 0.5%) atmospheres provide a beneficial supplement to proper temperature management for flavor quality maintenance and insect control. Exclusion of oxygen is usually accomplished by vacuum packaging or by packaging in nitrogen.

In-shell almonds and walnuts are much more stable than shelled nuts. The shell is an effective package to prevent oxidation during storage. The pellicle of a walnut kernel acts as a protective barrier and contains antioxidants.

Insect control

Fumigation treatments using methyl bromide or phosphine may be used on stored nuts before final processing and packaging. Reduced temperatures and oxygen levels (less than 0.5%) are effective for insect control as mentioned above; their use as fumigation substitutes is expected to increase in the future. The use of insect-proof packaging is essential to prevent reinfestation.

References

1. Beuchat, L. R. 1978. Relationship of water activity to moisture content in tree nuts. *J. Food Sci.* 43:754-55, 758.

2. Guadagni, D. G., E. L. Soderstrom, and C. L. Storey. 1978. Effects of controlled atmosphere on flavor stability of almonds. *J. Food Sci.* 43:1077-80.

3. Holmberg, D. 1978. Almond harvest. In *Almond orchard management*, ed. W. Micke, 143-45. Univ. Calif. Div. Agric. Sci. Publ. 4092.

4. Kader, A. A., C. M. Heintz, J. M. Labavitch, and H. L. Rae. 1982. Studies related to the description and evaluation of pistachio nut quality. *J. Am. Soc. Hortic. Sci.* 107:812-16.

5. Labavitch, J. M. 1978. Relationship of almond maturation and quality to manipulations performed during and after harvest. In *Almond orchard management*, ed. W. Micke, 146-50. Univ. Calif. Div. Agric. Sci. Publ. 4092.

6. Labavitch, J. M., C. M. Heintz, H. L. Rae, and A. A. Kader. 1982. Physiological and compositional changes associated with maturation of 'Kerman' pistachio nuts. *J. Am. Soc. Hortic. Sci.* 107:688-92.

7. Martin, G. C., G. S. Sibbett, and D. E. Ramos. 1975. Effect of delays between harvesting and drying on kernel quality of walnuts. *J. Am. Soc. Hortic. Sci.* 100:55-57.

8. Olson, W. H., G. S. Sibbett, and G. C. Martin. 1978. *Walnut harvesting and handling in California*. Univ. Calif. Div. Agric. Sci. Leaf. 21036. 7 pp.

9. Phillips, D. J., S. L. Purcell, and G. J. Stanley. 1980. *Aflatoxins in almonds*. USDA SEA ARM-W-20. 11 pp.

10. Sommer, N. F., J. R. Buchanan, and R. J. Fortlage. 1976. Aflatoxin and sterigmatocystin contamination of pistachio nuts in orchards. *Appl. Environ. Microbiol.* 32:64-67.

11. Wells, A. W., and H. R. Barber. 1959. *Extending the market life of packaged shelled nuts*. USDA Mark. Res. Rep. 329. 14 pp.

12. Woodroof, J. G. 1979. *Tree nuts: Production, processing, products*. Westport, CT.: AVI Publ. Co. 712 pp.

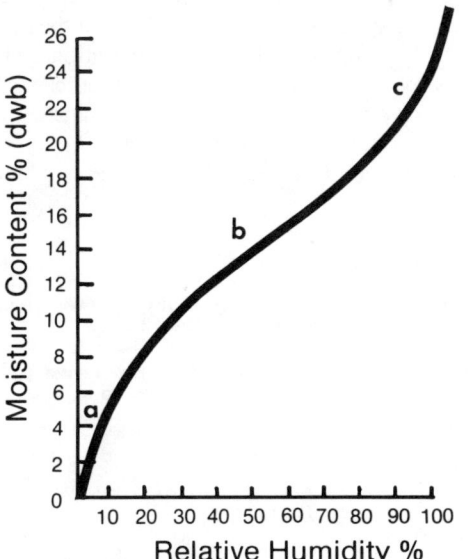

Fig. 28.2. Relationship between moisture content on a dry-weight basis (dwb) of dried products and ambient relative humidity.

29

Postharvest Handling Systems: Ornamentals

MICHAEL S. REID

Introduction

The international trade in ornamentals has greatly expanded in recent years, becoming an important source of overseas exchange in a number of countries, notably Israel, Colombia, Guatemala, Kenya, Thailand, and Mexico. While the bulk of this trade is in cut flowers, other ornamental commodities are important. These include cut foliage, rooted and unrooted cuttings of ornamental plants (particularly tropical foliage plants), bulbs, and seed materials.

Little information is available concerning the postharvest physiology of whole-plant ornamentals, or of other materials such as cuttings, nursery plants and turf. Despite lack of specific information, postharvest treatment of cut flowers (and indeed, fruits and vegetables) provides many guidelines for the care of other ornamental crops.

Factors in the Senescence of Ornamental Crops

Temperature

The rate of development and senescence of all perishables, ornamentals included, is strongly influenced by temperature. Table 29.1 shows that, between the proper storage temperature (0°C = 32°F) and room temperature, the Q_{10} (defined as the effect of a 10°C (18°F) rise on the rate of a process) for respiration and heat production of carnations is as high as 8—a dramatic effect of temperature on the metabolism of this commodity. Above room temperature, the Q_{10} drops to about 2. These effects mean that relatively short times at ambient temperature, even when the ambient is relatively low, can greatly reduce the overall storage or vase life of cut flowers (fig. 29.1). Proper temperature management is clearly the primary goal in upgrading the handling of cut flowers and other ornamental commodities. Their high respiration and heat production rates mean that failure to adequately cool cut flowers can result in the heating of packed products, leading to accelerated deterioration.

As with many fruits and vegetables, ornamental plants from tropical or subtropical regions may be deleteriously affected by temperatures below about 12.5°C (55°F) (fig. 29.2). Chilling injury in ornamental plants is most often seen in the nursery, because transport and storage temperatures are usually above the chilling threshold. The threshold chilling temperatures for ornamentals need to be determined so that losses of these sensitive commodities can be prevented.

Water Relations

Ornamental crops differ markedly from the other perishable commodities in their water relations. In most cases, they are exceedingly sensitive to desiccation, having a high surface area in relation to volume. Water that is lost during the postharvest period can normally be replaced from the vase solution when the commodity enters the retail distribution system. Nevertheless, desiccation is one of the most important postharvest problems in the handling of cut flowers. The prime example is "bent neck," a disorder of cut roses in which the water needs of the foliage and flower are provided at the expense of the relatively soft stem tissue just below the flower. A high relative humidity during storage and transportation can reduce the chances of water stress when ornamentals are removed from storage.

Present information on the movement of water in the stems of cut flowers is sketchy, but we do know that it can be strongly affected by the composition of the vase solution. Acidic solutions, for example, move much more readily through the stems of cut flowers than solutions which are neutral or alkaline. "Plugging" of the cut surface of the stem is considered a major limitation to the vase life of many cut flowers. Plugging can be caused by air occlusions or embolisms, microbes contaminating the vase solution, particulate or colloidal material in the water, or exudations from the cells surrounding the conducting tissues of the stem.

Table 29.1. Respiration rates and heat production of carnations at different temperatures

Temperature (Celsius)	Respiration rate (mg CO_2/kg-hr)	Heat production (Btu/ton-hr)	Q_{10}*
0°	10	89	*
10°	30	275	3
20°	239	2,191	8
30°	516	4,730	2.2
40°	1,053	9,653	2.0
50°	1,600	14,718	1.5

*$Q_{10} = \dfrac{\text{Rate of deterioration at T} + 10°C}{\text{Rate of deterioration at T}}$

Fig. 29.1. The optimum temperature for most common cut flowers is 0°C. Flowers kept for 7 days at 10°C deteriorate three times faster than those kept at 0°C.

Fig. 29.2. Anthurium flowers are sensitive to chilling injury as shown above for the flower kept at 0°C for 7 days. The optimum temperature is 10°C.

Carbohydrate Supply

Unlike fruits and vegetables, flowers can be cut before they are mature, at the bud stage. This is the normal commercial practice for some flowers (rose, gladiolus, see fig. 29.3); for others, flowers are normally cut almost fully open, but buds can be harvested to allow longer storage (carnation). The dry weight of a fully expanded rose flower is over twice that of the harvested bud. The flower stem cannot supply all the materials necessary to provide this increase in dry weight, so flowers opened in deionized (DI) water are generally of very poor quality and short vase life.

Fig. 29.3. Pulsing gladiolus flowers overnight in the cold room with a 20 percent sucrose solution enhances flower size, number, and vase life.

Because the supply of carbohydrates for development and respiration is important to the quality and vase life of flowers, the provision of adequate reserves is an important consideration in any strategy for handling these commodities.

Growing conditions

The most important preharvest factor governing the postharvest quality of cut flowers is the carbohydrate provided to the flower stem by the plant. For example, the effect of sucrose on the vase life of cut carnations is not observed at times of the year when light intensity during the day is high, and temperatures are relatively cool. Under these conditions, photosynthesis is optimal, and the flowers have high carbohydrate reserves at harvest. The high quality of imported Colombian flowers is thought to be due to their high carbohydrate content.

Storage conditions

The effect of temperatures on the respiration rate noted above indicates how higher storage temperatures might deplete a flower's carbohydrate reserves. Apart from reducing the rate of senescence, proper storage temperatures also conserve carbohydrates, allowing longer vase life at the retail level.

The use of floral preservatives to increase the quality and extend the vase life of cut flowers is now standard for florists. Preservatives contain sucrose as a carbohydrate source.

The effect of sucrose on the quality of gladiolus flowers is shown in table 29.2. The florets of spikes "pulsed" with a solution containing sucrose before storage opened better and were larger than those of control spikes.

Growth Regulators

Much information is available regarding the effects of growth regulators on the senescence of ornamental commodities. Ethylene can cause a dramatic acceleration of senescence for some flowers. Treating carnations with relatively low concentrations of ethylene for 2 days completely destroys them (table 29.3). The effect, like other physiological responses to ethylene, can be inhibited by elevated carbon dioxide atmospheres. When handling flowers sensitive to ethylene, the exclusion and removal of the gas, or inhibition of its action, is very important.

Cytokinins have been shown to extend the vase life of some cut flowers. For example, a 50 percent increase in the longevity of cut carnations can be obtained by adding 5 ppm kinetin to the vase solution. This material is an amendment in at least one floral preservative. Cytokinins are used in Israel's ornamental industry to prevent foliage senescence in cut flowers and other commodities.

Pathogens

Petal tissues are very susceptible to attack by fungal pathogens. Most important is the gray mold *Botrytis cinerea*, which grows readily when the humidity is high, even at low temperatures.

Postharvest Technology of Ornamental Crops

Maturity

Among ornamental crops, maturity is normally estimated based solely on external factors (size of plant, openness of flower), so there are few problems associated with assessing proper harvest maturity. Maturity may be important in the postharvest life of the commodity; for example, tight-bud roses are much more susceptible to "bent neck" than those that are harvested slightly more open.

Harvesting

Harvesting is usually a hand operation for ornamental crops. Mechanical aids are few, but they include mobile picking platforms, shears designed to hold the cut stem, and overhead conveyor systems for transporting flowers to the packing shed. Studies of "once-over" harvesting for single crops such as chrysanthemums suggest that this technique is economically feasible, but no practical application of these findings exists to date.

Grading

Though there are no official grade standards for cut flowers, the Society of American Florists has promulgated some standards for its members. Flower quality is judged by bud or bloom regularity, stem strength and straightness, foliage quality, and stem length. For roses, equipment that separates flowers according to stem length is in widespread use.

Table 29.3. Survival of carnations after treatment with 0.2 ppm ethylene (C_2H_4) and 2 percent carbon dioxide (CO_2) for 2 days

Treatment	Vase life (days)
C_2H_4	0
C_2H_4 + CO_2	6.1
Control	5.4
Least significant difference at the 5% level	0.4

Table 29.2. Effect of a sucrose pulse on gladiolus spikes held in deionized water for 2 days at 20°C after storage for 10 days at 2°C

Treatment	Floret diameter (cm)	Florets open per spike
Pulsed in water	6.2	10.8
Pulsed in 20% sucrose	11.1	14.8

Bunching

Flowers are normally sold in bunches of 10 to 25 stems each. Most bunching is done manually, sometimes with simple mechanical aids such as mirrors and electric-eye counters. A bunch is generally protected by a paper or polyethylene sleeve.

Packaging

Flowers are normally packed flat in long boxes, with the heads of alternate layers at alternating ends of the box (fig. 29.4). In California, a standard fully telescopic box with an end dimension of 51 cm by 30 cm (20 inches by 12 inches) is in wide use. For some crops such as gladiolus and marguerite daisies, special packages are still used (tall upright boxes and waxed cardboard hampers, respectively).

Fig. 29.4. A cutaway view shows the proper packing of chrysanthemum flowers. Flower heads are placed at both ends of the carton for maximum space utilization, and immobilized to minimize transport damage.

Postharvest temperature management

Forced-air precooling. Several large, forced-air cooling facilities are already used in California to remove field heat from flowers. It is likely that the proportion of the California cut flower crop handled in this way will increase rapidly. The standard box is provided with holes and flaps that permit proper air movement through the carton and around the commodity and can be closed after cooling.

Room cooling. The majority of the cut flower crop is still cooled by simply placing the flowers—packed, unpacked, in closed or open buckets—into a cooler. This is more than adequate for unpacked flowers.

Vacuum cooling. The high surface-to-volume ratio of ornamental commodities makes them good candidates for vacuum cooling. Some years ago the method was considered to have great potential for ornamentals, and one of the largest California growers still has a (inoperable) vacuum cooler at his facility. The ease with which the products can be cooled with forced air has limited further interest in this method.

Package icing. Traditionally, roses have been transported with flake ice, or, more recently, frozen package blocks in each box. The practice originated when air freight, which provided no external means of refrigeration, was the sole means of transporting flowers. Many shippers still consider this an important part of their operation, particularly for flowers that may not always be refrigerated during distribution. Proper placement of flake or gel ice is important to the efficient interception of incoming heat. Water from the melting ice weakens boxes and adds expensive weight. Icing should therefore be avoided if the flowers are precooled and transported under refrigeration throughout the distribution chain.

Water relations

After temperature management, adequate water is the most important consideration in the handling of ornamental crops.

The bucket syndrome. Without adequate refrigeration, it is impossible to handle cut flower crops "dry," so the whole industry is geared to the use of water buckets as containers for cut flowers. Now that proper temperature management is being implemented, these buckets—which are heavy, use a great deal of refrigeration space, and are often badly contaminated with bacteria, fungi, and algae—should become less important in the handling of cut flowers.

Water quality. The dramatic effects of varying water quality—pH, and dissolved particulate and colloidal matter and solids—on the vase life of cut flowers show that good-quality water is an important element in the postharvest handling of cut flowers. In many California greenhouses the water comes from deep wells, and has a high pH and a very high salt content. Many operators therefore use ion-exchange resin deionizers to provide water of an adequate quality for postharvest use.

Water amendments. Considerable improvement in the hydration of cut flowers can be achieved with any water supply, simply by adding sufficient acid to reduce the pH to between 3 and 3.5. In practice, citric acid is a good amendment because it provides this pH without the danger of the pH falling lower if too much is used. The salts of 8-hydroxyquinoline and aluminum sulphate, commonly used in commercial flower preservatives as bactericides, probably also benefit flowers because of their low pH.

Dry handling. In the future, we may see cut flowers change to a dry handling system where the flower is not put into water until it enters the retail distribution system—or even, perhaps, the consumer's home. The advantages would include longer storage life, more efficient use of refrigeration space, and reduced costs of postharvest handling. This goal will not be realized without modifying the harvesting, handling, and storage systems to keep water loss to a minimum. Specifically, reducing the time between harvest and establishment of the proper storage temperature through the use of forced-air precooling will play a major role. To prevent desiccation during storage, high-humidity storage systems, humidifiers, and containers with vapor barriers may all be important.

Carbohydrate supply

Carbohydrate supply is best in quality flowers grown under high-light-intensity conditions and harvested at proper maturity, but even these flowers will benefit from additional carbohydrates. Commercial preservatives (Floralife, Oasis, Florever, Viva La Fleur, and the like) contain sucrose or glucose as a carbohydrate source. Extra carbohydrates can be supplied in one or both of two ways.

Pulsing. It is possible to supply adequate carbohydrates for some flowers by standing them in concentrated sucrose solutions for up to 24 hours immediately after harvest. Table 29.2 shows the effects of such a pulse sucrose treatment on the quality of gladiolus.

Vase solutions. Sucrose and other sugars constitute an important part of most preservatives marketed for use in vase solutions, and are normally combined with a biocide. The problem with using proprietary preservatives as vase solutions is that the optimum sucrose concentration for different flowers varies widely. Concentrations above 1.5 percent cause severe foliage burn in cut roses, though they have little effect on the vase life of carnations. In practice, most formulations use relatively low sucrose concentrations to avoid the danger of phytotoxicity, but they fail to provide adequate carbohydrate for maximum benefit in many flowers. Moreover, the cost of prepared commercial preservatives far exceeds the cost of their ingredients. A range of simply formulated, useful vase solutions is shown in table 29.4.

Growth regulators

The technologies that are available to eliminate ethylene from storage rooms, handling areas, and packages are seldom utilized by the ornamental industry. Purafil pads are sometimes hung in cool rooms, but this amounts to a token gesture. Much is made of the use of hypobaric containers for extending the storage life of carnations, but those favoring their use are primarily their manufacturers. Adequate ventilation is probably the cheapest method of reducing ethylene concentrations in flower handling areas. Gross contaminations with ethylene are common in all phases of the distribution of ornamental products.

The antiethylene effects of the silver ion are being utilized now by some California growers and shippers of cut flowers, who pulse the flowers with a solution of the silver/thiosulfate complex after harvest. This treatment completely inhibits the action of ethylene in causing the flowers to "sleep."

As mentioned earlier, the effects of the cytokinins on vegetative and floral tissue senescence are in some use, particularly in Israel's ornamental industry.

Storage technology

At present, organized, long-term storage of flowers is uncommon, despite the obvious benefits of successful storage for an industry where demand changes dramatically through the year. Conditions in most cool storage areas are far from ideal, and the adoption of more sophisticated technologies, such as controlled atmosphere and hypobaric storage does not seem warranted until maximum efficiency has been achieved in common cool storage. Only minor benefits have been reported by studies on the use of newer storage techniques with ornamentals, although one grower does store his production of chrysanthemum cuttings in hypobaric conditions. Carnation buds have recently been shown to survive up to 6 months storage at 0°C (32°F) in sealed polyethylene packages that insure high humidity and a modest increase in carbon dioxide concentration. This successful storage depends entirely on the use of new fungicides (Rovral, Ornalin, and so on) to control *Botrytis cinerea*.

Table 29.4. Preservative solutions for cut flowers

Solution components	Uses
1.5% sucrose, 320 ppm citric acid	Vase solution, roses
1.5% sucrose, 320 ppm citric acid, 25 ppm silver nitrate	Vase solution (except roses)
1.5% sucrose, 250 ppm 8-hydroxyquinoline citrate	General vase solution
20% sucrose, 250 ppm 8-hydroxyquinoline citrate	Overnight pulsing of gladiolus
10% sucrose, 200 ppm physan	Bud opening of carnations
2% sucrose, 200 ppm physan	Bud opening of chrysanthemums

References

1. Halevy, A. H., and S. Mayak. 1979. Senescence and postharvest physiology of cut flowers, part 1. *Hortic. Rev.* 1:204-36.

2. _____. 1981. Senescence and postharvest physiology of cut flowers, part 2. *Hortic. Rev.* 3:59-143.

3. Kofranek, A. M., and M. S. Reid, eds. 1981. Second international symposium on postharvest physiology of cut flowers. *Acta Hortic.* 113:1-189.

4. Maxie, E. C., F. G. Mitchell, and N. F. Sommer. 1974. Postharvest handling of cut flowers. *ASHRAE Trans.* 80:350-55.

5. Rij, R. E., J. F. Thompson, and D. S. Farnham. 1979. *Handling, precooling, and temperature management of cut flower crops for truck transportation.* USDA SEA AAT-W-5, 26 pp. (Univ. Calif. Div. Agric. Nat. Resour. Leaf. 21058).

30

The Extension Link: University Research and the Horticultural Crops Industry in California

I. Cooperative Extension Methods

ROBERT F. KASMIRE

There is a great need for effective extension programming in postharvest technology to help solve the industry's problems and to disseminate research results and other information useful to the fresh market horticultural-crop shipping, marketing, and distribution industry. To fill this need extension personnel must be highly motivated professionals, well-trained and equipped. They must be located where they can readily and effectively interact with both research workers and the agricultural industry.

The Role of Extension

The overall objectives of an extension postharvest program are (1) to improve the quality and value of horticultural crops available to consumers, (2) to reduce marketing losses of horticultural crops, and (3) to improve marketing efficiency. Specific objectives focus on solving particular problems affecting one commodity or a group of commodities.

To plan a postharvest extension program, one must first identify the problem, and identify and understand the clientele being addressed. Sometimes the audience's composition, background, and distribution are themselves major constraints. The fresh produce shipping, marketing, and distribution industry is very heterogeneous and widespread, often including handlers at distant points in the same country or in more than one country. Initially, this clientele will know little, if anything, about extension's role and objectives, and may be highly suspicious of any attempts to change their handling practices, techniques, or facilities. To gain their attention and confidence, a program must first demonstrate that it can benefit them, most often economically. Once this is accomplished, they can be as eager and cooperative as any other extension audience. The clientele of the postharvest extension specialist consists of local (county) extension workers (farm advisors or county agents) and industry personnel involved in the preparation, shipping, marketing, and distribution of fresh-market horticultural products. It also includes students, via teaching, guidance, and counseling. The local extension worker's clientele consists primarily of industry personnel within a specific county.

After learning about your audience, the following steps can be helpful:

1. Identify problems to be resolved and determine their priorities. It is not possible to work on all problems at one time. Determine which are the most important and which can be realistically solved, and then assign priorities in both long- and short-range plans.

2. Develop the long- and short-range (1-year) objectives of the program. For example a long-range objective would be to improve the nutritional quality of fresh fruits and vegetables sold to consumers; a short-range objective would be to improve cooling practices and facilities used before loading.

3. Determine the elements needed to conduct the long- and short-range programs. Compare the manpower, equipment, and facilities that will be needed with those that are currently available. It is best to involve cooperators in programs whenever possible.

Extension Methods

Many methods are available for extension workers conducting useful programs. The relative effectiveness of these methods depends upon the involvement of research workers, extension specialists, local agents, and industry clients. Methods most commonly used in postharvest extension programs include:

Applied research. Many handling problems and practices in the industry cause increased product deterioration and marketing losses. Some problems are relatively simple and can be studied easily. Most applied research studies the causes or magnitude of deterioration or losses, and develops and evaluates possible corrective measures. While some research must be conducted in laboratories, other research is conducted in industry facilities. All applied research must use scientifically sound methods and procedures. It is often necessary, even desirable, to cooperate with other researchers from universities, government agencies, and industry. This makes researchers aware of industry problems and adds their scientific input, and it makes industry aware of the problem-solving assistance available from scientists.

Consultations. Requests for consultations most often originate with industry leaders who want to improve their operations. Consultations include individuals, companies, or other groups. Consultations deal with specific subjects (e.g., cooling methods) and broad subjects (e.g., commodity-day meetings to cover various aspects of producing and handling a given commodity). Repeat requests for consultations are sure indications of the client's satisfaction with the technique.

Group meetings. Meetings with groups allow the extension worker to present information to larger audiences than through consultations. Group meetings also encourage audience participation and discussion.

Demonstrations. Extension personnel can use demonstrations to show how to use a new practice, procedure, or facility, or to illustrate its results. Demonstrations are often used to extend the results of applied research. Careful attention to the equipment, facilities, and visual aids used in demonstrations can increase their effectiveness. Tutorial audio visual programs such as slide sets, video tapes, or movies, can also be used to demonstrate new procedures and facilities.

Short courses. An intensive, broad coverage of specific subjects can be administered to an audience in a short course, generally through classroom lectures, laboratory demonstrations, or tours. The Postharvest Technology Short Course offered annually by the University of California at Davis is an example. Other short courses and training workshops are conducted by individual marketing firms, for their own personnel, or by trade association specialists. Cooperative Extension postharvest specialists can expand the effectiveness of their programs by participating in other industry short courses. Short courses may be from 1 day to about 2 weeks in length. They may be used to refresh the audience with previously learned information, provide updated or new information on a subject, or both. Effective short courses require much planning, considerable professional involvement and input, proper facilities, and sufficient follow-up to evaluate effectiveness. Inadequate preparation reduces a course's effectiveness and can result in complete failure. Printed material, including a syllabus, is generally provided to each participant in a short course.

Workshops. Cooperative Extension workshops can improve the skills of individuals or groups. For example, in postharvest technology, one might conduct workshops on grading, sorting, packing, cooling, or careful handling. Workshops can last for one or more hours, and can meet only once or, as a series, over a period of time.

Tours. Facilities and operations are often easier to understand after a well-run tour. A tour can be a very effective way to introduce a new subject to an audience. Industry cooperation is essential to a successful tour.

Publications. A publication can be written in a variety of formats: technical, semitechnical, or popular. It is most important to publish information in a way that will reach its intended audience. Following are examples of types of publications:

Technical and semitechnical. These include articles on applied research published in professional societies' publications and in university and government technical reports. They are written primarily for the benefit of professional workers; very few industry representatives read these publications.

Progress reports. These publications extend current information to cooperators, research sponsors, and other interested persons—for example, observations or results of a just-completed preliminary study on the effects of a questionable, presently used industry practice, distributed to sponsors and industry personnel. They are usually brief reports, ranging from one to a few pages long. Their main advantages are timeliness, brevity, and directness, and they keep cooperators informed about the results.

Brief, single subject guides. This type of publication addresses a single specific subject or development. The University of Florida Fact Sheets are examples.

Newsletters. These periodical publications extend information to broad audiences on a regular basis, typically four to six times a year. A good newsletter is an effective route for extension of brief, pertinent reports and articles to the postharvest industry and to fellow Extension and research workers. Our *University of California Perishables Handling—Horticultural Crops Newsletter* has been a very effective informational tool since its founding in 1962. An issue usually contains a review article on a specific subject related to postharvest handling, brief articles on recent postharvest research results, and a list of recent postharvest publications and reports. The readership for this newsletter comprises about 50 percent professionals (extension, research, libraries, and government agencies) and about 50 percent industry recipients. The University of Florida postharvest newsletters for industry, *Packinghouse Newsletter* and *Handling Florida Vegetables*, and Washington State University's *Postharvest Pomology Newsletter* are other effective newsletters. The industry also has its own postharvest newsletters.

Trade publications. Trade articles are most effective for extending information to the industry handlers who regularly read such publications. Monthly magazines and bulletins produced by grower-shipper associations—*The Western Grower and Shipper*, published by the Western Growers Association; monthly *PMA Bulletin*, published by the Produce Marketing Association; and the *Fresh Outlook*, published quarterly by the United Fresh Fruit and Vegetable Association—and weekly newspapers of the fresh produce industry, *The Packer* and *Produce News*, are included in this group. Articles by postharvest extension workers that appear in trade publications extend information to a broader, often harder-to-reach audience. These trade associations and their publications can be very effective media for extension postharvest programs.

Results from specific studies or other relevant information are often published in two or more types of publication in order to extend the information to a broader audience and achieve the maximum desired effect. For example, the more scientific details of a study might be reported in a professional society journal, the semitechnical aspects published in a university or government report, and a popular report of the study, showing its relevance to the postharvest industry, might be published in an Extension newsletter and one or more trade publications.

Committees and programs. Extension postharvest technology programs can be effectively enhanced through participation in professional society and trade association programs and committees. This means more work for extension personnel, but it can also mean greater program effectiveness. In the University of California postharvest programs, we are all active in American Society for Horticultural Science work and we participate actively with several industry trade associations, both as program participants and as working committee members (e.g., United Fresh Fruit and Vegetable Association, Produce Marketing Association, California Grape and Tree Fruit League, Western Growers' Association, and the American Society of Heating, Refrigeration and Air Conditioning Engineers). We also work closely with various committees and advisory boards for fresh-market horticultural crops established by state marketing order programs. Trade associations and marketing order boards can help extend information more effectively and widely than is possible working alone.

Personal Program Development

Postharvest extension workers must maintain a high degree of professionalism if their programs are to be effective. They should continue to take courses and short courses when possible, and take sabbatical leaves to advance their capabilities. They must be familiar with appropriate modern techniques and instrumentation. Extension workers need to be creative and use initiative in program approaches, techniques, and skills. They also must be highly motivated. They must seek out most postharvest handling problems and the groups confronted with those problems because industry handlers will not reveal their problems until extension gains their confidence. Finally, extension workers must be able to communicate effectively in speech and in writing, and they must be good listeners, able to take advantage of any constructive criticism made about their program or skill. Criticism can simply mean that people are listening to what the speaker is saying and are observing extension programs. By listening to criticism, extension workers involve more people in their programs, and that helps program effectiveness.

Program Coordination

The key to success in extension programs is to maintain effective coordination of research and industry personnel. There are several ways to accomplishing this coordination, and one is discussed in the following section of this chapter.

30
The Extension Link—Continued

II. Extension and the California Fruit Industry

F. GORDON MITCHELL

The State University System

In the United States, extension and local research programs are conducted under the auspices of each state's land grant university. Additionally, the U.S. Department of Agriculture operates a research program designed to study problems of national or regional scope. It also maintains offices to coordinate research and extension activities among the states.

In California, agricultural research and extension responsibilities rest with the University of California. Agricultural faculties on the various campuses have dual responsibilities for research and teaching, and Cooperative Extension personnel are responsible for all extension activities.

The California Deciduous Fruit Industry

Total fruit and nut production in California covers about 2 million acres, with an annual product value of about $3 billion. The program described here involves the pomology portion and excludes grapes (over 600,000 acres) and subtropicals (over 300,000 acres), which have their own programs. Pomological crops involve over 1 million acres in California, with an annual product value exceeding $1.5 billion, including 13 major crops and 7 important minor crops. California produces 90 percent to 100 percent of the U.S. total for 11 of these crops.

Some deciduous fruit production occurs in most of California's 58 counties, although the biggest part of production is centered in the large, fertile Central Valley and in several smaller coastal valleys. Although typically of a Mediterranean climate, considerable climatic differences exist between fruit-growing areas within the state.

Research Responsibility

The pomological research responsibility rests with the Pomology Department of the University of California, Davis. Considerable fruit research activities are also conducted by other University departments in such areas as pest management, engineering, food science, soils, and water. The academic staff of the Pomology Department is now at about 20, and researches various areas of plant improvement, production physiology, and postharvest handling. Additional research responsibilities rest with extension personnel.

Extension Responsibility

The agricultural portion of the University of California Cooperative Extension is essentially composed of farm advisors and statewide extension specialists. The farm advisor offices are located in most California counties, and the specialists work with research department personnel on University campuses or at field stations. About 30 farm advisors have part- or full-time responsibility for pomological crops, and 5 specialists are attached to the Pomology Department.

Farm Advisor Programs

As academic staff members of the University of California, all farm advisors are expected to develop their own programs, using those techniques that are best suited to the needs of the fruit industry within their area. Thus, program emphasis varies considerably up and down the state. Most local programs involve some combination of newsletters, local publications, radio broadcasts, short courses, field meetings, field demonstrations, demonstration plots, problem-solving research, and individual problem diagnosis.

In recent years, there have been some "area" or "cross-county" assignments for farm advisors. This has sometimes allowed the farm advisors to specialize in narrower commodity or subject orientations, and has eliminated duplication of effort within an area. Usually, cooperative agreements between counties are established so that each county program receives its full allotment of extension resources.

Research activities by farm advisors have also received increasing emphasis. Awareness of the value of extension involvement in research has grown over the years, and now most farm advisors conduct some local research. Depending on the advisor and the needs of the county or area, this can involve a substantial dedication of time.

Research can be conducted jointly with departmental research personnel, other advisors, and specialists. Such activities provide important local solutions to pressing industry problems and help the advisors to become recognized authorities in pomology.

Extension Specialist Programs

Over the years, the specialist program has changed dramatically in California. Originally, specialists' offices were all together with the extension administrators', sometimes far from the campus of the subject matter department. At present, virtually all specialists are attached to and housed with the relevant subject matter department, and they often function essentially as members of the department. Many have researcher and lecturer titles within the department, especially if they are involved in graduate student training, and some have joint appointments in extension and the department.

Extension specialists typically maintain active programs in cooperation with farm advisors throughout the state. This involves meetings, conferences, training sessions, publications, and individual consultation as needed. The specialist coordinates the field research being conducted by a number of farm advisors, maintains contact with statewide organizations within the fruit industry, and makes the research department aware of industry research needs. The pomology production specialists have divided their responsibilities along commodity lines.

In recent years another type of extension specialist has evolved in California, the *clientele specialist*. The clientele specialist works in less traditional areas that are often of limited concern to the farm advisor, and typically working with segments of the agricultural industry or agribusiness that do not fit into county lines. This specialist's major activities involve direct contact with industry clientele and more limited contact with farm advisors.

Each extension specialist is now expected to maintain a personal research program. The research emphasis of the specialist depends on the needs of his or her program. Research may be conducted jointly with departmental research personnel and farm advisors. Projects are expected to involve problem solving or adaptive studies, the results of which are directly applicable to industry needs.

The Extension Professional

The strength of the extension program is in large measure determined by the quality of its personnel. In California, extension personnel are full academic staff members of the University of California. While salaries may not match some in the private sector, they are high enough to keep employee turnover to a minimum. Considering how many years it takes for an extension worker to reach peak effectiveness, this is important. Extension personnel are entitled to sabbatical leave privileges and are expected to develop sabbatical study programs designed to improve their effectiveness. They are also expected to participate in professional societies and other programs to maintain their professional competence. The trend toward closer contact with research personnel further helps to maintain that professionalism.

Pomology-Extension Continuing Conference

During the past several years, pomology extension personnel and Pomology Department researchers have coordinated their efforts through the Pomology-Extension Continuing Conference in an attempt to improve their overall response to the California fruit industry's needs. As the name implies, this is a continuing program and involves farm advisors, extension specialists, and Pomology Department academic personnel. The conference has several parts and several purposes. About once each year all personnel meet for 2 to 3 days to report, review, and plan. Smaller commodity working groups coordinate statewide research and extension activities within individual commodity areas. These groups may involve people from other departments or agencies, and may call upon the fruit industry for participation. The groups assign research priorities, discuss plans, and project the needs for publications or other activities. New cooperative projects have emerged from some of the discussions.

While the entire program is voluntary, participation has been good. Many participants have responded to group decisions by starting, stopping, or modifying research projects. Coordination of the varied elements of the pomology program has been facilitated, and the conferences's deliberations have helped project future staffing needs for both organizations.

Pomology Short Courses

While certainly not a new concept, the short course has emerged as an important activity closely linking extension, research, and the fruit industries. Pomology short courses are scheduled through the Pomology-Extension Continuing Conference, with generally no more than two scheduled each year. To date, pomology short courses have been held on walnut, almond, and prune production; kiwifruit production and handling; and stone-fruit handling.

These short courses combine classroom lectures and field or laboratory demonstrations in a 3- to 5-day program. This format allows more in-depth coverage of the subject matter than is possible in shorter meetings. The courses are intended to train and refresh industry personnel who have active responsibilities in the subjects covered.

Short courses provide a special opportunity for research personnel and extension personnel (both farm advisors and specialists) to work together in planning, preparation, and presentation. Faculty selection for a short course is based entirely on subject-matter competence and teaching ability. Faculty are asked to prepare written material for a course syllabus, and whenever possible, syllabi are reworked into permanent form as comprehensive published manuals on the subject covered. To date, these short courses have been well-attended and well-received by fruit industry personnel.

One Program

From the point of view of the California fruit and nut industries, the University activities in research and extension are a single program. The agricultural audience does not distinguish between research and extension personnel; it is simply concerned with obtaining the help and information it needs from an appropriate authority. The University of California's Pomology-Extension Continuing Conference is an attempt to effect a single, coordinated program that will serve these needs.

Index

A

Almond
  handling systems, 170-71
  harvesting, 13, 170
  insect control, 102
  maturity indices, 170
  quality standards, 130
  storage
    recommendations, 172-73

Anise, sweet
  quality standards, 126

Anthurium
  chilling injury
    susceptibility to, 175

Apple
  anthocyanin development, 4
  bruising, 29
  chilling injury
    susceptibility to, 6
  cooling, 144
  culls and cull utilization, 27
  disease
    control, 89
    defense mechanisms, 77, 84
    organisms, 86, 91
  energy use, 115
  ethylene
    effects on, 35
    production rate, 4
  handling systems, 144
  harvesting, 13
  maturity indices, 8, 10
  maturity
    prediction of, 11
  modified atmospheres, 58-59, 61, 144
  physiological disorders, 5, 143-44
  quality factors, 119, 120
  quality standards, 124, 128
  respiration
    pattern, 4, 76
    relative rate, 3
  shipping containers, 30-32, 144
  storage
    recommendations, 144
  temperature effects, 6
  transport
    mixed loads, 110
  water loss, 56

Apricot
  carotenoid development, 4
  chilling injury
    susceptibility to, 6
  disease
    organisms, 86
  ethylene
    production rate, 4
  harvesting, 13
  modified atmospheres, 59
  preparation for market, 20
  quality standards, 123-24
  respiration
    pattern, 4, 76
    relative rate, 3
  temperature effects, 6

Artichoke
  chilling injury
    susceptibility to, 6
  cooling, 133
  culls and cull utilization, 26
  ethylene
    production rate, 4
  handling systems, 134
  harvesting, 12
  maturity indices, 9
  modified atmospheres, 60
  quality standards, 126
  respiration
    relative rate, 3
  storage
    recommendations, 134
  temperature effects, 6

Asparagus
  chilling injury
    susceptibility to, 6
  cooling, 133
  ethylene
    effects on, 72
    production rate, 4
  geotropism, 4
  handling systems, 133
  harvesting, 12, 131
  maturity indices, 8
  modified atmospheres, 60
  preparation for market, 22
  quality factors, 118
  quality standards, 126, 129
  respiration
    relative rate, 3
  storage
    recommendations, 134
  temperature effects, 6
  texture, 4

Avocado
  chilling injury
    susceptibility to, 6, 153
  culls and cull utilization, 26
  disease
    control, 90, 95, 153
    defense mechanisms, 85
    organisms, 87, 92, 153
  ethylene
    production rate, 4
    threshold concentration, 69
    treatment, 154
  handling systems, 154
  harvesting, 13, 154
  maturity indices, 10, 152
  maturity
    prediction of, 11
  modified atmospheres, 59, 156
  packinghouse operations, 154
  quality standards, 124
  respiration
    pattern, 4, 76
    relative rate, 3
  ripening, 152
  storage
    recommendations, 156
  temperature effects, 6
  transport
    mixed loads, 110

B

Banana
  chilling injury
    susceptibility to, 6
  disease
    control, 95, 160
    defense mechanisms, 84
    organisms, 87-88, 92, 160
  ethylene
    inhibition of action, 73
    production rate, 4
    threshold concentration, 69
  fungicide treatment, 160
  handling systems, 158-61
  harvesting, 13, 157-58
  maturity indices, 9, 10, 158
  modified atmospheres, 58-59, 63
  quality standards, 123
  respiration
    pattern, 4, 76
    relative rate, 3
  ripening facilities, 161-62
  shipping containers, 161
  temperature effects, 6
  transport
    marine, 161

Beans
  chilling injury
    susceptibility to, 6
  cooling, 52
  harvesting, 12, 139
  maturity indices, 9
  modified atmospheres, 60
  quality factors, 121
  quality standards, 126, 129
  respiration
    relative rate, 3
  storage
    recommendations, 142
  temperature effects, 6
Beets
  chilling injury
    susceptibility to, 6
  harvesting, 12
  modified atmospheres, 60
  quality standards, 126, 129
  respiration
    relative rate, 3
  temperature effects, 6
Blackberry
  chilling injury
    susceptibility to, 6
  harvesting, 13
  quality standards, 125
  respiration
    pattern, 4
    relative rate, 3
  temperature effects, 6
Blueberry
  ethylene
    production rate, 4
  harvesting, 13
  quality standards, 124, 128
  respiration
    pattern, 4
Brazil nut
  quality standards, 130
Breadfruit
  harvesting, 13
  respiration
    pattern, 4
Broccoli
  chilling injury
    susceptibility to, 6
  cooling
    methods, 42, 133
  handling systems, 131-32, 134
  harvesting, 12, 131
  maturity indices, 8, 10
  modified atmospheres, 60
  preparation for market, 22
  quality factors, 118
  quality standards, 126, 129
  respiration
    relative rate, 3
  storage
    recommendations, 134
  temperature effects, 6
  transport
    mixed loads, 110
    modified atmospheres, 109
Brussels sprouts
  chilling injury
    susceptibility to, 6
  cooling
    methods, 39, 133
  handling systems, 131-32
  harvesting, 12, 131
  maturity indices, 10
  modified atmospheres, 60
  preparation for market, 23
  quality standards, 126
  respiration
    relative rate, 3
  storage, 134
  temperature effects, 6
  transport
    modified atmospheres, 109
Bush berries
  anthocyanin development, 4
  chilling injury
    susceptibility to, 6
  cooling, 39
  harvesting, 13
  modified atmospheres, 63
  quality standards, 128
  temperature effects, 6
  transport
    air, 108
    mixed loads, 110

C

Cabbage
  chilling injury
    susceptibility to, 6
  cooling, 133
  ethylene
    effects on, 72
  handling systems, 132
  harvesting, 12, 131
  maturity indices, 8, 10
  modified atmospheres, 60
  preparation for market, 23
  quality factors, 118
  quality standards, 126, 129
  respiration
    relative rate, 3
  storage
    recommendations, 134
  temperature effects, 6
Cabbage, Chinese
  cooling, 133
  storage
    recommendations, 134
Cantaloupe
  cooling, 39, 43-45, 140-41
  ethylene
    effects on, 35
    production rate, 4
    threshold concentration, 69
  handling systems, 140-41
  harvesting, 12, 139-40
  modified atmospheres, 60
  packing
    field, 139
  quality standards, 126
  storage
    recommendations, 141
  transport
    mixed loads, 141
    rail, 141
    truck, 141
  waxing, 140
Carnation
  carbohydrates
    supply, 178
  ethylene
    effects on, 36, 72
  modified atmospheres, 61
  preservative solutions, 178
  respiration
    relative rate, 174
  storage
    recommendations, 175-76, 178
  temperature effects, 174
Carrot
  chilling injury
    susceptibility to, 6
  cooling, 42
  disease
    defense mechanisms, 82
  ethylene
    effects on, 36, 72
  handling systems, 135
  harvesting, 12, 135
  maturity indices, 8, 135
  modified atmospheres, 60
  preparation for market, 23
  quality standards, 126, 129
  respiration
    relative rate, 3
  temperature effects, 6
  transport
    mixed loads, 110
Cassava
  curing, 136
  harvesting, 12
  quality factors, 121
  storage
    recommendations, 138
Cauliflower
  chilling injury
    susceptibility to, 6
  cooling, 39, 42, 133
  ethylene
    effects on, 72
    production rate, 4

handling systems, 131-32, 134
harvesting, 12, 131
maturity indices, 8, 10
modified atmospheres, 60
preparation for market, 23
quality standards, 126, 129
respiration
   relative rate, 3
storage
   recommendations, 134
temperature effects, 6

Celeriac
harvesting, 12

Celery
chilling injury
   susceptibility to, 6
cooling, 42, 133
handling systems, 131-32
harvesting, 12, 131
maturity indices, 8
modified atmospheres, 60
preparation for market, 22-23
quality standards, 126
storage
   recommendations, 134
temperature effects, 6

Cherimoya
chilling injury
   susceptibility to, 6
ethylene
   production rate, 4
respiration
   pattern, 4
temperature effects, 6

Cherry
anthocyanin development, 4
chilling injury
   susceptibility to, 6
cooling, 36, 44
disease
   control, 80, 89
   organisms, 86
ethylene
   production rate, 4
harvesting, 13-15
maturity indices, 10
modified atmospheres, 58-59, 63
quality factors, 118
quality standards, 124, 128
respiration
   pattern, 4, 75
   relative rate, 3
temperature effects, 6

Chrysanthemum
harvesting, 176
preservative solutions, 178
shipping containers, 177
storage
   recommendations, 178

Citrus fruits
carotenoid development, 4
chilling injury
   susceptibility to, 6, 153
degreening, 153
disease
   control, 96-97, 153-54
   defense mechanisms, 84-85
   organisms, 87, 93, 153
ethylene
   effects on, 35-36
   production rate, 4
   treatment, 153
handling systems, 154-56
harvesting, 154
insect control, 102, 154
maturity indices, 8, 10
modified atmospheres, 59, 156
quality standards, 124
respiration
   pattern, 4, 76
   relative rate, 3
storage
   recommendations, 156
temperature effects, 6

Corn, sweet
carbohydrates
   changes, 4
chilling injury
   susceptibility to, 6
cooling, 42, 44, 141
handling systems, 140
harvesting, 12, 139-40
maturity indices, 8, 10
modified atmospheres, 60
quality standards, 126, 129
respiration
   relative rate, 3
storage
   recommendations, 142
temperature effects, 6
transport
   mixed loads, 110

Cranberry
ethylene
   production rate, 4
harvesting, 13
quality standards, 124, 126

Cucumber
chilling injury
   susceptibility to, 6
ethylene
   effects on, 36, 71, 141
   production rate, 4
handling systems, 140
harvesting, 12, 139-40
maturity indices, 8
modified atmospheres, 60
quality standards, 126, 129
respiration
   pattern, 4
storage
   recommendations, 142
temperature effects, 6
transport, 141
   mixed loads, 110
waxing, 140

D

Date
harvesting, 13
maturity indices, 10
quality standards, 125
respiration
   relative rate, 3

Dried fruits and vegetables
insect control, 101-02
modified atmospheres, 59
respiration
   relative rate, 3

E

Eggplant
chilling injury
   susceptibility to, 6
cooling, 141
ethylene
   effects on, 141
   production rate, 4
handling systems, 140
harvesting, 12, 140
quality standards, 126
respiration
   pattern, 4
storage
   recommendations, 142
temperature effects, 6
transport, 141
waxing, 140

F

Feijoa
chilling injury
   susceptibility to, 6
ethylene
   production rate, 4
maturity indices, 10
respiration
   pattern, 4
temperature effects, 6

Fennel
handling systems, 133
harvesting, 12
storage
   recommendations, 134

Fig
chilling injury
   susceptibility to, 6
disease
   organisms, 87

Fig *(continued)*
  ethylene
    production rate, 4
  harvesting, 13
  modified atmospheres, 59
  respiration
    pattern, 4
    relative rate, 3
  temperature effects, 6

Filbert
  harvesting, 13
  quality standards, 130

G

Garlic
  chilling injury
    susceptibility to, 6
  cooling, 137
  curing, 136
  handling systems, 136
  harvesting, 12, 135-36
  maturity indices, 136
  preparation for market, 22
  quality factors, 118
  quality standards, 126
  respiration
    relative rate, 3
  sprouting, 4
  storage
    recommendations, 137-38
  temperature effects, 6
  transport
    mixed loads, 110

Ginger
  harvesting, 12
  storage
    recommendations, 138

Gladiolus
  carbohydrates
    supply, 175-76, 178
  geotropism, 4
  handling systems, 177
  preservative solutions, 178

Grape
  chilling injury
    susceptibility to, 6
  cooling 37, 151
  disease
    control, 149-51
    organisms, 86
  ethylene
    effects on, 35
    production rate, 4
  handling systems, 149-51
  harvesting, 13-14, 149-50
  insect control, 101
  maturity indices, 10, 149
  modified atmospheres, 59
  preparation for market, 16
  quality standards, 125, 128
  respiration
    pattern, 4, 76
    relative rate, 3
  shipping containers, 30-31, 150
  temperature effects, 6
  transport, 151
  water loss, 149

Grapefruit
  chilling injury
    susceptibility to, 153
  degreening, 153
  handling systems, 155
  harvesting, 13, 154
  modified atmospheres, 59, 156
  respiration
    pattern, 4
  storage
    recommendations, 156

Guava
  chilling injury
    susceptibility to, 6
  disease
    control, 90
    organisms, 88
  ethylene
    production rate, 4
  harvesting, 13
  respiration
    pattern, 4, 76
  temperature effects, 6

H

Honeydew melon
  cooling, 37
  ethylene
    production rate, 4
    threshold concentration, 69
  handling systems, 140
  harvesting, 140
  modified atmospheres, 60
  preparation for market, 23
  quality standards, 126-27
  ripening, 141
  storage
    recommendations, 141

Horseradish
  harvesting, 12
  quality standards, 127

J

Jerusalem artichoke
  harvesting, 12

Jujube
  chilling injury
    susceptibility to, 6
  ethylene
    production rate, 4
  respiration
    pattern, 4

K

Kiwifruit
  chilling injury
    susceptibility to, 6
  culls and cull utilization, 27
  disease
    control, 95
    organisms, 87, 92
  ethylene
    effects on, 36
    production rate, 4
  harvesting, 13
  maturity indices, 9-10
  maturity
    prediction of, 11
  modified atmospheres, 59-61
  quality standards, 125
  respiration
    pattern, 4
    relative rate, 3
  temperature effects, 6
  transport
    mixed loads, 110

Kohlrabi
  handling systems, 133
  harvesting, 12
  storage
    recommendations, 134

L

Leafy vegetables
  cooling, 39, 42, 44, 133
  ethylene
    effects on, 71
    production rate, 4
  handling systems, 131-32
  harvesting, 12
  quality factors, 121
  quality standards, 126-27
  storage
    recommendations, 134
  transport
    mixed loads, 110

Leeks
  cooling, 133
  modified atmospheres, 60
  storage
    recommendations, 134

Lemons
  chilling injury
    susceptibility to, 153
  degreening, 153
  ethylene
    threshold concentration, 69
    treatment, 153
  handling systems, 154-55
  harvesting, 13
  modified atmospheres, 59, 156

quality standards, 124
respiration
  pattern, 4
seed germination, 4
storage
  recommendations, 156

Lettuce
chilling injury
  susceptibility to, 6
cooling, 45, 47, 133
ethylene
  effects on, 72
  production rate, 4
handling systems, 131-32
harvesting, 12, 131
maturity indices, 8, 10, 131
modified atmospheres 58-60, 63
preparation for market, 23
quality factors, 118
quality standards, 127
storage
  recommendations, 134
temperature effects, 6
transport
  mixed loads, 110

Lime
chilling injury
  susceptibility to, 153
harvesting, 13
modified atmospheres, 59, 156
quality standards, 124
respiration
  pattern, 4
storage
  recommendations, 156

M

Mammee apple
ethylene
  production rate, 4

Mandarin and tangerine
chilling injury
  susceptibility to, 153
harvesting, 13
modified atmospheres, 156
quality standards, 124
respiration
  pattern, 4
storage
  recommendations, 156

Mango
chilling injury
  susceptibility to, 6
disease
  control, 97-98
  defense mechanisms, 84-85
  organisms, 94
ethylene
  production rate, 4

threshold concentration, 69
fumigation, 166
fungicide treatment, 165
handling systems, 165-66
harvesting, 13, 165
maturity indices, 10, 165
modified atmospheres, 59
respiration
  pattern, 4, 76
ripening, 166
shipping containers, 165

Melons
chilling injury
  susceptibility to, 6
cooling, 141
handling systems, 140
harvesting, 13, 139-40
maturity indices, 10
preparation for market, 23
quality standards, 127
respiration
  pattern, 4
ripening, 141
storage
  recommendations, 141
temperature effects, 6
transport, 141
  mixed loads, 110

Mushroom
harvesting, 12
modified atmospheres, 60
quality standards, 127, 129
respiration
  relative rate, 3

N

Nectarine
chilling injury
  susceptibility to, 6
disease
  control, 89
  organisms, 86, 92
ethylene
  production rate, 4
harvesting, 13
modified atmospheres, 59
quality standards, 125
respiration
  pattern, 4, 76
  relative rate, 3
temperature effects, 6

Nuts
insect control, 101
maturity indices, 170
modified atmospheres, 59
quality standards, 130
respiration
  relative rate, 3

O

Okra
chilling injury
  susceptibility to, 6
ethylene
  production rate, 4
handling systems, 140
harvesting, 12, 140
maturity indices, 8
modified atmospheres, 60
quality standards, 127, 129
temperature effects, 6
storage
  recommendations, 142

Olive
chilling injury
  susceptibility to, 6
ethylene
  production rate, 4
harvesting, 13
modified atmospheres, 59
quality standards, 125
respiration
  pattern, 4

Onion, dry
chilling injury
  susceptibility to, 6
cooling, 137
curing, 136
handling systems, 135
harvesting, 12, 135-36
maturity indices, 8, 136
modified atmospheres, 60
preparation for market, 22-23
quality factors, 118
quality standards, 127, 129
relative humidity, 6
respiration
  relative rate, 3
rooting, 4
sprouting, 4
  inhibition, 137
storage, 52
  recommendations, 137-38
temperature effects, 6
transport
  mixed loads, 110
  rail, 105-106
  truck, 104

Onion, green
cooling, 39, 133
handling systems, 131-32
harvesting, 12
modified atmospheres, 60
quality standards, 127
respiration
  relative rate, 3
storage
  recommendations, 134

Orange
  chilling injury
    susceptibility to, 153
  culls and cull utilization, 26
  ethylene
    threshold concentration, 69
  handling systems, 154-55
  harvesting, 13, 154
  modified atmospheres, 59, 156
  quality standards, 124
  respiration
    pattern, 4
  storage, 49
    recommendations, 156
  transport
    rail, 106

Ornamentals: bulbs
  ethylene
    effects on, 72

Ornamentals: cut flowers
  carbohydrates
    supply, 175-76
  chilling injury
    susceptibility to, 174
  cooling, 39-40, 177
  disease, 176
  ethylene
    effects on, 72
    inhibition of action, 74
    production rate, 4
  growth regulators, 178
  handling systems, 176-78
  harvesting, 176
  maturity indices, 176
  quality factors, 118
  respiration
    relative rate, 3
  storage
    recommendations, 178
  transport
    air, 104
    truck, 104

Ornamentals: potted plants
  cooling, 177
  ethylene
    effects on, 72
    inhibition of action, 74
  transport
    air, 108
    mixed loads, 110
  water relations, 174, 177

P

Papaya
  chilling injury
    susceptibility to, 6
  culls and cull utilization, 26
  disease
    control, 96, 163-64
    defense mechanisms, 84-85
    organisms, 88, 93, 162
  ethylene
    effects on, 35
    production rate, 4
  fumigation, 164
  handling systems, 163-64
  harvesting 13, 163
  insect control, 101, 163-64
  maturity indices, 10, 163
  modified atmospheres, 59
  quality standards, 123
  respiration
    pattern, 4
  storage
    recommendations, 165
  temperature effects, 6
  transport, 165

Parsley
  cooling, 133
  handling systems, 132
  harvesting, 12, 131
  quality standards, 127
  storage
    recommendations, 134

Parsnip
  harvesting, 12
  quality standards, 127

Passion fruit
  chilling injury
    susceptibility to, 6
  ethylene
    production rate, 4
  harvesting, 13
  respiration
    pattern, 4
  temperature effects, 6

Pea
  carbohydrates
    changes, 4
  chilling injury
    susceptibility to, 6
  cooling, 141
  harvesting, 12, 139-40
  maturity indices, 10
  quality standards, 127, 129
  respiration
    relative rate, 3
  storage
    recommendations, 142
  temperature effects, 6

Peach
  carotenoid development, 4
  chilling injury
    susceptibility to, 6
  cooling, 37
  disease
    control, 78-79, 81-82, 89
    defense mechanisms, 78, 82
    organisms, 86
  ethylene
    effects on, 35
    production rate, 4
  harvesting, 13
  modified atmospheres, 59
  quality standards, 123, 125, 128
  respiration
    pattern, 4, 75-76
    relative rate, 3
  temperature effects, 6

Pear
  bruising, 14-15, 28-29
  chilling injury
    susceptibility to, 6
  cooling, 36, 38, 43, 144
  disease
    control, 89
    defense mechanisms, 84-85
    organisms, 86, 91
  ethylene
    production rate, 4
  handling systems, 144-45
  harvesting, 13, 144-45
  maturity indices, 8, 10
  modified atmospheres, 58-59
  physiological disorders, 143-44
  preparation for market, 14-17
  quality standards, 125, 128
  respiration
    pattern, 4, 76
    relative rate, 3
  shipping containers, 30, 144
  storage
    recommendations, 144
  temperature effects, 6
  transport
    mixed loads, 110

Pecan
  harvesting, 13
  quality standards, 130

Pepper
  chilling injury
    susceptibility to, 6
  cooling, 42, 141
  ethylene
    production rate, 4
  handling systems, 140
  harvesting, 12, 140
  modified atmospheres, 60
  quality factors, 118
  quality standards, 127, 129
  respiration
    pattern, 4
    relative rate, 3
  seed germination, 4
  storage
    recommendations, 141
  temperature effects, 6
  waxing, 140

Persimmon
  chilling injury
    susceptibility to, 6
  ethylene
    production rate, 4

harvesting, 13
maturity indices, 10
modified atmospheres, 59
quality standards, 125
respiration
  pattern, 4
temperature effects, 6

Pineapple
chilling injury
  susceptibility to, 6, 168
disease
  control, 95
  defense mechanisms, 84
  organisms, 88, 93
ethylene
  production rate, 4
fungicide treatment, 167
handling systems, 166-68
harvesting, 13, 166
maturity indices, 166
modified atmospheres, 59
physiological disorders, 168
quality standards, 123, 125
respiration
  pattern, 4, 76
storage
  recommendations, 168
temperature effects, 6

Pistachio
aflatoxin, 172
drying, 171
handling systems, 171
harvesting, 13, 170
maturity indices, 170
quality standards, 123
storage
  recommendations, 172

Plantain
chilling injury
  susceptibility to, 6
disease organisms, 87-88
ethylene
  production rate, 4
respiration
  pattern, 4
temperature effects, 6

Plum and prune
chilling injury
  susceptibility to, 6
disease
  control, 89
ethylene
  production rate, 4
modified atmospheres, 59
quality standards, 125
respiration
  pattern, 4, 76
  relative rate, 3
temperature effects, 6

Pomegranate
chilling injury
  susceptibility to, 6
ethylene
  production rate, 4
harvesting, 13
maturity indices, 10
quality standards, 125
respiration
  pattern, 4
temperature effects, 6

Potato
carbohydrates
  changes, 4
chilling injury
  susceptibility to, 6
cooling, 137
culls and cull utilization, 26-27
curing, 136
energy use, 115
ethylene
  effects on, 72
  production rate, 4
handling systems, 135-36
harvesting, 12, 135-36
light exposure, 6
maturity indices, 8, 135
modified atmospheres, 58, 60
preparation for market, 23
quality factors, 118-21
quality standards, 127, 129
respiration
  relative rate, 3
rodent control, 137
sprouting, 4
  inhibition, 137
storage, 49, 52
  recommendations, 137
temperature effects, 6
transport
  rail, 105-106

Pumpkin
chilling injury
  susceptibility to, 6
ethylene
  production rate, 4
handling systems, 140
harvesting, 12, 140
quality standards, 127
relative humidity, 6
storage, 52
  recommendations, 141
temperature effects, 6

Q

Quince
quality standards, 125

R

Radish
chilling injury
  susceptibility to, 6

handling systems, 135-36
harvesting, 12, 135
maturity indices, 135
modified atmospheres, 60
quality standards, 127
temperature effects, 6

Raspberry
harvesting, 13
quality standards, 125, 128

Rhubarb
harvesting, 12
quality factors, 121
quality standards, 127

Root vegetables
cooling, 39, 137
ethylene
  production rate, 4
modified atmospheres, 58
preparation for market, 23, 136
quality standards, 126-28
relative humidity, 6
rooting, 4
sprouting, 4
storage
  recommendations, 137-38
texture, 4

Rose
carbohydrates
  supply, 175
maturity indices, 176
preservative solutions, 178

Rutabaga
harvesting, 12
preparation for market, 22

S

Sapote
chilling injury
  susceptibility to, 6
ethylene
  production rate, 4
respiration
  pattern, 4
temperature effects, 6

Shallot
quality standards, 127

Snapdragon
ethylene
  inhibition of action, 74
geotropism, 4

Spinach
chilling injury
  susceptibility to, 6
cooling, 133
handling systems, 12, 131
modified atmospheres, 60
quality factors, 121
quality standards, 127, 129
respiration
  relative rate, 3

Spinach *(continued)*
  storage
    recommendations, 133
  temperature effects, 6

Squash, summer
  chilling injury
    susceptibility to, 6
  cooling, 141
  ethylene
    effects on, 71
    production rate, 4
  handling systems, 139-40
  harvesting, 12, 140
  quality standards, 127
  storage
    recommendations, 142
  temperature effects, 6
  transport
    mixed loads, 110

Squash, winter
  handling systems, 140
  harvesting, 12, 140
  quality standards, 127
  storage, 52
    recommendations, 141

Stone fruits
  cooling, 41
  culls and cull utilization, 26
  disease
    control, 81, 89
  harvesting indices, 10
  insect control, 102
  preparation for market, 16, 18-19
  quality factors, 119

Strawberry
  anthocyanin development, 4
  bruising, 145
  chilling injury
    susceptibility to, 6
  cooling, 36, 40, 146
  disease
    control, 80, 90, 145, 147
    organisms, 86, 92, 145
  energy use, 117
  ethylene
    effects on, 35
    production rate, 4
  handling systems, 146-48
  harvesting, 13, 146
  maturity indices, 146
  modified atmospheres, 58-59, 63, 147
  quality standards, 125, 128
  respiration
    pattern, 4, 76
    relative rate, 3
  shipping containers, 146
  temperature effects, 6
  transport
    air, 148
    mixed loads, 110, 147
    truck, 147
  water loss, 145

Sweet potato
  chilling injury
    susceptibility to, 6
  curing, 136
  fungicide treatment, 136
  handling systems, 135
  harvesting, 12, 135-36
  maturity indices, 136
  preparation for market, 22
  quality standards, 127, 129
  respiration
    relative rate, 3
  rodent control, 137
  storage
    recommendations, 137-38
  temperature effects, 6

T

Tamarillo
  ethylene
    production rate, 4
  respiration
    pattern, 4

Tangelo
  harvesting, 13
  quality standards, 124

Taro
  harvesting, 12
  maturity, 136
  storage
    recommendations, 138

Tomato
  carotenoid development, 4
  chilling injury
    susceptibility to, 6
  cooling, 140-41
  disease
    control, 98
    organisms, 88, 94
  energy use, 116
  ethylene
    effects on, 35
    production rate, 4
    threshold concentration, 69
  handling systems, 139-40
  harvesting, 12, 139
  maturity indices, 8, 10
  modified atmospheres, 58, 60, 141
  physiological disorders, 5
  preparation for market, 23
  quality factors, 119
  quality standards, 123, 128-29
  relative humidity, 141
  respiration
    pattern, 4, 76
    relative rate, 3
  ripening, 141
  seed germination, 4
  storage
    recommendations, 141
  temperature effects, 6, 142
  transport
    mixed loads, 110, 141
    truck, 141
  waxing, 140

Turnip
  chilling injury
    susceptibility to, 6
  harvesting, 12
  quality standards, 128
  temperature effects, 6

W

Walnut
  drying, 171
  handling systems, 171
  harvesting, 13, 170
  maturity indices, 170
  quality standards, 130
  storage
    recommendations, 172-73

Watermelon
  chilling injury
    susceptibility to, 6
  ethylene
    effects on, 141
    production rate, 4
  handling systems, 140
  harvesting, 140
  maturity indices, 10
  quality standards, 128
  respiration
    pattern, 4
  storage
    recommendations, 141
  temperature effects, 6

Y

Yam
  curing, 136
  storage
    recommendations, 138

WARNING ON THE USE OF CHEMICALS

Pesticides are poisonous. Always read and carefully follow all precautions and safety recommendations given on the container label. Store all chemicals in their original labeled containers in a locked cabinet or shed, away from food or feeds, and out of the reach of children, unauthorized persons, pets, and livestock.

Recommendations are based on the best information currently available, and treatments based on them should not leave residues exceeding the tolerance established for any particular chemical. Confine chemicals to the area being treated. THE GROWER IS LEGALLY RESPONSIBLE for residues on his crops as well as for problems caused by drift from his property to other properties or crops.

Consult your County Agricultural Commissioner for correct methods of disposing of leftover spray material and empty containers. **Never burn pesticide containers.**

PHYTOTOXICITY: Certain chemicals may cause plant injury if used at the wrong stage of plant development or when temperatures are too high. Injury may also result from excessive amounts or the wrong formulation or from mixing incompatible materials. Inert ingredients, such as wetters, spreaders, emulsifiers, diluents, and solvents, can cause plant injury. Since formulations are often changed by manufacturers, it is possible that plant injury may occur, even though no injury was noted in previous seasons.

To simplify information, trade names of products have been used. No endorsement of named products is intended, nor is criticism implied of similar products which are not mentioned.

The University of California in compliance with the Civil Rights Act of 1964, Title IX of the Education Amendments of 1972, and the Rehabilitation Act of 1973 does not discriminate on the basis of race, creed, religion, color, national origin, sex, or mental or physical handicap in any of its programs or activities, or with respect to any of its employment policies, practices, or procedures. The University of California does not discriminate on the basis of age, ancestry, sexual orientation, marital status, citizenship, nor because individuals are disabled or Vietnam era veterans. Inquiries regarding this policy may be directed to the Affirmative Action Officer, Division of Agriculture and Natural Resources, 2120 University Ave., University of California, Berkeley, California 94720 (415) 644-4270.

Issued in furtherance of Cooperative Extension work, Acts of May 8 and June 30, 1914, in cooperation with the U.S. Department of Agriculture. Jerome B. Siebert, Director of Cooperative Extension, University of California.